APPLICATIONS OF NUCLEAR TECHNIQUES

International Conference Proceedings

APPLICATIONS OF NUCLEAR TECHNIQUES

Heraklio, Crete, Greece, June 1990

Editors

George Vourvopoulos
Western Kentucky University, USA

Themis Paradellis
NCSR Demokritos, Greece

World Scientific
Singapore • New Jersey • London • Hong Kong

Published by

World Scientific Publishing Co Pte Ltd
5 Toh Tuck Link, Singapore 596224
USA office: 27 Warren Street, Suite 401-402, Hackensack, NJ 07601
UK office: 57 Shelton Street, Covent Garden, London WC2H 9HE

British Library Cataloguing-in-Publication Data
A catalogue record for this book is available from the British Library.

APPLICATIONS OF NUCLEAR TECHNIQUES
Proceedings of the International Conference

ISBN-13 978-981-02-0374-0
ISBN-10 981-02-0374-8

PREFACE

In the recent years, the bulk of the experiments in basic nuclear physics has shifted from small or modest size laboratories to major regional facilities with large, inclusive, and multi-parameter detection systems. Such a shift could have spelled the demise of most of the nuclear facilities around the world, and this would have been announced as the price to pay for progress. Such a demise, fortunately, did not occur. The versatility, breadth, and overall usefulness of nuclear science prevailed, and a large number of nuclear scientists discovered that their knowledge and their equipment were still valuable. One laboratory after another started dedicating more and more time in applications of nuclear techniques in various fields ranging from medicine to the cement industry.

Transitions from a pure research environment to an industrial or mission oriented environment are not easy to accomplish. It is not a matter of simply changing the accelerator target from a thin foil to a ball bearing. Furthermore, since each nuclear facility has been built or is being built with certain goals in mind, applications are chosen conforming to the specifications of these facilities.

We initiated the series of the conferences on applications of nuclear techniques aiming to facilitate such a transition. The aim has been to discuss topics which can be transplanted to the industry or of major interest to other disciplines.

The topics discussed during the Second International Conference on Applications of Nuclear Techniques, which took place at Heraklion, Crete, were chosen with two purposes in mind:

a. The in-depth discussion of three particular areas, namely applications of nuclear techniques in the life sciences, industrial applications, and applications in materials. These three areas were chosen because nuclear methodology is playing an important role to their development, existing nuclear facilities can still be utilized for developmental purposes, and nuclear scientists can still actively participate and share into their improvement.

b. The presentation of reports from a large number of nuclear laboratories detailing their facilities and the range of chosen applications. Some laboratories still continue to perform traditional nuclear physics experiments, thereby applications constitute only a portion of their effort. Other laboratories were constructed with specific applications in mind. These reports constitute a useful resource to those seeking to contact laboratories of similar size and range of applied activities.

The approach taken was interdisciplinary, and this is reflected in the selection of the speakers to this conference. In most of the cases, the authors drew from their considerable experience from working with others who did not share the same knowledge of nuclear physics, and for whom the particular nuclear technique was only a tool. Considerable emphasis was placed by many authors in detailing those ideas that were successfully transplanted into other fields. The first report, however, of Brian England on technology transfer, illustrated in a very vivid way the pitfalls and considerable effort required for such a transplant.

The coverage of a diverse topic can be meaningful only with a judicious choice of speakers. We believe that the presentations included in the proceedings reflect the high quality of the speakers. For their choices, we wish to express our thanks to the members of the advisory committee Drs. P. Assimakopoulos, E. Cereda, J.L. Debrun, J.B. England, J. Keinonen, H. Morinaga, I.L. Morgan, B. Rosner, J.P.F. Sellschop, and V. Valkovic. A large number of details were taken care of through the untiring efforts of the members of the organizing committee Drs. C. Kalfas, V. Katselis, and E. Kossionides. The burden of dealing with all the elements that scientists usually avoid (hotel, transportation, participants' needs etc.) fell again on the shoulders of our trusted collaborator F. Vlachos and his Tourist and Travel Advisor group. For his unwavering through the years support we owe him our gratitude.

G. Vourvopoulos
T. Paradellis

CONTENTS

Preface v

Conference Overview 1
G. Vourvopoulos

Technology Transfer from Nuclear Research to Applications in Industrial, Medical and Other Fields 10
J. B. A. England

Applications of Charged Particle Beams in Life Sciences 30
V. Valkovic

Nuclear Microscopy in Medical Diagnosis and the Impact on Trace Element Biology 66
U. Lindh

Studies of Metabolism using Stable Isotopes as Tracers and Charged Particle Activation as Analytical Method 87
N. Molho

The Role of Protons in Radiotherapy 109
D. E. Bonnett

Activities of the Medical Physics Laboratory at the Physics Dept. of Milan University, Italy 131
M. C. Cantone et al.

Multielement Analysis of Human Tooth Enamel and Dentine using Proton-induced X-ray Emission Spectroscopy 137
M. C. Buoso et al.

Investigation of Biological Nitrogen Fixation via Nuclear Techniques 143
C. E. Douka et al.

Comparison of XRF and PIXE Methods for Measurement of Extracellular Water in Human Body 148
H. R. Shao et al.

Detection of Elements and Trace Elements in Endoscopy Biopsies of Colonic Mucosa in Normal and Ulcerative Colitis Patients 153
P. E. Lecis et al.

Trace and Ultratrace Elements Determined by PIXE in Natural Waters and Sediments to Study Pollutant Distribution and their Toxicological Effects 159
R. Cecchi & G. Ghermandi

Lateral and Depth Profiling in Al Alloy Implanted with Nitrogen using $^{27}Al(p,\gamma)^{28}Si$ and $^{15}N(p,\alpha\gamma)^{12}C$ Resonant Nuclear Reactions with Proton Microprobe 166
S. Lucas et al.

Quantitative Analysis of Aluminium in Bauxites using PIGE 174
A. Savidou et al.

On the Applicability of PIXE Microanalysis at 10^5 eV Bombarding Energies 179
S. K. Biswas et al.

Application of PIGE Method for the Determination of Sodium and Boron in Aerosol Samples 185
S. K. Biswas et al.

Proton Induced Monoenergetic Photon Beams for Analytical Purposes 192
A. G. Karydas & T. Paradellis

Precision of Neutron Activation Analysis in Archaeometric Studies of Ceramics 197
H. Mommsen et al.

Nuclear Fuel Analysis by Neutron Activation 202
N. N. Papadopoulos

Applications of Muons in Condensed Matter Physics 206
E. J. Ansaldo

Ion Beam Analytical Techniques at the Cise 3.5 MV Tandem VDG Accelerator Laboratory 211
C. Boni et al.

Development and Use of Nuclear Methods at the C.E.R.I. for Research and Applications 216
G. Blondiaux & J. C. Debrun

Applications of Nuclear Techniques in the Accelerator Laboratory of the University of Helsinki 223
J. Keinonen

Applied Nuclear Techniques at C.R.N. Strasbourg 228
G. J. C. Costa

The Applied Nuclear Physics Program at the Institute of Nuclear Physics of the NCSR "Demokritos" 232
X. Aslanoglou

Application of Nuclear Techniques in the Chemistry Department of the University of Thessaloniki, Greece 238
P. Misaelides

Ion Beam Facilities at Catania University 243
G. Foti

Activities of the Radionuclides Metrology Group of the Central Bureau for Nuclear Measurements 249
T. Altzitzoglou et al.

Process Monitoring using Gamma-rays in the Power Industry 254
I. L. Morgan & T. Thompson

Potential of Nuclear Techniques for On-line Bulk Analysis in the Mineral Industry 268
J. P. Baron & L. Debray

Industrial Applications of Positron Emission Tomography 297
G. M. Field et al.

Ion Beam Techniques for Materials Modification and Analysis 307
F. D. McDaniel et al.

Sealed Tubes Neutron Generators for Detection and Analysis 321
P. Bach

"Minerals Physics ... and Ion Probes" 326
J. P. F. Sellschop

Nuclear Technologies for Finding Clandestine Explosives 338
L. Grodzins

Radiometry of Heavy Mineral Sands as a Tracer of Coastal Erosion 360
M. B. Greenfield et al.

The Neutron Spectrum of the Hiroshima A-Bomb and the Dosimetry System 1986 379
W. Rühm et al.

A Universal Method of Producing Ultra-fine Microbeam 397
A. Isoya

On Neutron Transmutation Doping of Silicon 414
K. Dyrbye

Structural Characterization of II–VI Semiconductors by Ion Channeling 427
B. Daudin et al.

Muonic X-rays as a Tool for Nondestructive Materials Research 447
H. Daniel

The Design of the Israeli X-ray Telescope 472
B. Rosner

Conference overview

G. VOURVOPOULOS
Department of Physics and Astronomy
Western Kentucky University
Bowling Green, Kentucky 42101, USA

This is not an exhaustive, all inclusive review of everything that was presented at this conference. I apologize beforehand if some people find that their work will not be mentioned. This summary, in spite of my efforts to avoid it, includes my biases. Perhaps more seriously, it reflects my inadequacy to assimilate quickly such a wealth of information in a variety of nuclear activities.

The theme of the conference was not on nuclear techniques but on applications of nuclear techniques. This point was made by a number of speakers. As one contemplates moving a laboratory nuclear technique to an industrial setting, a key point to consider is whether this technique adds something new to the existing methodology. Does it improve reliability, precision, accuracy, throughput, cost, etc.. As B. England put it very succinctly in his presentation, "If it does not sell, Industry does not want it. If it does not improve the life expectancy or well-being of patients, Medicine does not want it." Let me bring as an example the nuclear microprobe. When it came out a few years ago, researchers were presenting their results by comparing them with electron microscope pictures. They were indicating that a nuclear microprobe can approach the resolution of the electron microscope. It would be extremely hard to persuade a biologist or a materials scientist to give up their standard analysis technique for a more expensive one but with poorer resolution. A nuclear microprobe, however, is a very useful tool in trace element biology. As U. Lindh told us in his presentation, a nuclear microprobe has a detection limit of 0.0001% or better and it can detect all elements of the periodic table. Scanning electron microscopy on the other hand, by using window-less detectors can detect the elements lithium through uranium. There are advantages therefore in using a nuclear microprobe in clinical research. An interesting example on the use of the nuclear microprobe was the investigation of the Chronic Fatigue Syndrome in Sweden. U. Lindh told us of the analysis of individual blood cells of patients that revealed measurable concentrations of mercury. A bulk analysis would not have revealed anything significant because of the small fraction of the blood cells involved in the mercury accumulation. The Chronic Fatigue Syndrome was tied to the release to the human body of mercury used in the dental amalgams.

In trying to promote nuclear techniques, we do so with a built-in handicap: fear or reluctance of the public in general to use anything that includes the word nuclear. B. England brought to us the example of food irradiation. After using a series of logical questions and answers in order to establish the usefulness of the method, the last question and answer "will Joe Soap or Fred Bloggs accept it? The answer generally is NO" sealed the fate of the method. Other fields have simply removed any traces of the world nuclear e.g. changing Nuclear Magnetic Resonance (NMR) to Magnetic Resonance Imaging (MRI).

Irrespective of the method that one wishes to use, let me add two points that were a recurring theme in a number of talks:

1. Quality Assurance/Quality Control
2. Sampling

Trace element investigation seems to be one of the strong points of the nuclear techniques. A number of presentations dealt either with the detection of trace elements or the insertion of them in materials. Handling of a large number of samples and measuring trace element concentrations in them require well established QA/QC procedures. Errors can be introduced at any phase from sample preparation to data reduction.

Sampling is another issue whose importance can best be manifested by referring to some of the results presented. In the use of Positron Emission Tomography (PET), G. Field presented preliminary data of a mixing process. A single 3 mm radioactive tracer placed in a liquid slurry (Fig. 1) remained in a certain region within the slurry, indicating a non-uniform distribution.

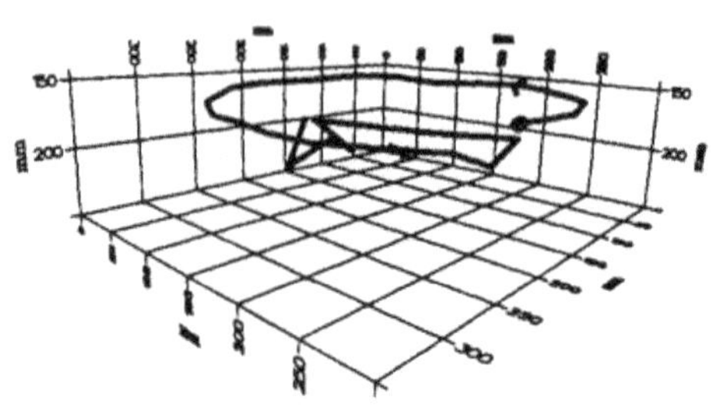

Fig. 1 Mixing process in a liquid slurry.

Such result would indeed be very important in the cases of trace element determination using the Inductive Coupled Plasma (ICP) technique. In ICP, the material to be analyzed is placed in solution and then is aspirated in an ion source. There has been an effort to use coal slurries instead of dissolving the various elements present in a coal sample. G. Fields's results indicate that one should be concerned about the measurements obtained with a coal slurry. Studying the distribution of toxic elements in fly ash, V. Valkovic reported on their distribution in particulate matter. Some toxic elements were found to reside mostly on the surface, others were distributed inside the particulate. From my own experience examining trace elements in coal, the concentration of some of them showed a dependency on the size of the coal sample. By starting with three distinctive groups of coal (up to 9 mm, between 9 mm and 25 mm and between 25 mm and 75 mm) and subsequently grinding them to a uniform powder and looking at the trace elements with XRF, elements such as V, Cr, Mn did not show any size dependency, while Zn and Ti showed a strong one. Finally, let me repeat here a sampling method with a light side in it, as told by Pierre Gy, author of a book on sampling theory. A pine apple processing plant developed an elaborate computerized sampling of the incoming trucks laden with pine apple. A mechanical arm would dip into the pine apple load at random positions and the pine apples would be analyzed for sugar content etc. The producer would then be paid accordingly. The truck drivers soon discovered that the elaborate mechanical arm, although it would scoop pine apples from random positions, it would never dip below a certain point, possibly for the fear of damaging the arm due to variations of truck bed heights. The trucks soon started arriving laden with stones up to the dipping point of the arm. The stones were collected from the pine apple fields. The pine apple producers were accomplishing a double feat. They were getting rid of the stones from the fields and they were paid for it.

All of the above only mean to emphasize the importance of sampling especially in elemental analysis.

In the **life sciences** applications, a point that was shown in a number of presentations was the inter-dependence of trace elements. In an example discussed by U. Lindh, Se was found to exert a protecting effect against Hg toxicity as well as Cd toxicity. This realization leads to the conclusion that multi-element analytical techniques are important to life sciences. From the presentations, a more useful definition of the toxicity of elements emerged. Our knowledge of the essentiality of most of the elements increased considerably in the last few years. Thus, we do not talk about toxic elements but about toxic concentrations. Let us look at Se in particular, since a number of authors spoke about it. U. Lindh indicated that deficiency of Se in the human diet results in heart disease. There is a region in China where such deficiency exists, and the indigenous population receives Se supplements. Large concentrations of Se on the other hand, create a toxic effect which results in liver degeneration. Se was also found, as reported by P. Lecis, to be an effective inhibitor of colon tumorigenesis in rats. One can surmise from the case of Se that indeed toxic concentrations are the issue and not toxic elements.

V. Valkovic gave us a very informative overview of the nuclear analytical methods used in life sciences. You will find his article very useful containing a large number of references (74 in total). V. Valkovic always give credit to the researchers for their work. He also presented some results of PIXE analysis of keratin material and of bone. Both analyses show that the concentration of the trace elements is mostly on the surface, with a subsequent slow diffusion into the bulk of the material. N. Molho presented some interesting results in metabolic studies using charged particle activation analysis. In order to study the metabolism of a certain element, he introduces two stable isotope tracers and subsequently identifies both of them by using a nuclear reaction. Let us look at the case of iron metabolism. In order to study the iron transfer through plasma, two iron stable isotopes were used (Fig 2.). ^{58}Fe was administered intravenously and ^{57}Fe orally.

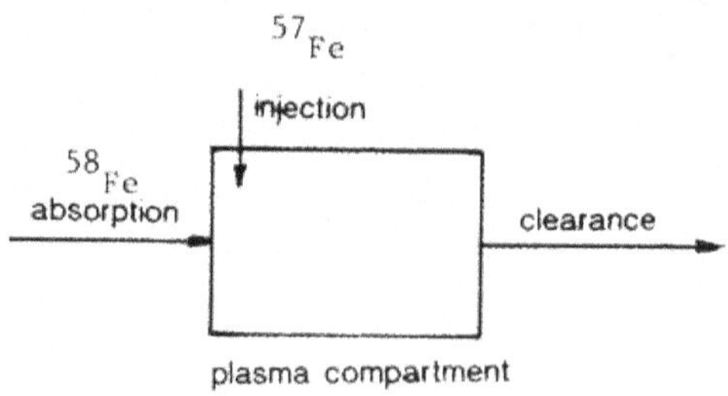

Fig 2. Use of two stable isotope tracers for the study of iron transfer in plasma.

Blood samples were drawn at various time intervals and the ^{58}Fe and ^{57}Fe concentrations were determined via (p,xn) reactions. The results of the measurements allowed the study of the iron clearance and the iron intestinal absorption. Is PNA the only method for metabolic studies? The answer is no. Radioactive tracers can and have been used, as well as single stable isotope tracers with subsequent analysis via mass spectroscopy or neutron activation. PNA is thus competing with these methods which are more established. The uniqueness of the method however is exhibited when one wishes to study two uptake paths simultaneously as in the case of iron in plasma. In that case PNA is a powerful method.

Proton radiotherapy as presented in the review talk of D. Bonnett, becomes important in the case of tumors that cannot respond to other forms of radiotherapy because of their position. He brought the example of chordoma and chondrosarcoma which are tumors that occur very close to the spinal cord, optic nerve, or brain stem. Because of that proximity and the high radiation required, they are difficult to treat. An 82% effective treatment has been reported after treatment with 160 MeV protons at the Harvard/Massachusetts General Hospital facility. This is to be compared with a 33% local control expected with conventional treatment. We are looking forward to the results from the first dedicated proton radiation facility at Loma Linda University Medical Center in California. It is a facility with a 250 MeV proton synchrotron accelerator designed and built at Fermilab, and scheduled to begin treatments this year.

In **industrial applications**, what has acquired importance in the last few years is the on-line or real time determination of various parameters. These parameters could be the ash content of coal, the silica content of cement, or the degradation of high pressure steam pipes. Problems such as the above are amenable to solutions based on nuclear techniques. In L. Morgan's work as presented by F. McDaniel, we learned about the utilization of gamma rays to detect defects or degradation of components used in the power industry. Although other non-destructive techniques such as x-ray radiography, ultrasonics etc. could accomplish the same result, they usually require the shutdown of the power unit. Plant shut-downs for inspection cost to the industry hundreds of millions of dollars in lost revenues. By developing a high efficiency multi-detector gamma ray system the inspection can be performed on real time bases. The system that he described (Fig. 3) utilizes a 50 Ci ^{60}Co source. The photons are transmitted through the power plant component and are detected by an array of 128 high speed scintillation counters.

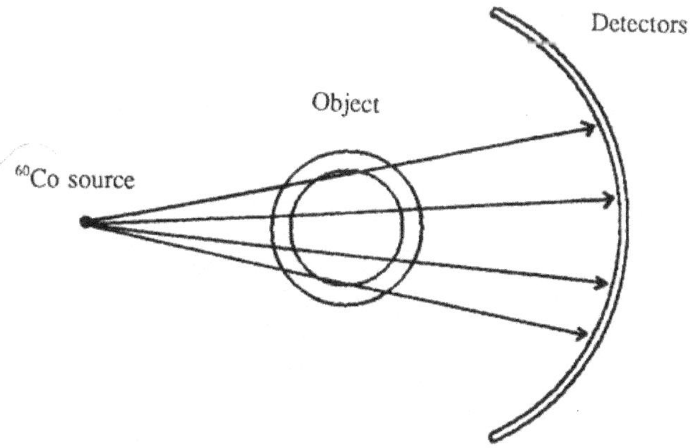

Fig. 3 Schematic diagram of a multi-detector gamma ray system.

The system of the source and the detectors can rotate around the object as well as translate along the object's axis with a speed of 30 cm/min. A seam in a pipe can be scanned in 30 s with a spatial resolution of 0.5 mm. J.P. Baron described the nuclear techniques used in the mineral industry for on-line analysis. These are natural radioactivity methods, γ,γ methods, γ,n methods and pulsed neutron methods. The γ,γ and γ,n methods have been around for a number of years. They have been successful enough to be commercialized and to be used routinely. For example, gamma ray backscattering is used in the coal industry for the on-line determination of the ash content of coal. An important development for on-line analysis is the utilization of pulsed neutron generators. The fast neutrons generated via the D-D and D-T reactions can be used for neutron activation and for (n,n′ γ) reactions. They can also be thermalized and be used for thermal neutron capture reactions. The system that J.P. Baron utilizes for the on-line quality control in the cement industry, is composed of a sealed tube neutron generator (described in this conference by P. Bach) and a NaI(Tl) detector. The time structure of the neutron pulses is shown in Fig. 4. The neutrons are on for 100 μs followed by a 200 μs pause. Every 100 neutron pulses there is a longer, 20 ms pause. The measuring periods are those between neutron pulses. During the 200 μs pauses gamma rays from thermal neutron capture and from activation are measured. During the 20 ms pauses, neutron activation is measured exclusively. He is thus able to determine the concentration of Si, Ca, Fe, and O, which are of primary importance to the cement industry. A very useful addition to the above measurements would be the measurement of gamma rays during the neutron pulse, emanating from (n,n′ γ) reactions. Such measurements are currently difficult because the detector is virtually saturated during the neutron pulse, unless the neutron yield is limited at controlled intervals to allow in-beam measurements.

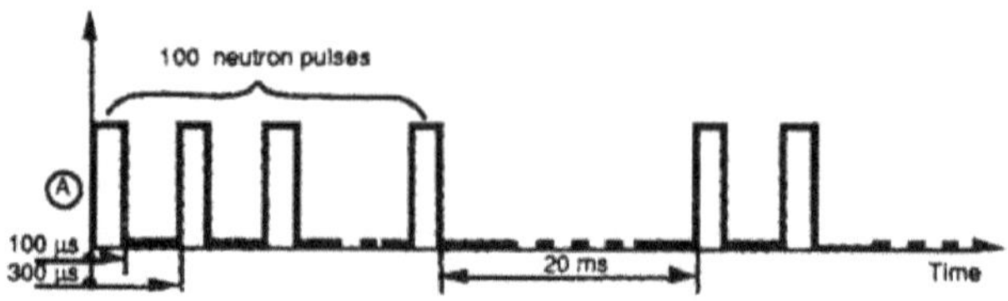

Fig. 4 Time structure of the neutron pulses .

Positron Emission Tomography (PET) has been around for a number of years. This two-photon method has been promoted as a significant improvement over the one photon Computer Assisted Tomography for diagnostic purposes. Because of the short half-lives of the positron emitters used in medical diagnosis, only medical centers with in-house accelerators for isotope production could use PET. G. Field, however, presented to us some very interesting results from the industrial applications of PET. She showed us tracing of ball bearing lubrication, particle tracking in a fluidized bed of a coal-fired power plant and various stages of a mixing process, by using a 50-100 μCi ^{22}Na source.

In the recent hearings in the US Congress about the on-line detection of explosives hidden in suitcases to be loaded on airplanes, thermal neutron activation (TNA) was the method mentioned most often as the appropriate method to be employed. L. Grodzins gave us a very nice description of the key elements needed to be determined in order to minimize false alarms. These elements are C, N, and O. He reviewed briefly a number of techniques utilizing nuclear probes for the detection of explosives. Although the first detectors of explosives currently under installation are based on thermal neutron capture, a number of other promising methods are developed including those using gamma ray resonances.

F. Sellschop showed us the power of the nuclear methods. In mineral physics, the nuclear methods are more than analytical. They can also perform surface studies, 2-D analysis, and depth profiling. Methods such as scanning microbeams, ion implantation and ion channeling play an important role in mineral physics. Combination of these methods allows detailed investigations. As an example, ion implantation of ^{13}C coupled to ion channeling can locate precisely the implanted ion. As F. Sellschop put it, nuclear methods can be yet even more than just analytical. They can tell us about bonding of surface layers and hyperfine interactions.

In talking about surface phenomena, let me take exception on the hydrogen profiling work presented by F. Rauch (University of Frankfurt, Germany). His work dealt with reactions between silicate glasses and aqueous solutions, or humidity. For the analysis of the hydrogen content, the reaction $H(^{15}N,\alpha\gamma)^{12}C$ was utilized at the resonance energy of 6.5 MeV. The method was chosen because it a) has a good depth resolution and b) hydrogen profiling is not influenced by matrix effects. Results of this type have an impact on the optical industry. His findings indicate that :

1. In coatings, the hydrogen content influences the optical, mechanical, thermal, and long term properties of the lens
2. The hydrogen content of the polished surface prior to coating can change. This change depends on the manufacturing process and the mode of storage.

In seeking a solution to a problem amenable to nuclear techniques, depending on the nature of the problem, the technique can be a simple one or a very sophisticated one. Two such problems discussed in this conference are a shore erosion problem and the calculation of the neutron spectrum of the Hiroshima A bomb. Both make use of the analytical capabilities of the nuclear techniques. The shore erosion problem studied by M. Greenfield uses a simple hand held scintillation counter with which shore surfaces are scanned. By looking at the natural radioactivity emitted by the heavier constituents of sand, he finds out that there is a selective removal and deposition of light sand fractions. Such findings are of importance to areas were erosion is a major problem (Florida) or where land is too precious to have it eroded (Netherlands). E. Nolte used the very sophisticated accelerator mass spectrometry method to study the concentrations of the isotopes ^{36}Cl and ^{41}Ca produced in a gravestone granite located 107 m from the hypocenter of the Hiroshima explosion. Since ^{36}Cl and ^{41}Ca are produced via the n,γ reaction, an estimate could be made of the neutron spectrum of the A bomb. His findings indicate that the neutron spectrum which activated the stable isotopes ^{35}Cl and ^{40}Ca is much harder than the neutron spectrum used in the standard dosimetry system.

In **materials-semiconductor** applications, ion implantation remains a major industrial technique. It was with interest then that we received A. Isoya's presentation for a method for producing ultra-fine microbeams. He described a double acceleration lens designed to produce ion beams of 0.1 μm diameter. Such beams, combined with PIXE or RBS are a powerful tool for surveying micro-structures on surfaces. To obtain a beam of 0.1 μm diameter starting with a 0.1 mm diameter beam from an ion source, would require a magnification of 10^{-3}. Magnifications of the order of 10^{-2} are hard to obtain with one acceleration step. He thus proposes a composite lens (Fig. 5) utilizing a 1 kV and a 100 kV acceleration potentials. Starting with a 0.1 mm beam, it is focused to an approximately 10 μm size. A 10 mm accelerating gap follows with a 1 kV potential difference. The beam focuses down to a 0.1 μm size at a distance of 350 mm from the first focal point and is then accelerated with a 100 kV einzel lens. A large series of calculations on trajectories, magnifications, beam dispersion, etc. can be found in A. Isoya's paper.

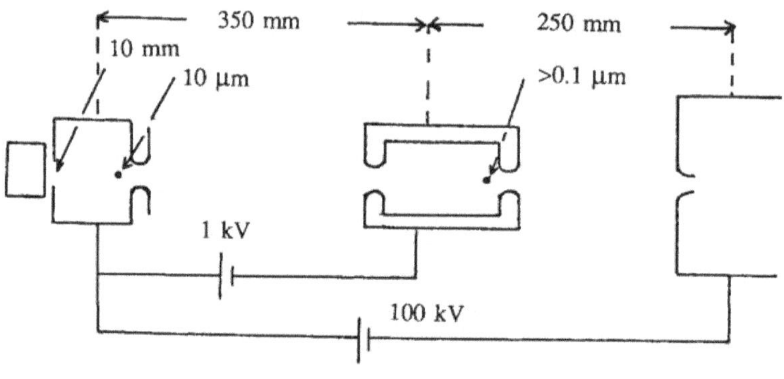

Fig. 5 Double acceleration lens.

A unique silicon doping method was described by K. Dyrbye. Pure silicon crystals are irradiated in a reactor core with doses of 10^{16}-10^{17} neutrons/cm^2. The crystals have diameters up to 12.5 cm and a length of 60 cm. Neutron capture by ^{30}Si nuclei creates unstable ^{31}Si which beta decay to ^{31}P. The crystals created through neutron transmutation doping have an irradiation inhomogeneity of only ±4%. Such technique is the only method that can produce large semiconductor devices with homogeneity.

In the characterization of semiconductors, B. Daudin told us about II-VI semiconductors such as CdTe and ZnTe. He uses channeling for the determination of quantum wells. Advantages of the channeling method are that is non-destructive and it does not require sample preparation.

Characteristic x-rays emitted from celestial objects are proposed to be detected by an x-ray telescope as described by B. Rosner. An innovative way for surface analysis with muonic x-rays was presented by H. Daniel. Although they are difficult to produce since muon facilities are not that plentiful, there are advantages in using them : the analysis is nondestructive, the radiation risk is low, and the method easily yields a bulk analysis. H. Daniel presented some interesting results of experiments with muonic x-rays performed on organic materials from animals and on archaeological artifacts.

A number of laboratories presented a short review of their activities on applications of nuclear techniques. Although their reports were short due to time limitations, they were eloquently given, and the detailed description of the activities as presented in this volume will be a very useful addendum.

I apologize that due to space limitations I will not review the very informative work presented by colleagues in the form of contributed papers. Their work showed ingenuity, precision and an excellent working relation in an interdisciplinary environment.

Let me finish by reiterating my belief that nuclear physics has offered us a large number of tools to work with. As shown by the material presented during this conference, these techniques are in many cases unique, we can draw information from muonic x-rays to nuclear reactions, and colleagues have been most ingenious in using them in a productive and meaningful manner. It was a priviledge to be in the company of such a distinguished gathering.

TECHNOLOGY TRANSFER FROM NUCLEAR RESEARCH TO APPLICATIONS IN INDUSTRIAL, MEDICAL AND OTHER FIELDS.

J.B.A. ENGLAND

Q–Par Angus Limited, Barons Cross Lodge, Barons Cross, Leominster, HR6 8RS. United Kingdom.

ABSTRACT

In the past twenty years research in nuclear physics and in particle physics has pushed the frontiers of certain areas of technology to new limits. However, the benefits of these new precise nuclear techniques of measurement and analysis and of the new instruments have been disappointingly slow to be realized in areas outside pure research.

Part of the problem is the often over–optimistic claims of the scientists involved in the new development and part is an over–cautious attitude to "ultra–new technology" on the part of industry and elsewhere. Where the new technology of nuclear science has been successfully transferred to industrial, medical and environmental areas, the benefits have been very considerable.

Some examples of successful transfer will be discussed and a notable failure analysed. Areas where new nuclear techniques may find applications in the future will be mentioned. The problems which can be encountered in taking a well proven research instrument and turning it into a viable industrial product are discussed from the point of view of the personal experience of the speaker.

There is currently a great deal of pressure on Universities and other Research Laboratories world–wide to transfer their techniques from the research bench to applications in industry, in medicine and in studies of the environment. This pressure is in some ways felt more keenly in nuclear research laboratories of all types, mainly because they are seen, rightly or wrongly, as the "big spenders". However, the histories of the attempts to transfer new ideas and technologies from the nuclear research laboratories to, in particular, industry over the past twenty years have not been particularly happy ones. Industrialists are often rather suspicious of Academia, particularly British Industrialists of British Academia, and one has a hard time trying to get new ideas and techniques accepted.

One must ask the question, "Is Industry, or Medicine for that matter, always to blame for not taking up new technology?". The answer is quite emphatically "Certainly not!". All too often the idea, invention, technology coming from the research laboratory, including nuclear research laboratories, is inadequately developed. They may involve costly and highly complicated manufacturing or operational techniques which do not easily transfer from the small scale laboratory project to the production line or to the hospital treatment room. Even if this transfer can be achieved, the new technique, production method or instrument, by the time it has been completely "de–bugged", may have little advantage over less costly traditional competitors. If it does not sell, Industry does not want it. If it does not improve the life expectancy or well–being of patients, Medicine does not want it.

It is difficult to get over to the enthusiastic scientist that his idea, technique, instrument must work 99% of the time and that there must be an opening, need or market for it. It is no use if only 1 in 3 of the instruments, for example, work properly. That is a sure way to failure. However, on the research laboratory scale, 1 in 3 may be a very good success rate for a new product or technique. It may even be a big improvement in, for example, the medical field where a 30% increase in successful cancer therapy would be a very important result. However, it is no use going to industry until you know why the other 2 fail and can do something about it.

Similarly, there is no point in transferring a brilliant new idea or instrument into industry or medicine if there is no real outlet or use for it, either because of its expense or in the complexity of its use. Industry, and to a lesser extent medicine, is littered with such "white elephants". A recent minor industrial example of such a white elephant is the "Electric Corkscrew". It works, it takes corks out of bottles which are difficult to do by hand, but it costs $40 while the traditional corkscrew costs about $2.

The resistance to accepting new ideas and technology is largely a 20th century phenomenon. Writing in the early 19th Century, Emerson was stating an obvious truth of the time when he wrote:

> "If a man ... make a better mousetrap than his neighbour, though he build his house in the woods, the World will make a beaten path to his door" (Ralph Waldo Emerson, 1803–1832)

Things have changed a lot since Emerson's day.

Not all reluctance on the part of industry and medicine to take up new "revolutionary" ideas is without foundation. The number of plain cases of fraud, or of deliberate laundering of statistics, in the field is surprisingly large. Some scientists work on the Barnum Principle:

> "There's a sucker born every minute" !

It is against such a background that we must look at the criteria by which ideas, technology and instruments from the nuclear field can be judged as suitable for application in more general fields in industry, medicine and the environment. These criteria are set out in Table 1. Such criteria can be used, together with others specific to a given new technique, instrument, application, to judge whether it has been successfully transferred to a new area.

When we use such a set of test criteria on an instrument, idea or technique transferred from the nuclear field to some other, the result is often a resounding "yes", it is successful. In other cases it is a cautious "maybe". In a few cases it is a resounding "NO".

Let us now look at some of the applications which are in use today from the nuclear field. The lists presented here are by no means exhaustive and I apologize in advance to anyone whose particular speciality is not included. A good many of these techniques will be discussed in more detail over the next few days, so for illustration, I shall choose those which are least represented in our Conference for comments on their success or otherwise. Tables 2(a) to 2(e) group these techniques under general activity headings.

Now let us look at a few of these in more detail. I make no apology for choosing amongst these ones in which I myself have been actively engaged.

My first candidate is Surface Analysis by ion beam techniques. These are now widely used to examine doping of semiconductor surfaces, hydrogen permeation of specially prepared metal surfaces and impurity concentrations brought about by surface adsorption. They can also be used to examine the composition of sputtered and evaporated films of compounds where preferential sputtering may have occured. Two distinct energy regimes are currently in use.

(1) Low energy techniques such as Rutherford Back Scattering (RBS) where a few tens of nanometres up to 2–3 micrometres depth into the surface is examined. These are the depths involved in ion implantation in semiconductor devices and in the drifting of light element impurities into semiconductor substrates during annealing. Another powerful low energy technique is Proton Induced X-ray Emission (PIXE) where the depth examined may be up to 30 micrometres.

Figure 1(a) shows an RBS spectrum of sputtered $YBa_2Cu_3O_7$ on a magnesia substrate [1]. This spectrum was taken using a beam of 2 MeV He^+ ions. Figure 1(b) shows the computer fit to the various components of this spectrum. The sputtered surface layer is one of the new high T_c superconductor materials and the interest is to see if the sputtered layer has the same composition as the original. Thus by careful measurement and calibration an accurate quantative determination of both the depth profile and the depth density can be obtained.

My second example from the low energy regime is from an environmental study using PIXE [2]. The sample is sediment taken from the Severn Estuary near Bristol. Figure 2(a) is the induced X-ray spectrum following 2.5 MeV proton bombardment with no absorber before the SiLi X-ray detector. Figure 2(b) is the same spectrum but now with a 100 micrometre Al absorber interposed to reduce the low energy count and allow the higher energy X-rays from heavier elements to be more easily observed. The interest here is in the heavy metals which are thought to originate from a smelting plant near the mouth of the river Avon.

(2) High energy (10 - 30 MeV) techniques using beams of accelerated complex projectiles such as helium, nitrogen and flourine. Here the techniques are Elastic Recoil Analysis (ERA) and Reaction Product Analysis (RPA). The ERA method is particularly suitable for the assay of diffused concentrations of hydrogen and helium isotopes in metals. The RPA method can also be used for hydrogen assay but is more suited to the assay of very low impurity concentrations on or near the surface. Examples illustrating these come from ref [3].

Figure 3 shows the spectrum of recoil hydrogen ions obtained by a glancing angle bombardment of the surface of a thick silicon substrate with 25 MeV alpha-particles. The depth profile is now measureable to approximately 100 micrometres. In the case illustrated here, the depth profile of the hydrogen extends to only 20 micrometres inside the surface.

Using the RPA method we can look at surface trace elements and Figure 4 shows a surface trace of fluorine on silicon at a concentration of 1 in 10^6 atoms obtained via the (α, t) reaction. By using pure elastic scattering, it is possible to measure deeply diffused concentrations of light elements. Figure 5 shows a comparison of a freshly etched silicon surface and one which has been nitrided in 1 bar of ammonia gas at 400°C for 1 hour in radiation hardening studies. The nitrogen peak, of width 2.5 micrometres, can be seen very clearly. This sample also contained a considerable concentration of hydrogen in the surface layer which was measured simultaneously with the nitrogen and was shown in Figure 3.

Since all of these measurements can in principle be done by other methods, the criteria on which the success or otherwise of these nuclear methods has to be judged are those against competing processes. Using, therefore, the criteria of Table 1(a) we can say:

(i) Yes it is safe (provided standard procedures are followed)

(ii) It is more efficient and more accurate than non-nuclear methods.

(iii) It takes about 4 minutes for the hydrogen in silicon assay shown in Figure 3, for example, compared with 12 hours by the best non-nuclear method.

(iv) Yes it is more cost effective by a factor of between 2 and 10.

(v) Provided the accelerator is available, it is easy to use.

(vi) Here we have to use "General Industrialist" rather than "General Public" and the answer is certainly "YES".

From this comparison, it is clear that these techniques of ion beam surface analysis are very successful and they have now gained a firm footing in the assay field.

Now let us look at a completely different application of a nuclear technique, Food Irradiation. This technique has been in use in various countries for over 30 years. The principle is fairly simple, a sufficiently high dose of gamma radiation is given to the food to produce the following effects:

(1) Reduce the incidence of food-borne disease and pathogen contamination – Salmonella in poultry products, for example.

(2) Reduce post harvest loss so that more food is available – grain-mite infestation of stored wheat and barley, sprouting of potatoes, for example.

(3) Prolong the life or "Sell by date" of food by inhibiting bacterial growth and fungoid growth which cause food to perish – fungoid rotting of fruit such as apples and pears, for example.

There are certain well defined limits to this irradiation.

(i) The maximum gamma-ray or X-ray energy must be below 7 MeV to avoid activation of the material by (γ, n) reactions.

(ii) The Codex General Standard for the Irradiation of Food set out by the IAEA must be followed.

(iii) The implementation of this standard and its interpretation must be agreed and followed in all national procedures.

When these conditions are satisfied, the stored quality and safe lifetime of most foods can be very considerably enhanced. Wastage and loss due to deterioration in storage is reduced to near zero in many cases. Let us look at this technique in terms of our criteria. We should probably use the criteria of Table 1(b) since no general method of food preservation or sterilization competes with this process.

(i) Is it Safe? All scientific tests show that it is safe. In fact almost all major organ transplant patients are fed on irradiated food while in intensive care because of the very much improved safety standards of this food.

(ii) Is it successful? It is demonstratively successful in prolonging the useable life of food and in reducing pathogen contamination.

(iii) Does it allow us to do something not previously possible? Yes it does for a large number of foods.

(iv) Is it complicated to use? No.

(v) Is it costly to use? Not more than other very much less effective methods such as deep refridgeration.

(vi) Is it reliable? Provided a standard volume dose is given, yes it is reliable.

(vii) Will "Joe Soap" or "Fred Bloggs" or "Willi Schmidt" accept it? The answer generally is "NO" !

What are the reasons for this non-acceptance? They vary from country to country. In the United Kingdom, for example, there has been a blaze of publicity recently in the popular press, full of pseudo-scientific "facts", half truths and pure "science fiction".

"Radioactive Food" screamed one tabloid over an article which implied that the irradiation made the food radioactive, which it very clearly does not.

"Radiation poisoned food" shouted another tabloid over a report which claimed that "scientists" had proved that proteins were broken down into free radicals which were harmful. The hard scientific fact is that, yes, some proteins are broken down, but only in a similar way to normal cooking and the radicals so-formed are harmless – they give the "flavour" to many cooked foods!

There were further highly controversial statements by fringe members of the U.K. Anti-nuclear lobby which purported to be scientifically proven facts. No matter what respected food scientists said to counter these statements, a large fraction of the population believed them and opinion polls showed that over 80% of the U.K. population rejected any form of irradiated food.

So for the forseeable future, irradiation of food stands no chance of being accepted as a successful technique in the United Kingdom, except in the medical application mentioned above.

Now let us try to use a crystal ball and see if there are any areas where nuclear techniques may be successful in the future and to try to see if any new techniques may be available to extend areas where nuclear techniques are already in use.

In a later contribution to this Conference, we will be hearing about the uses of Positron Emission Tomography (PET). This technique relies on the use of positron-emitting nuclides such as ^{18}F, ^{11}C, ^{13}N etc. However, for many cases, it is not possible to get a suitable positron emitting nuclide to tag or mimic the process we want to study. Powell [4)] has proposed that a system can be devised by which any nuclide that emits two or more ultra-prompt gamma-rays can be used to obtain a localization without the need for the collinearity inherent in the positron decay. His technique involves the use of one detector which is collimated to observe a line of sight through the object under study. The collinearity requirement is replaced by a precise measurement of the time–of–flight difference of the two or more gamma-rays. This defines a hyperbolic surface in space which is cut by the line focus of the collimated detector, this intersection being the locus of the decay. The method requires picosecond timing and is therefore at present restricted to the use of plastic scintillators and low energy radiation for reasonable efficiency and localization. However, if it can be developed successfully, it opens up a much wider field of study than currently open to us using PET.

Are there any other areas where nuclear techniques might be successful in the future? One possibility which has been mooted for the future is the application of nuclear techniques to detect forged bank notes. The idea which looks most feasible is to "dope" some part of the note or of the materials of which it is made with a particular isotope finger–print, that is, a particular set of rare stable isotopes which are sufficiently far from the stable mean that their X–ray isotope shift is easily detected. This set would have a characteristic intensity and spectral distribution under XRF at a particular energy which would be almost impossible, and certainly very costly, to mimic, yet be easily identifiable. The detector for the fluorescent X-rays could possibly by that time be one of epitaxially grown diamond, which at room temperature would give equal or better performance than a current SiLi detector at liquid nitrogen temperature. Such a system is already being contemplated and the diamond detectors are already within reach of being a reality.

My final illustration is in the transfer of instrumental technology developed for nuclear physics into a commercially viable product which has applications well beyond the nuclear field. Here I draw on my own experience in taking silicon radiation detector development from a University research laboratory and turning it into the basic product of a successful commercial Company.

My experience of what happens when you attempt to do this is very similar to playing the childrens game of "Snakes and Ladders". So to illustrate the problems, I show here as Figure 6 my "Technology Transfer Snakes and Ladders" board. I would recommend a study of this board as a dire warning to anyone thinking of trying to start their own Company to exploit a research product.

The start of the game is the "idea". This may be an already proven instrument or technique. In my case it was a new technique enabling more versatile silicon surface barrier detectors to be manufactured. When the decision was taken to set up the Company, after some "arm twisting" by the University I should add, I had already passed the first few hurdles. Thus the first real part of the game for me was to research the market. Some of you sitting here today received questionnaires from me in connection with this.

Funding is the next step. It is essential to have a good market projection for the instrument or technique if you want venture capital from a bank. You may not get all you require, so a good plan is to find a backer in already established industry. This is the route which I took and so avoided the "snake" which can occur if you get venture capital from a "partner" who eventually takes over the business and pushes you out.

Finding suitable manufacturing premises – University Science Parks are to be treated with caution – and finding an acceptable Company name and logo are the next steps. These things may seem trivial, but with the new EEC regulations over trade markes and trade names, it is not always easy to get a Company name and trade logo accepted for registration.

The most important next step is to get your first customer and to start manufacturing. In my case, since I had deliberately tested the export potential, my first customer was from New Zealand, my second from Australia and only then did I get a U.K. customer.

If you really want to sell a new high tech product, especially one from the nuclear field, you must take it to the potential customer, and sales tours are a very effective method for doing this. If you can talk direct to the customer in his own scientific language and convince him or her that you know what you are talking about, you will very probably get an order, with repeat orders in the future. Very occasionally a new market suddenly opens up for you. This happened to my Company following the Chernobyl accident when continuous monitoring of the environment for alpha-particles became important. Such a sudden, large, market can be a potential disaster for a new Company because the temptation is to expand rapidly to meet it. In my case, we had to expand but we did this in such a way that we used part-time "consultants" and also overtime working, rather than a very large expansion. The danger is that once the new market is saturated, sales drop and you find yourself with staff with nothing to do. This is a very good way to become bankrupt!

By hard work and very careful financial managment we have now passed the main pitfalls or "snakes" and are operating at a steadily increasing rate. Eighteen months ago we merged our activities with those of the original small industrial backer which gives us a much broader based scientific market. Semiconductor technology developed for detector manufacturing provides the link for the two parts of the present Company. We are always on the lookout for new markets and new products. As some of you are aware, we are planning, subject to a successful market research response, to bring out a range of new detectors based on monocrystalline carbon in the form of epitaxially grown diamond in about eighteen months from now. The starting material for these new detectors is already available, we now have to prove the manufacturing process for the detectors.

Those of you who know me in the United Kingdom will, I hope, appreciate the meaning of the final "home" square on my "snakes and ladders" board.

REFERENCES.

1) Sokhi, R.S., Gilberd, P.W. and Earwaker, L.G., Nuclear Instruments and Methods, B50 140 (1990)

2) Sokhi, R.S., Davis, A., Randle, K., West, J.R., and West, M.S., Nuclear Instruments and Methods, B49 328 (1990)

3) England, J.B.A. and Sokhi, R.S. Private Communication (1989)

4) Powell, M.D. Medical Physics (USA) 16 809 (1989)

TABLE 1

1(a) CASES WHERE A NUCLEAR TECHNIQUE IS COMPETING WITH AN ESTABLISHED NON-NUCLEAR TECHNIQUE

(i) Is it safe?
(ii) Is it more accurate/efficient/successful?
(iii) Is it faster?
(iv) Is it more cost efficient?
(v) Is it more, or less, complicated to use?
(vi) Will the general public, or general industrialist accept it?

1(b) CASES WHERE A NUCLEAR TECHNIQUE ALLOWS A NEW FIELD TO BE OPENED UP

(i) Is it safe or are there side effects?
(ii) Is it accurate/efficient/successful?
(iii) Does it allow one to do something previously not possible or to treat something previously not treatable?
(iv) Is it complicated to use?
(v) Is it costly to use?
(vi) Is it reliable?
(vii) Will "Joe Soap" accept it?

TABLE 2

2(a) TECHNIQUES INVOLVING BEAMS OF CHARGED PARTICLES.

TECHNIQUE	FIELD OF APPLICATION
Surface analysis by techniques such as RBS, ERA, RPA and PIXE	Mechanical wear, Hydrogen in metals, semiconductor research, Environmental studies, Biological Sciences, Archaeology, Medicine
Microprobe Analysis	Medicine, Agriculture, Archaeology, Metallurgy
Component activation for wear studies	Industry, Medicine
Charged particle activation analysis for trace elements	Industry, Environment, Archaeology, Fine Art
Accelerator-based dating techniques for ^{14}C, ^{36}Cl, ^{10}Be	Archeaology, Environment, Fine Art
Charged particle cancer therapy using electrons, protons, alphas and pions	Medicine

2(b) TECHNIQUES INVOLVING BEAMS OF NEUTRONS.

TECHNIQUE	FIELD OF APPLICATION
Neutron radiography	Industry
Neutron scattering analysis for water (hydrogen)	Industry
Neutron activation analysis	Archaeology, Fine Art, Metallurgy, Medicine
Neutron cancer therapy	Medicine
Environmental monitoring for heavy metals, e.g. Zn, Hg, Pb	Environment, Medical, Industry, Archaeology

2(c)TECHNIQUES INVOLVING X-RAYS OR GAMMA-RAYS

TECHNIQUE	FIELD OF APPLICATION
X-Ray Fluoresence (XRF)	Mining, Geology, Industry, Environment
Hard X-ray radiography	Industry
Food irradiation	Food industry, Farming
Equipment sterilization	Medicine, Vetinary Science
Industrial level/thickness monitoring and measurement	Industry
Cancer therapy	Medicine

2(d) TECHNIQUES INVOLVING RADIOISOTOPES

TECHNIQUE	FIELD OF APPLICATION
Radiography of welds	Industry
Tracer techniques	Industry, Medicine, Environment, Agriculture
Positron Emission Tomography	Industry, Medicine
Smoke alarms	Industry, Home

2(e) TECHNIQUES INVOLVING INSTRUMENTATION

TECHNIQUE	FIELD OF APPLICATION
PET (multiwire proportional detectors and electronics)	Industry, Medicine
Solid state detectors of Si and Ge	Environment (Radon in buildings) Medicine, Industry
Scintillation detectors of BGO, LiF etc	Industry, Medicine
Radiation monitors and electronics	Environment, Industry, Medicine
Compact electron linear accelerators	Industry, Medicine
Compact cyclotrons	Medicine

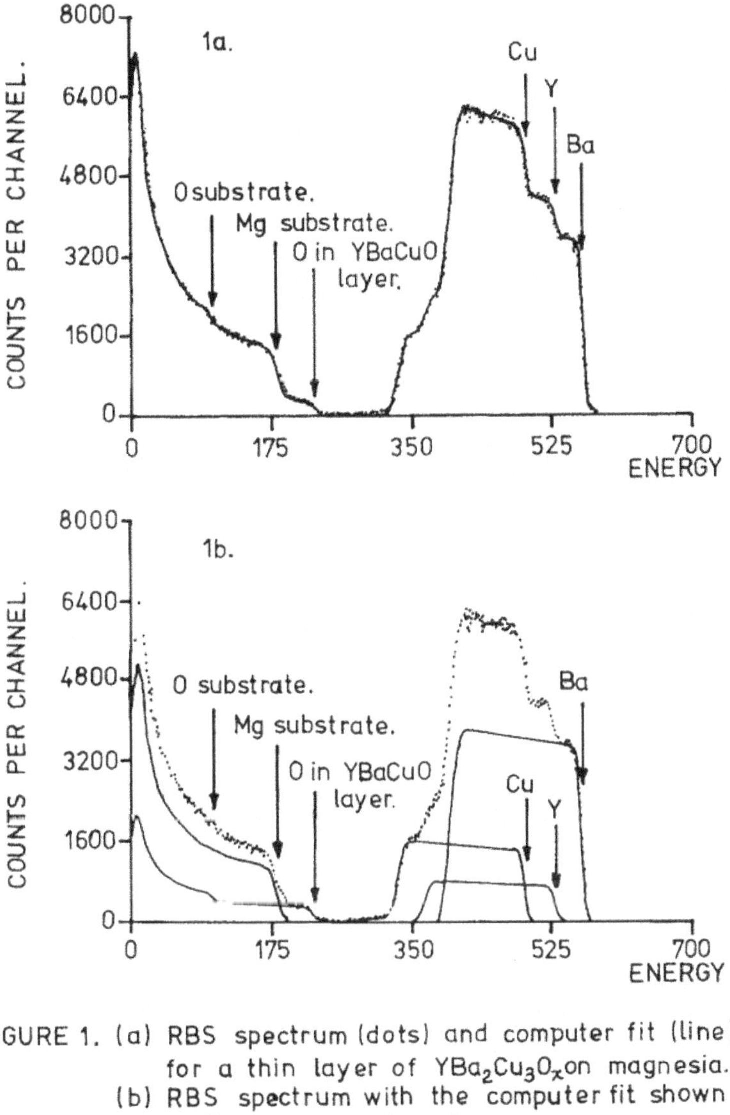

FIGURE 1. (a) RBS spectrum (dots) and computer fit (line) for a thin layer of $YBa_2Cu_3O_x$ on magnesia.
(b) RBS spectrum with the computer fit shown in the component parts.

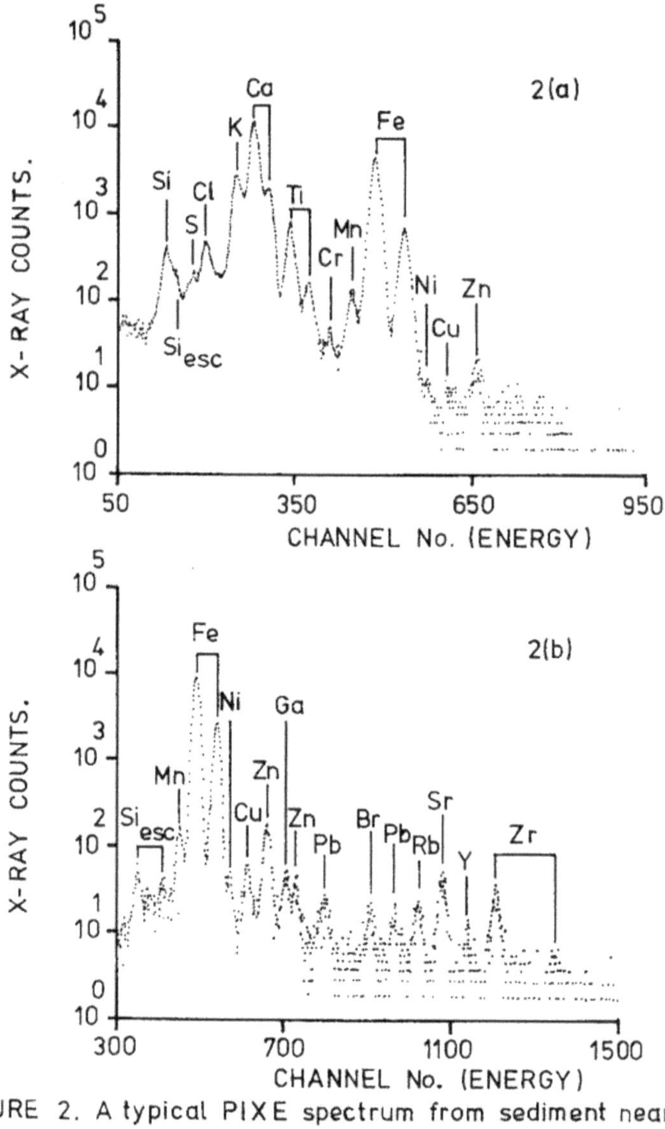

FIGURE 2. A typical PIXE spectrum from sediment near Portishead. (a)without,(b)with 100um Al absorber.

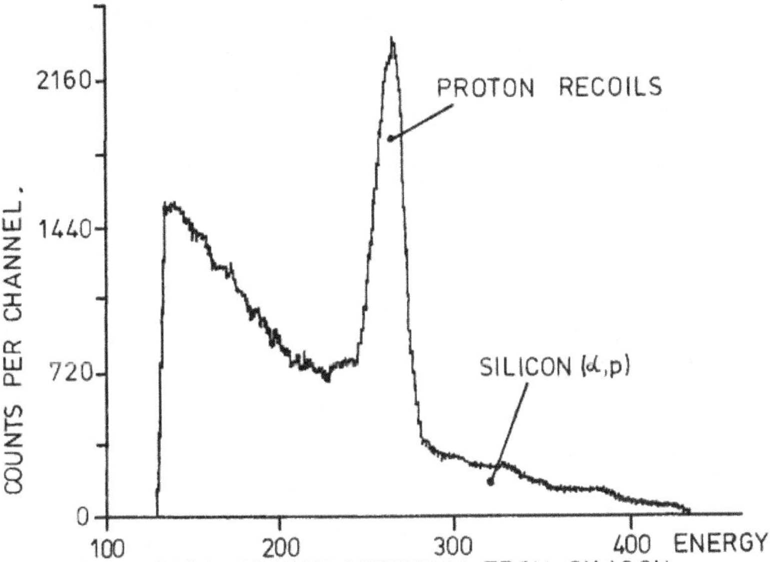

FIGURE 3. RECOIL PROTON SPECTRUM FROM SILICON.

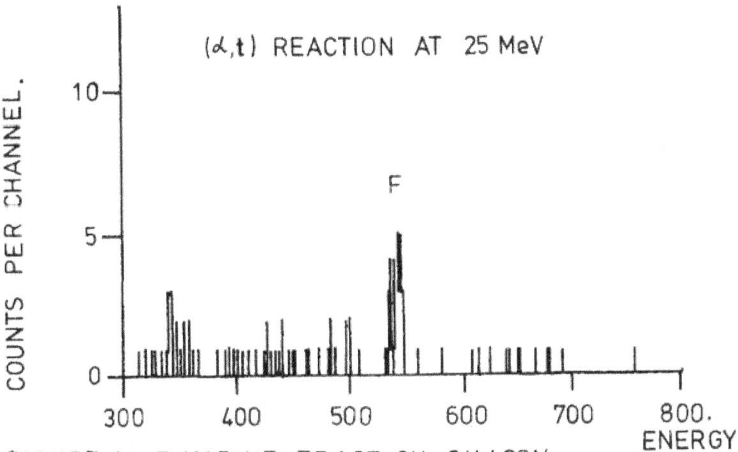

FIGURE 4. FLUORINE TRACE ON SILICON.

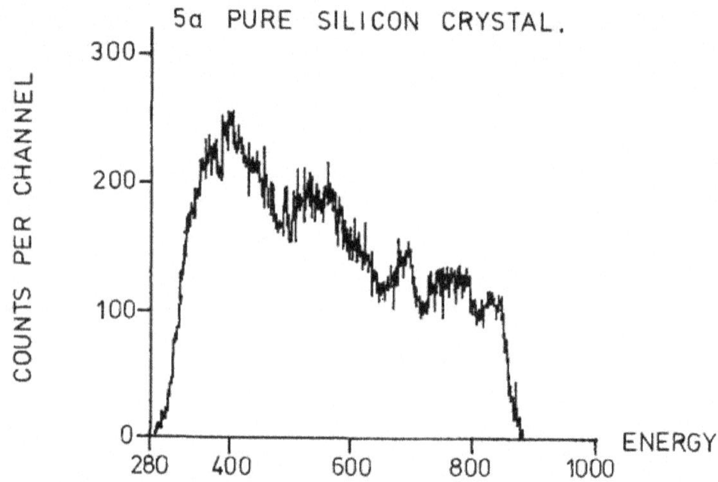

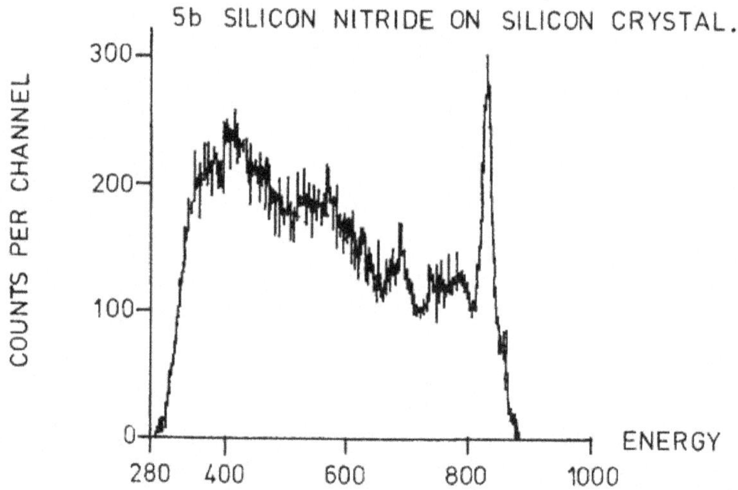

FIGURE 5. ELASTIC SCATTERING OF 25 MeV α-PARTICLES.

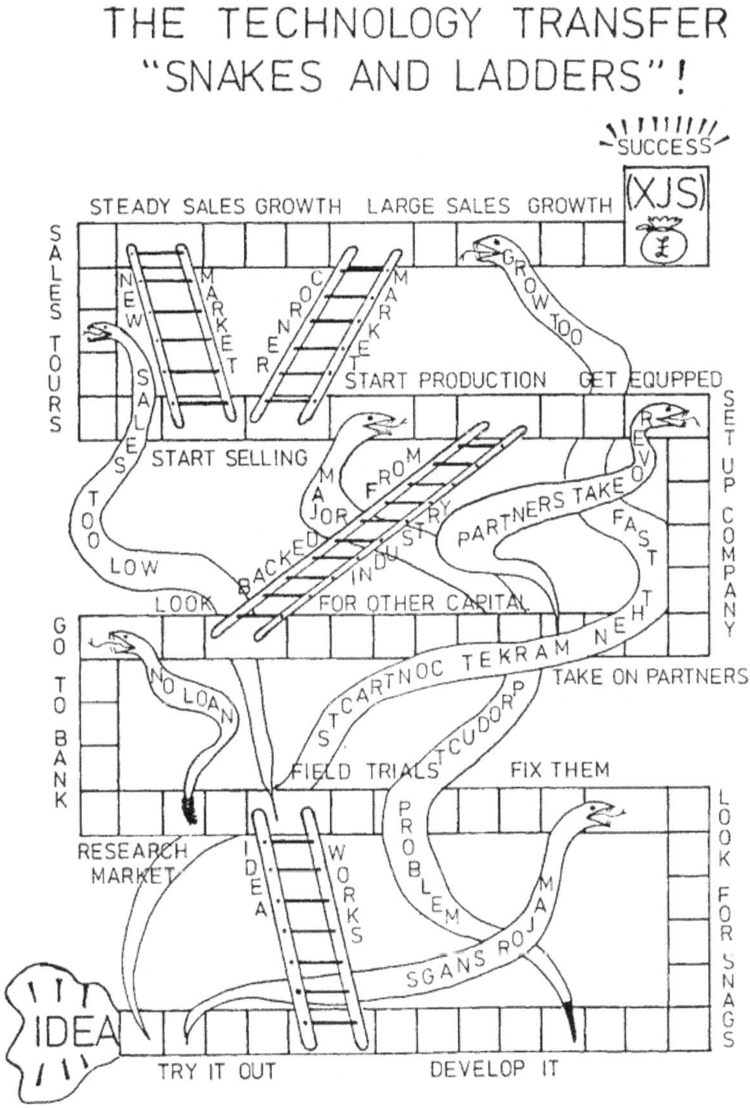
THE TECHNOLOGY TRANSFER
"SNAKES AND LADDERS"!
SUCCESS
(XJS)
STEADY SALES GROWTH
LARGE SALES GROWTH
SALES TOURS
NEW MARKET
RENROC TEKRAM
GROW TOO
START PRODUCTION
GET EQUPPED
SET UP COMPANY
SALES TOO LOW
START SELLING
MAJOR FROM
BACKED INDUSTRY
PARTNERS TAKE OVER
FAST
LOOK FOR OTHER CAPITAL
GO TO BANK
NO LOAN
STCARTNOC TEKRAM NEHT
TAKE ON PARTNERS
STCUDORP
FIELD TRIALS
FIX THEM
RESEARCH MARKET
IDEA WORKS
PROBLEM
MAJOR SGANS
LOOK FOR SNAGS
IDEA
TRY IT OUT
DEVELOP IT

FIGURE 6.

APPLICATIONS OF CHARGED PARTICLE BEAMS IN LIFE SCIENCES

Vlado Valkovic*
Institute Ruder Boskovic, Zagreb, Croatia, Yugoslavia

ABSTRACT

An overview of the use of charged particle beams in life sciences is presented. Special attention is given to recent developments in nuclear analytical techniques . Progress in accelerator based nuclear analytical techniques is associated with developments in areas of

- Beam focusing and construction of microprobes;
- sample preparation methods;
- simultaneous use of several detection methods;
- hardware and software for data collection and analysis

In addition, ion induced material modification, accelerator mass spectrometry, positron emission tomography, radiotherapy and other uses of accelerators in biology and medicine are briefly discussed.

* Present address: IAEA, Agency's Laboratories, A-2444 Seibersdorf, Austria.

TABLE OF CONTENTS

1. INTRODUCTION

2. NUCLEAR ANALYTICAL TECHNIQUES

2.1 Analytical Requirements

2.2 Experimental Arrangements

2.2.1 Light element measurements
2.2.2 PIXE
2.2.3 Concentration profile measurements
2.2.4 Microprobes

2.3 Sample Preparation Procedures

2:4 Examples of Applications

2.4.1 Body fluids
2.4.2 Soft tissues
2.4.3 Keratin
2.4.4 Hard tissues
2.4.5 Plant material
2.4.6 Cell cultures and microorganisms
2.4.7 Microanalysis of biological material

2.5 Material Modifications

2.6 Positron Emission Tomography

2.7 Radioisotope Production

2.8 Interaction of Charged Particles with Living Matter

2.9 Radiotherapy with Charged Ions

3. CONCLUSIONS

REFERENCES

1. INTRODUCTION

In this report the applications of charged particle beams in life sciences, biology and medicine in particular, are described. Charge particle beams can be obtained by a variety of accelerators, and in different energy range. Our discussion will be mainly limited to energy interval of few MeV to few tens of MeV. Accelerators producing ions in this range fall into the category of so-called "small accelerators". The continuing growth of the field of using small accelerators in different applications is best followed through a series of conferences on these subjects (see e.g. ref. 1-4). In addition, many different applications os small accelerators have been elaborated in monographs and books. Here we shall only briefly describe some uses of charged particle beams produced by small accelerators - most of them being cyclotrons or electrostatic type.

2. NUCLEAR ANALYTICAL TECHNIQUES

2.1 Analytical Requirements

Biological systems require a number of trace elements for their functioning. The great majority of trace elements serve chiefly as key components of the enzyme systems of proteins with vital functions. The metal is usually embedded deep inside the structure of the protein. The knowledge accumulated so far on elements essential for life can be summarized as follows. The elements which form the bulk of living matter: H, C, N, O, Na, Mg, P, S, Cl, K and Ca; essential trace elements (for warmblooded species): F, Si, V, Cr, Mn, Fe, Co, Ni, Cu, Zn, As, Se, Mo, Sn, and I. This list should not be considered as final; with the improvements in experimental methodology and analytical techniques it will probably be extended. The common characteristics for all essential trace elements is a narrow range of suitability, and toxicity or deficiency effects for deviation in concentration. In addition, a number of elements (Cd, Hg, Pb, etc.) have only toxic effects even when present in very low concentrations.[5]

Studies of trace elements in biological materials have resulted in significant improvements of our understanding of the functioning of biological systems. Much information has been accumulated about essential trace elements and the response of living systems to their concentration levels.

What is usually measured are whole body, organ, or blood trace element levels. However, many tissues have distributions of concentration patterns reflecting function or metabolic activity as well as body stores of some elements. In the case in which tissue has an inherent variation in element concentration, the whole-tissue concentration value will not contain essential information. It is expected that the knowledge of the distribution of chemical elements on a microscopic scale in biological systems may be of help in understanding basic phenomena in these systems.

2.2 Experimental Arrangements

When a sample of some material is bombarded with a beam of charged particles a number of processes takes place. All of these processes can be used to obtain analytical information about the sample investigated. Some of them have been developed into accepted nuclear analytical techniques, for example the use of nuclear resonant reactions (NRR), backscattering (RBS), production of characteristic X-rays (PIXE), proton elastic scattering (PESA), coincidence measurements and time-of-flight measurements of atoms and molecular fragments. By proper combination of detection of reaction products one can obtain information on total elemental composition related to the morphology of sample (optical information).

Elemental concentration profiles can be measured with a charged particle beam spot size of 1-10μm on the scale of 1-2mm with a microprobe. Microbeam analysis of light elements in different samples can be accomplished by the use of nuclear reaction properties. The information on spatial distribution of heavier elements can be achieved by detection of reaction products excited by focused beams of charged particles.

2.2.1 Light element measurements

Determination of light elements in biological material is usually done by the use of nuclear reaction analysis (NRA), Rutherford backscattering (RBS) and charge particle activation analysis (CPAA). The differentiation between these methods is made upon the detected radiation: in the case of NRA this is usually of the outgoing charged particles, in the case RBS this is scattered beam particle, while in the case of CPAA this is γ-radiation (either prompt or delayed).

Recent advances in activation analysis with charged particles are summarized in the paper by Hoste and Vandecasteele.(6) The main difference with respect to neutron activation analysis is the fact that in CPAA simultaneous irradiation of standard and sample is generally impossible.

Charged particle activation analysis allows the determination of light elements such as Li, B, C, N and O with good accuracy and high sensitivity in a variety of materials. When the concentrations of these elements are sufficiently high, i.e. of the order of μg/g or higher, CPAA offers independent assessment of the accuracy compared with more classical chemical analyses.

The reactions which can be used are as follows:

Lithium can be determined either by proton or deuteron activation. However, the most suitable reaction is $^{7}Li(p,n)^{7}Be$ reaction at an energy below 11.3 MeV so as to avoid interferences from other light elements.

Boron can be determined by the use of many reactions. The most sensitive determination is obtained from the $^{11}B(p,n)^{11}C$ reaction but interference from the $^{14}N(p,\alpha)^{11}C$ reaction cannot be avoided.

The $^{10}B(p,\alpha)^{7}Be$ and the $^{10}B(d,\alpha n)^{7}Be$ reactions have the advantage of yielding a long-lived species and allow a chemical separation if required. However, the Li interference cannot be avoided. By irradiating with deuterons below 5.9 MeV the $^{10}B(d,n)^{11}C$ reaction is interference-free.

Carbon can be determined by a wide variety of reactions, but the most selective one is the $^{12}C(d,n)^{13}N$ reaction at an energy not higher than 4.9 MeV. In many cases ^{13}N has to be separated from the matrix by dissolution of the samples and distillation of the nitrogen as NH_3 followed by coincidence counting of the annihilation radiation.

Nitrogen can be also determined by a large number of reactions. The most interesting ones are, however, the $^{14}N(p,n)^{14}O$, $^{14}N(p,\alpha)^{11}C$ and $^{14}N(d,n)^{15}O$ reactions, each with its own advantages and disadvantages. The $^{14}N(p,n)^{14}O$ reaction is usually interference-free and ^{14}O is measured through the gamma at 2313.0 keV. Low-Z elements are, however, also activated and limit the sensitivity. The $^{14}N(p,\alpha)^{11}C$ reaction is interfered by the $^{11}B(p,n)^{11}C$ reaction, the induced activities being approximately equal at 12 MeV. Moreover, the ^{11}C activity has usually to be separated from the matrix. The $^{14}N(d,n)^{15}O$ reaction has the advantage of being interference-free below 10.6 MeV. Due to its short half-life a separation of ^{15}O is not practical.

Oxygen can be most selectively analyzed by the $^{16}O(^{3}He,p)^{18}F$ and the $^{16}O(t,n)^{18}F$ reactions, the latter being specific if low energy tritons are used. As the number of accelerators where triton activation is possible is rather limited, ^{3}He activation is the most widely applied.

2.2.2 Proton induced X-ray emission spectroscopy (PIXE)

PIXE has been successfully applied to multielemental analysis of different samples. PIXE has the advantage of being applied with focused proton beams (~1μm), this gives the possibility to x-y maping of elemental distribution on any surface. The bulk concentration measurements can be done by using X-ray fluorescence (XRF). In addition, PIXE can be combined with NRA or ERDA in the single measurement.

The accumulation of some metals in some organs in several diseases has been known for some time. Much attention has been paid to the accumulation of early measurable trace metals in organs during the growth of tumors. This has been suggested as a diagnostic tool. Since Cu, Fe, and Zn concentrations in blood serum (normal values: 1 to 2 ppm) hold much promise as a clinical test, several methods have been developed for determining their concentrations. Charged-particle induced X-ray emission spectroscopy offers a very interesting approach to these problems because of its ability to detect simultaneously several trace elements. In addition the technique has the advantage that only very small quantities of serum are needed in target preparation; good results can be obtained with a drop of blood serum on an appropriate backing.

The specific aims of such investigations should be: to determine the clinical usefulness of multiple trace element concentration measurements for the definitive relationships to disease activity and efficiency of therapy; to explore further the preliminary observations that changes in trace element levels precede other symptoms or disease activity; to determine trace element concentrations in normal vs. diseased tissues.

In the coming years, the presence of traces of heavy elements in the human organism is going to be of great interest in the detection and probably in the treatment of human diseases. Particle-induced X-ray emission (PIXE) analysis has already become a very effective means of detecting trace elements in biomedical samples and sensitivities from 0.1 to 1.0 ppm wet weight can now easily be obtained. These sensitivities can generally be achieved in approximately 15 min, using as little as 100 μg of material. If necessary, an analysis can also be easily performed on biopsy tissues, utilizing even less material.

Another advantage of PIXE analysis stems from the fact that 10 to 15 elements can simultaneously be detected in a human organ in a single analysis.

In all these measurements samples have to be introduced into the vacuum (scattering chamber) in order to perform measurements. This is a limitation, especially for liquid samples, large samples etc. In order to avoid these difficulties external beam PIXE is being used in many laboratories.

2.2.3 Concentration profile measurements

Measurements of depth profiles can be accomplished by the use of reactions with a sharp resonance at a given bombarding energy. This technique takes advantage of the fact that resonance reactions occur only at a specific energy and the energy of a charged particle beam progressively decreases as it penetrates the sample. The depth at which a particular reaction occurs is varied by changing the energy of the incident beam. This then provides an indication of the local variation in concentration. The precision with which the depth of a resonance can be located depends on the straggling (this is the principal cause of uncertainty), the width of the resonance, the initial energy distribution of the beam and the energy resolution of the system. Let us mention only some examples:

The penetration of fluorine compounds in teeth has been measured by several authors using the $^{19}F(p,\alpha\gamma)^{16}O$ reaction. Fluorine uptake from different fluorine compounds is usually compared. The F concentrations are measured at intervals of approximately 0.1 μm to a depth of several μm. The resonance is excited by 672 keV protons and 6.13 MeV γ-rays are emitted. The next resonance level is at 838 keV, so that by varying the incident energy from 835 keV to 672 keV the fluorine content can be measured from 0 to 2.4 μm.

The $^{11}B(p,\alpha)$ reaction is utilized by several authors to obtain the boron distribution in different materials. Concentration profiles of carbon in different samples have been determined using $^{12}C(p,\gamma)^{13}N$ reactions.

2.2.4 Microprobe

Electron microprobe offers a very powerful method for the analysis of microstructures. The limitation is the strong bremsstrahlung background from the electrons. This gives a rather poor sensitivity, of the order of $1:10^4$, so that only the major constituents of the sample can be determined. An obvious way to improve this situation is to replace the electrons by protons. Proton microprobe is based on the use of a well focused proton beam for the excitation of target characteristic X-rays. Using only collimators from the electron microscope the dimensions of the beam from the Van de Graaff accelerator can be reduced to small dimensions (25 μm).

Experimental arrangements for focusing a proton beam to the dimensions of 1 μm x 1 μm have been developed in several laboratories. Most of them use systems of several quadrupole lenses to focus the proton beam. Combinations of two, three and four lenses are in use in microbeam systems. Doublets have the advantage of simplicity, but because of the difference in focusing action in the two planes, the demagnification in one direction is low (~ 5), which means that to achieve small spot diameters, a very small object aperture is required which may be difficult to fabricate. The operation of multipoles is simplified by connecting lenses in pairs to the same power supply, so that only two adjustments are required.

One of the most serious problems associated with the use of microprobes concerns the sample alteration induced by the strongly focused beam impact on an approximately one square micrometer area. In the case of more sensitive organic materials, the damage caused can be very important and sometimes erroneous analytical results are observed, chiefly when intensities greater than 100 pA/μm^2 are used. Very little work has been carried out on this subject but it is now well established that effects which severely complicate classical analysis are induced, namely local temperature increase on the target, elemental migration, atomic displacements, shrinkage and sputtering. An excellent review of these phenomena was recently published by Cookson(8), while Watt et al.(9) studied damages on a single pollen grain under impact of a 1 μm section beam. The effects of the mass loss on the accuracy of the X-ray microanalysis are discussed in details by Themner et al.(10) Some of these effects are avoidable. For a given total collected charge, atom displacement and radiolytic effects remain present, while damage induced by local temperature increases (losses of volatile elements, melting and shrinkage) can be reduced by an increase of the scanning speed.

Advantages of high speed scanning for microprobe analysis of biological samples is discussed by Llabador et al.(11)

Quantitative analysis of micro-PIXE results is additionally complicated by many factors including surface roughness (see Orlic et al.(12)). Lövestom and Swietlicki(13) have described an external beam line for the Lund microprobe. This set-up has been added as an option to current scanning proton microprobe set-up for the analysis of fragile samples, such as art objects, and for X-ray imaging of larger areas of a sample.

The summary of recent microprobe techniques and instrumentation has been presented by Leppe and Saint.(14)

2.3 Sample Preparation Procedures

Sample preparation in elemental analysis of biological materials represents a very critical step. For any trace element analysis preconcentration procedures need to be applied when nuclear analytical methods are used. For X-ray emission spectroscopy, the sample excitation mode determines to some extent the sample preparation procedure. The possibility of microbeams is the unique feature of nuclear analytical techniques, however this results in specific requirements on sample characteristics.

For most nuclear analytical techniques, thin targets are required. In order to fabricate thin targets, the sample must be prepared as a homogeneous solution or as a finely divided powder (homogeneous on the scale of μg). Matrix reduction is an essential step in biological sample preparation; it can be accomplished by:

- high temperature ashing;
- low temperature (oxygen excited) ashing;
- wet oxidation procedures.

Ash from low temperature ashing can be dissolved in hydrochloric acid or in nitric acid with or without the aid of measured pressure and temperature. A number of wet-oxidation procedures for destroying organic matrices has been developed which require either a mixture of acids or a single acid.

Sometimes thick targets are used for analysis by nuclear analytical methods. Biological and biomedical samples are then first converted into a fine powder by means of "brittle fracture" techniques.[15] If a thick target is to be made from powders then the sample should be mixed with graphite before pressing. Another possibility is the evaporation of Al onto the surface of the pellet.

A number of problems is encountered in microanalysis of biological material including:

- lack of suitable preparatory procedures;
- specimen stability during analysis;
- lack of appropriate standards for quantification.

It should be kept in mind that localization of an element is possible only if the structure is preserved. Chemical information should be directly related to a well defined morphological structure, therefore correlation between image and measured concentrations are required. The substance to be analyzed should not be altered in composition either during sample preparation or under proton or X-ray bombardment. Rapid freezing of fresh tissues is considered to provide minimum ion movement. Target preparation for nuclear analytical methods should follow rapid freezing by sectioning in a cryostat.

Sample preparation for nuclear microprobe studies requires often the use of staining techniques. For example Larson et al.[16] used the following procedure for preparation of samples of senile plaques:

- freeze-drying the frozen section;
- toluidin blue staining with the spraying technique;
- optical identification of possible senile plaques
- analysis of half of each possible senile plaque;
- silver staining of the other halves;
- pathological confirmation of the senile plaques.

After that, the sample was mounted on electron microscopy grid. The grid (made of copper) was glued over a small hole (2 mm diameter) in a carbon disc (about 2 mm thick) and above a formvar film was mounted. The tissue was mounted on this film of formvar to avoid cracking the tissue section. This is a well known technique in electron microscopy. With this technique the beam can be positioned easily and accurately at the area which has been pre-determined.

It should be kept in mind that these are several routine histochemical procedures, however, the quantification of histochemical reactions by the micro-PIXE method is possible only if reaction products contain the element(s) which are not normal constituents of the investigated tissue or, at least, are present in trace amounts. The list of selected histochemical procedures which may be combined with the micro-PIXE method is presented in a paper by Rokita et al.[17]

2.4 Examples of Applications

Possibilities of nuclear analytical techniques in the trace element analysis of biological material are illustrated by presenting only some of many examples.

2.4.1 Body fluids

Concentrations of the essential trace elements in body fluids are rather low. For example, blood samples contain a large number of trace elements, however, only Cu, Fe, Zn and Sr are found in concentrations higher than 1 ppm in healthy adults. These elements can be easily determined by X-ray emission spectroscopy using samples of some 60 µl dried on an appropriate backing material. However, many elements of biochemical interest (some of them holding a promise to be used as a clinical test in pathological conditions) occur in much lower concentrations. The same is true of the other body fluids as well. An example of essential trace elements that occur at low concentrations in body fluids is selenium (human blood serum concentration 30-100 ppb). It has been shown[18] that with a suitable preconcentration procedure (acid digestion) both PIXE and T-XRF could be used in the analysis of Se in human blood serum. Much promise is given by the use of SR-XRF in total reflection geometry[19]. Because of the small source size and natural collimation of synchrotron radiation exiting the storage ring, X-rays can be directed with high precision at small glancing angles to the specimen and are particularly effective for use in the total reflection geometry. In addition, the high degree of polarization, high intensity, and continuous tunability of the beam energy are contributing factors for obtaining good analytical sensitivity. Such an arrangement is described by Spal et al.[20]. An MDL of 8 ppb was determined for selenium in human blood serum and in proposed NBS-SRM 1598 bovine serum. The results show that this method is sufficiently sensitive for analysis of Se in blood serum.

Let us mention that micro-PIXE allows the measurement of elemental profiles in individual blood cells. For example, Johansson and Lindh[(21)] have observed altered elemental profiles in erythrocytes and neutrophil granulocytes in patients with chronic lymphatic leukemia, chronic and acute myeloid leukemia. In another report, Johansson et al.[(22)] observed that the elemental profiles of erythrocytes and neutrophil granulocytes in breast cancer patients displayed significant alteration in some essential and nonessential elements.

Other body fluids have been also measured by nuclear analytical techniques. For example, Moeda et al.[(23)] have reported PIXE analysis of human spermatozoa free from contaminations of seminal plasma.

Trace element concentrations in urine have been widely used to monitor occupational exposure and intoxication with heavy metals, such as chromium, nickel, cadmium, and vanadium. The high content of salts in the urine causes disturbing background in PIXE analysis of trace metals. Pokarinen et al.[(24)] have used cation exchange resin, Chelex 100, to separate and preconcentrate trace metals in urine prior to PIXE analysis.

Extracellular water (ECW) is a component of body composition and an indicator of nutritional status. Oral bromide can be used for determination of ECW. In the work reported by Shao et al.[(25)] the variation of bromine concentrations in blood plasma were determined by PIXE, and ECW was calculated. The samples were mixed with graphite, in order to overcome charging problem.

In another work Nwosu et al.[(26)] have reported the use of assessment of ^{18}O enriched water as a marker of total body water. The method, which can be used to determine the total body water (TBW) in humans, has been tested on rabbits and mice. A measured amount of ^{18}O enriched water was administered and the ^{18}O concentration of the system was essayed using the $^{18}O(p,\alpha)^{15}N$ reaction. The rate of the distribution of ^{18}O injected into a rabbit was measured and the ^{18}O exchange between intracellular and extracellular water volume was determined. The targets for the nuclear reaction analysis were prepared by anodic oxidation of the tantalum at constant voltage directly from serum and plasma without the addition of a buffer.

2.4.2 Soft tissues

Tissue analysis is often performed both for essential and toxic trace elements. Many elements accumulate in different organs as a consequence of environmental exposure. Also, the accumulation of some metals in some organs in several diseases has been known for some time. Much attention has been paid to the accumulation of easily measurable trace metals in organs during the development of the disease. This has been suggested as a diagnostic tool. The specific aims of such investigations should be: to determine the clinical usefulness of multiple trace element concentration measurements for the definitive relationships to disease activity and efficiency of therapy; to explore further the preliminary observations that changes in trace element levels precede other symptoms or disease activity; to determine trace element concentrations in normal vs. diseased tissues.

X-ray emission spectroscopy has been used in a number of studies for the determination of elemental concentration levels in both biopsy and autopsy samples. Important data on metabolic processes have been obtained by using proton microprobe for elemental mapping of different tissues: brain[(27)], cancerous tissue[(22)], liver lobules[(28)], arteris[(29)] and others[(30)]. Sample preparation presents a serious problem in these studies. Some researchers have analyzed tissue sections while some others have looked into individual cells. Representativity of a sample is a crucial problem when individual cells are being analyzed.

For example, Pinheiro et al.[(31)] have used PIXE to determine K, Ca, Mn, Fe, Cu, Zn, Se and Rb in five different brain structures. Another interesting problem studied by proton microprobe (Vaux et al.[32], Larsson et al.[16]) is the analysis of senile plaques from Alzheimer patients.

Alzheimer's disease, a common form of senile dementia, has become a major problem in the Western world due to an increase in aging population. Neuropathologically, the disease is characterized by the presence of neurofibrillary tangles and senile plaques. Plaque sizes are about 100 μm and cores about one tenth of this. It has been shown that aluminium and silicon are present in plaque cores in the form of aluminosilicates (Candy et al.[33]). Results about the aluminium content of senile plaques from microprobe analysis have varied considerably.[33-37] There remain some technical problems of sample preparation and localizing of the interesting biological structures that have to be investigated before the nuclear microprobe technique can be applied in a large-scale investigation of Alzheimer's disease (Larsson et al.[16]).

2.4.3 Keratin

The analysis of keratin material (hair, nails) has attracted attention because of its possible use as a diagnostic tool. Both thin and thick targets are easily prepared for the analysis by X-ray emission spectroscopy. PIXE and μ-PIXE, as well as SR-XRF and T-XRF, can be used to measure longitudinal and radial concentration profiles. From such measurements information on element incorporation mechanisms into keratin structure may be obtained. Progress in this area has been recently summarized[38].

Hair is the unique biological material which, because of its growth, reflects the biomedical and environmental history of the subject. Because of convenience in handling and sampling and relatively high concentrations of metals, trace-element analysis of human hair has been applied widely for different purposes. Being highly individualistic, hair drew particular attention in forensic applications connected with crime detection in order to match or identify the common origin of a hair specimen.

Hair may offer an answer to the nutrient regime of the body since it accumulates trace elements over a period of time and is relatively inert metabolically. The concentration in human hair may be a reliable index of a total body concentration, at least for some metals. Efforts to relate the trace-element content of human hair to some diseases that are known to influence trace elements in some other parts of the body produce different results; such investigations are still being done.

Pronounced differences in the trace-metal content of hair related to geographical locations as a result of environmental effects have been established. Some investigators even suggested that trace-element analysis of human hair could be used for geochemical reconnaissance of large areas for possible ores. Trace-element properties of human hair may be used in archeological investigations to trace human population movements from one region to the other and to understand the eating habits, diseases, etc., of different civilizations, which may influence the contents of trace elements in human hair.

Poisoning by heavy metals is accompanied by their buildup in the growing hair. For example, the detection of Pb in hair provides an additional means of confirming the diagnosis of chronic plumbism in children. Poisoning by heavy metals has recently obtained much attention of the widespread use of poison in everyday life. Some of them, such as As, Cd, Pb, Te, and Hg, lead the list.

Hair analysis has unique advantages in the present context over analysis of other tissues or blood. This follows from the durability of hair, which, together with bone, is that part of the human body which survives for considerable periods after death. Some hair samples exist which are thousands of years old; a custom of preserving hair as a memento of the deceased has resulted in hair samples from the last century being fairly common. Thus, hair analysis offers the possibility of detecting poisoning at extended times after the internment of the victim.

Trace element incorporation routes into keratin structure through root section can also be studied by micro-XRF. So far only PIXE measurements have been reported, although there are radiation damage effects due to proton beam interaction with matter. It should be pointed out that trace elements incorporation routes into hair structure are only partially understood.

2.4.4 Hard tissues

An important part of both animal and human organisms is the skeleton, which is also a body store for some essential and toxic elements. Several authors (Ref. 39 and references therein) have discussed the concentration of trace metals in bones and their dependence on sex, age, and environmental factors. Trace element analysis of bone is a difficult analytical problem mainly because of inhomogeneous bone structure and inherent element variability within this structure. The extensive study of human skull bones[39-42] showed considerable differences in values among different persons as well as among different bones from a single individual. The use of the proton microprobe[25] clearly indicated the enrichment of metals on the surface of bones, resulting from the increased metabolic activity or exposure to the environmental contaminants.

Tadic et al.[43] have explored the diffusion of metals from crown into tooth interior by measuring concentration profiles for metal constituents of crowns on human teeth using PIXE spectrometry. For all elements (Au, Cu, Ag, Zn, Pd) the gradient in the distributions, towards the crown, indicates the presence of a diffusion process. Copper and gold are found to have a bigger mobility than silver or palladium. Fluoride distribution in enamel is often studied by the use of $^{19}F(p,\alpha\gamma)$ reaction; other elements (i.e. Ca and P) can be studied by the use of micro-PIXE (see report by Tros et al.[44]).

Trapper et al.[45] have demonstrated the use of external beam PIXE and PIGE methods for elemental analysis (Na, P, Ca, Mn, Fe, Cu,

Zn, Br, Sr and Pb) of excavated bones.

2.4.5 Plant material

The species of plants collected and studied often differ as widely as the kinds of soil on which they are grown. The necessity for sampling specific parts of plants to identify trace element deficiencies or toxicities affecting growth of agricultural crops is well recognized. Sample preparation of soil plants, and similar material for analysis by X-ray emission spectroscopy depends mainly on the element concentrations in the samples investigated. The uptake of the elements from the environment depends upon many factors including the elements' abundance in lithosphere, the form of the elements, the pH of the soil, the physical conditions of the soil, and the genetic constitution of the plant. The trace element composition of plants is a modified selection of the trace minerals present in the geochemical environment in which the plants are grown. There are many plants that accumulate elements to an anomalous degree; these are often used as biological monitors of pollution. Examples are mosses, lichens, grasses, tree needles, tree rings, etc.[(52)]

The X-ray emission spectroscopy is often used for tree ring analysis. PIXE with both internal and external beam is often used for the analysis of raw samples as well as samples treated by the preconcentration procedures. Tree ring sample preparation procedures for XRF measurements are discussed in Ref. 53. Carbonization of tree ring materials has been shown to be an adequate sample preparation technique for trace element analysis. However, when comparing the XRF and the PIXE results one has to keep in mind the effects of the charged particle beam on the sample matrix.

Sample preparation methods of tree-rings for PIXE-PIGE multi-elemental analysis, is discussed in details by Hall[(54)] (see also Pernestal et al.[(55)]).

Biological methods of prospecting for mineral include geobotany (visual examination of the vegetation cover), geozoology (use of animals for mineral prospecting) and biogeochemistry (chemical analysis of vegetation), and have been used for many years. Biological methods of prospecting have over the years become much more sophisticated and quantitative. One such effort is described by Mirzai et al.[(56)] where PIXE was used in leaf analysis serving as an explanatory tool in mineralogy.

Another interesting problem is protein content determination. Protein content of seeds and other plant material can be calculated from the determination of nitrogen concentration. A word of caution is needed when using total or crude nitrogen as an index of protein content. This follows from the fact that all nitrogen present in organic matter is not protein nitrogen. Nevertheless, total nitrogen per cent remains strongly correlated with protein content, but instead of using 6.25 as the correlation factor between them, one has to look for the correct factor according to the species in matter, e.g., this turns out to be 5.7 for wheat, 5.95 for rice, etc.

Nitrogen is easily measured by using a number of nuclear reactions with charged particles. In addition, MeV proton beams can be used to produce ^{11}C and ^{13}N isotopes in the plant material. The position decay allows determination of the N/C ratio (see work by Morales et al.[(57)]). This gives simultaneous measurement of proteins and organic material. In such a way protein distribution in the individual seeds can be determined (see for example work by Yukowa et al.[(58)]).

The two most often used charged particle nuclear reactions for nitrogen determination in organic matter: the (p,γ) reaction, first introduced by E. Ricci[(59)] and the (d,p), (d,α) reactions proposed and established by B. Sundqvist et al.[(60)]

In Ricci's work, the $^{15}N\ (p,\alpha\gamma)^{12}C$ reaction is used to determine nitrogen content, since the corresponding reaction for ^{14}N, namely, $^{14}N(p,\gamma)^{15}O$, has a rather small cross section at the energies of 0.7 and 1.0 MeV employed in the experiment; besides, the ^{15}N reaction is more adequate, since it has almost no interference problems in the prompt γ spectrum. Ricci was also able to obtain the ^{12}C and ^{13}C isotopic abundances at the same time, by the corresponding reactions: $^{12}C(p,\gamma)^{13}N$ and $^{13}C(p,\alpha)^{14}N$. All these reactions have positive Q values and suitable resonances in the energy interval from 0.5 to 1.7 MeV.

The $^{14}N/^{12}C$ ratio is deduced via the $^{15}N/^{12}C$ ratio, and since the $^{15}N(p,\alpha\gamma)^{12}C$ reaction has a very strong resonance at 898 keV, precise measurements of this ratio are attainable. The author used a pair of NaI(Tl) detectors to obtain the spectra, since the use of a Ge(Li) detector turned out to be unnecessary.

This technique has certain advantages: it is simple, fast, non-destructive (negligible radiation exposure) and in addition it requires a small accelerator facility.

Several other authors have used the $^{14}N(d,p)^{15}N$ and $^{14}N(d,\alpha)^{12}C$ reactions for nitrogen quantification. These two reactions have positive Q values bigger than the possible deuteron reactions on C and O isotopes, feature which simplifies the nitrogen analysis, since the corresponding yields for ^{14}N, for same sample thickness, appear isolated on the high energy part of the energy spectra.

2.4.6 Cell cultures and micoorganisms

X-ray emission spectroscopy can be successfully used for trace element analysis of cell cultures, microorganisms or any other similar material. The simultaneous determination of the concentrations of several elements can be used to provide an "elemental fingerprint" of the sample investigated. Several different methods of preparing samples from microorganisms are discussed in Ref. 5.

It is believed that gap junctions (multimetric protein cylinders) provide a mechanism for the movement of ions from one cell to another. Ionic permeability is regulated by variations in intra and extracellular Ca levels. Kurten et al.[46] have reported the preparation of intact and isolated gap junction specimens for transmission electron microscopy and X-ray spectroscopy by ultramicrotomy. Uniformly thick plastic sections displaying a blue interference color on water (approximately 200 nm thick) were cut with glass knives and collected on 300 mesh Cu support grids.

2.4.7 Microanalysis of biological material

The knowledge of elemental concentration distribution on μ-meter scale may yield information on metabolic processes involving essential or toxic elements. However, the microanalysis of biological material faces the following problems:

- lack of suitable sample preparation procedures,
- specimen stability during the analysis,
- lack of appropriate standards for quantification.

The preparation of sections of biological samples for X-ray microanalysis requires a structural and chemical preservation of tissue constituents. Localization of an element is possible only if the structure is preserved. Chemical information should be directly related to a well-defined morphological structure.

Another problem arises from the fact that in many cases biological specimens contain a high water content with diffusable elements. Specific osmotic and electrochemical potentials maintain ions in subcellular components.

In order to satisfy the requirement on correlation between

information on the elemental composition and the morphological structure, an image from the secondary electrons emitted during the μ-XRF measurement could be constructed.

Substances to be analyzed should not be altered in composition neither during the specimen preparation nor under charged particle or photon beams. It is considered that the rapid freezing of fresh tissues provides the minimum of ionic movement.

For example, targets for μ-XRF of cells having a diameter of 10 μm can be prepared by freeze-drying onto a formvar coated electron microscope grid.[47] Tissue samples from histological examinations removed by surgical cholecystectomy or needle biopsy should be immediately snapfrozen to the temperature of LN_2. Frozen blocks should be cut with cryomicrotome into slices of 4-8 μm thickness and mounted onto presolved Formvar backing film and freeze dried afterwards. By this procedure any migration of elements across the sample slice is prevented.[48]

Elemental concentration profiles for $Z>15$ elements can be measured by μ-PIXE on the scale of 1-2 mm with a proton beam focused to a spot size of 1-10 μm. The heat and radiation damage induced by such high current density beams on targets made from biological material are not yet well understood. The effects are minimized by rapid movements of either beam or target during the scanning. It has been shown[49] that the radiation damage from the XRF methods is several orders of magnitude less than for PIXE, for comparable trace element detection sensitivity. The study of radiation damage induced by synchrotron radiation is presented by Slatkin et al.[50] Damage to unfixed, air-dried human blood cells and tissues caused by synchrotron radiation was studied by scanning electron microscopy (SEM) and by light microscopy (LM). Morphological changes were found in unstained cell specimens after irradiation with photon fluences around 2×10^{19} photons/cm^2 at an average energy of 15.2 keV. At lower fluences (7×10^{18} photons/cm^2), morphological changes in irradiated cells are seen by LM after, but not before, staining by a standard hematological technique.

A series of typical applications of nuclear microprobe has been recently presented by Lindh.[51] Applications in cell biology, neurology, immunology, toxicology, trace element biology, and internal medicine are presented.

2.5 Material Modifications

The use of accelerators in material modification and subsequent analysis has been a fast growing area of accelerator based technologies.

Numerous are applications of ion implantation technology which have been transferred from the research laboratories to the industry. Two outstanding examples are the development of ion implantation for improving the wear resistance of artificial prostheses and the development of an ion beam treatment for improved bearing life. The utility of a surgical alloy for use in artificial joints is a function of its strength, biocompatibility, and corrosive wear resistance. Ion implantation technology developed at research laboratories dramatically increased the wear resistance of an otherwise highly suitable Ti alloy. Ion implantation is now being used on approximately 20% of the artificial hips and knees sold in the United States. The potential savings from the prevention of reworks of failed joints is $100 M per year.

A fairly recently development in the use of accelerator based technologies has been the combination of coating and ion implantation technologies, broadly known as ion assisted coating (IAC). Ion assisted coating promises to combine the best of both methods as a singularly effective means of engineering the surface properties of a very wide variety of materials. This technique permits the development of a new type of coating with advantages over both standard coating and implants. Ion assisted coating adhere strongly to the substrate due to an intermixed interface and are often denser than standard coating. Coating ranging from simple to complex in composition and structure can be grown on nearly any substrate material since the required energy is supplied by the ion beam, rather than an external heat source. These improvements are similar to those produced by plasma aided depositions, and arise from similar processes. However, ion beams are far more controllable than plasmas and produce a better defined and more reproducible process. Furthermore, because the energies are higher, intermixing is more extensive, resulting in better film adhesion. Because both metallic and gaseous ion species can be introduced by the ion beam, a much wider variety of modifications can be imposed. Ion assisted coating techniques even permit direct chemical modification of the coating and substrate during or after deposition, if required. Yet, because of the low ion doses which can often be used, the costs can be considerably less than those of implantation alone.

The future of ion beam technology appears to lie in two, related areas: direct ion implantation and ion assisted coating. For many applications, such as improving wear of bearings or prostheses or fracture strength of ceramics, ion implantation alone is the treatment of choice. For many other applications ion assisted coating processes have "built-in" advantages when compared with ion implantation. Perhaps the foremost of this is that coating technology is already generally familiar to all industries; in many applications type coatings are already part of manufacturers' specifications.

2.6 Positron Emission Tomography

Positron emission tomography is based on the fact that the activity distribution within a cross section of an object can be reconstructed from the measurements of its projections. Large number of short lived positron emitters can be produced either in reactors or by accelerators (usually cyclotron). However, for research and clinical studies in a hospital environment the most often needed are positron-emitting radionuclides of carbon, nitrogen, oxygen and fluorine.

For example, with compact cyclotron which delivers 16 MeV deuterons and 8 MeV deuterons carbon-11, nitrogen-13, oxygen-15 and fluorine-18 may be produced in useful quantities in the following chemical forms:

Radionuclide	Chemical Form
Carbon-11	$^{11}CO_2$ ^{11}Co $H^{11}CN$
Nitrogen-13	$^{13}N_2$ $^{13}N_2$(solution) $^{13}NH_3$
Oxygen-15	$^{15}O_2$ $C^{15}O$ $C^{15}O_2$ $H_2^{15}O$
Fluorine-18	$^{18}F_2$ $H^{18}F$

These precursors may be used directly, or in synthesis for the labelling of other compounds. Protons are used for producing carbon-11, and ^{13}N-labelled ammonia, while nitrogen-13 gas, oxygen-15 and fluorine-18 are produced from deuteron bombardment reactions. Simple and inexpensive target gases (N, Ne, and CO_2) are used with no need for isotopically enriched materials.

Precursors and reaction data for their production are listed in Table 2 and 3. Proton induced reactions are shown in Table 2, while deuteron induced reactions are indicated in Table 3.

Table 2

Precursors and reaction data for low energy protons

Parent precursor	Target	Nuclear reaction	Reaction threshold MeV	Energy at cross-section maximum (MeV)
$^{11}CO_2$	$N_2(+O_2)$	$^{14}N(p,\alpha)^{11}C$	3.1	4.3
$^{11}CH_4$	$N_2(+H_2)$	$^{14}N(p,\alpha)^{11}C$	3.1	4.3
^{13}NN	$^{13}CO_2$(ice)	$^{13}C(p,n)^{13}N$	3.2	
$^{13}NO_3$	H_2O	$^{16}O(p,\alpha)^{13}N$	5.5	17.5
$^{13}NH_3$	H_2O	$^{16}O(p,\alpha)^{13}N$	5.5	17.5
^{15}OO	^{15}N(liquid)	$^{15}N(p,n)^{15}O$	3.7	
^{18}F	^{18}O(liquid)	$^{18}O(p,n)^{18}F$	2.6	
$H^{18}F$	^{18}O(liquid)	$^{18}O(p,n)^{18}F$	2.6	
^{18}FF				

Table 3

Deuteron and ^{3}He induced reactions

Parent precursor	Target	Nuclear reaction	Reaction threshold MeV	Energy at cross-section maximum (MeV)
$^{11}CO_2$	B(solid)	$^{10}B(d,n)^{11}C$	0	5.4
$^{11}CH_4$	B(solid)	$^{10}B(d,n)^{11}C$	0	5.4
^{13}NN	CO_2	$^{12}C(d,n)^{13}N$	3	
$^{13}NH_3$	CH_4	$^{12}C(d,n)^{13}N$	3	
^{15}OO	N_2	$^{14}N(d,n)^{15}O$	0	3.8
^{18}F	O_2 or H_2O	$^{16}O(^{3}He,n)^{18}Ne\text{->}^{18}F$	3.8	7
$H^{18}F$	O_2 or H_2O	$^{16}O(^{3}He,n)^{18}Ne\text{->}^{18}F$	3.8	7
^{18}FF				

There is a large number of possible applications in diagnostic medicine and clinical studies.

2.7 Radioisotope Production

Let us remember that two-thirds of all radioactive nuclei were discovered via accelerator induced reactions. Perhaps in no other area has accelerator development had such a marked impact on mankind as in medicine - in nuclear medicine and in radiation therapy. Radioisotopes used for diagnostic purposes include ^{131}I, ^{57}Co, ^{51}Cr, ^{60}Co, ^{59}Fe, ^{68}Ga, ^{72}As, ^{82}Rb, ^{83m}Kr, ^{123}I, ^{172}Lu, ^{44}Sc, ^{42}K. This vast field cannot be reviewed within the scope of this article. Some of the useful radioisotopes can be produced only with special beams (like tritons or π).

2.8 Interaction of Charged Particles with Living Matter

Aspects of interactions of charged particles with living matter have been studied by many research groups. for example, Geard[(61)] has reported that chromosomal alterations are responsible for a significant proportion of cellular lethality and of genetic mutations, as well as being specifically associated with human cancers. For ionizing radiations the principal determinant of aberration likelihood is the linear energy transfer (LET). For radiation protection considerations one energy deposition event per cell nucleus is the ultimate low dose and such an approach bypasses traditional concepts of absorbed dose and relative biological effectiveness. Well characterized charged particles (protons, deuterons, ^{3}He, ^{4}He) generated by a Van de Graaff accelerator have been used to irradiate mammalian cells at LETs varying from 10 to 150 keV/µm. Measurements of the nuclear cross-sectional areas of mammalian cells at different stages of the cell cycle and known fluences of charged particles (per absorbed dose) at each LET enable relationships to be determined between charged particle incidence and both frequencies and types of chromosomal changes. Nuclear cross-sectional areas of cells tend to be log-normally distributed with overlapping distributions for the different stages of the cell cycle. There is a clear and profound effect of charged particle LET on cellular progression through the cell cycle and the induction of chromosomal change, two confounding endpoints. While there are variations in sensitivity to aberration induction (and cell cycle progression) for the different stages of the cell cycle, as LET per particle increases linearly, aberration frequency increases quadratically to a LET of about 100 keV/μm. Analysis of chromosomal changes as a function of LET has implications for the attribution of uniformity to alpha particles of different effective LETs following radon daughter breakdown.

The radiobiological effects of heavy charged-particles are of fundamental interest for the understanding of radioprotection problems in manned space flight and for the increasing use of heavy ions in radiotherapy. Therefore, an increasing number of radiobiological experiments are performed at heavy-ion accelerators. In these experiments biological targets as for instance cultures of living mammalian cells have to be irradiated under sterile conditions and atmospheric pressure. The problems studied in this field of activities include:

- neoplastic cell transformation induced by heavy ions:[62,63]
- biological effectiveness of low energy particles;[64]
- mutation induction by charged particles;[65]
- dose-effect relationship for small doses and many others.

2.9 Radiotherapy With Charged Ions

Heavy ion beams exhibit a favourable depth dose distribution for tumor therapy compared to conventional radiation like gamma- or X-rays. For ion beams less heavy than carbon, a low dose is deposited in the entrance channel which can be compared in the relative biological effectiveness (RBE) to sparsely ionizing radiation. Towards the end of the particle track, the energy deposition increases and a high dose is deposited by each single particle. Therefore, at the end of the range of the ion beam a high total dose combined with a high biological effectiveness is available for tumor therapy. The biological reaction mechanism in the regime of elevated efficiency (RBE>1) is not fully understood and a field of intense research activity at various heavy ion accelerators.

Several facilities have been used for proton radiation therapy. Since the first clinical studies began at the Lawrence Berkeley Laboratory in 1954, more than 5,000 patients have been treated with protons, using accelerators designed for physics laboratories. Superior results are reported for the control of selected diseases by the ten facilities which are currently investigating proton radiation therapy. The findings have resulted in expansion plans in several of these facilities, and in the formulation of plants for two new facilities.

Plans for improved radiation facilities have been described by Slater et al.[(66)], Alonso[(67)], Martin[(68)], Blosser et al.[(69)] and others. A very interesting concept of Light Ion Biomedical Research Accelerator (LIBRA) is discussed by Gough[(70)]. This is a light-ion, charged particle facility in a hospital environment dedicated to applications in biology and medicine. A similar proposal has been brought out by Bewley[(71)] described as European Light Ion Medical Accelerator (EULIMA). This European light-ion medical accelerator is a project to accelerate ions in the range hydrogen to neon to energies adequate for a penetration of 20 cm in tissue. The machine would have two sections, an initial cyclotron similar to the one recently installed at Clatterbridge for neutron therapy, followed by a second cyclotron to increase the energy to the required level.

Facilities for heavy ion treatment will soon be available at several locations, including GSI, Darmstadt, Germany (Gademann et al.[(72)]) and some others. This gives also the possibility of neurosurgical applications (Fabrikant et al.[(73)]).

The Bragg peak therapy of cancer patients with protons and heavier charged particles requires a very careful treatment planning involving the precise localization of the peak on a tumor volume. The present day techniques depend upon CT measurements which may not be as reliable as desired, especially when there is a large amount of thick bone or air path. Chatterjee et al.[(74)] have described a technique which utilizes high energy radioactive beams and a sensitive positron camera and provides a direct way of measuring the water equivalent path-length as traversed by a treatment-beam. The most commonly used radioactive beam has been ^{19}Ne so far; however, the technique is not limited to this particular isotope.

3. CONCLUSIONS

Charged particle beams are assuming an ever-more-prominent role in our everyday life. Today it is difficult to envision a research center in biomedical sciences or a contemporary hospital without the access to a particle accelerator.

With future development in technology and our better understanding of charged particle interaction with living matter the role of accelerators in life sciences will increase even more.

REFERENCES

1. J.L. Duggan and I.L. Morgan, Eda., Application of Accelerators in Research and Industry '86, Nuclear Instruments and Methods in Physics Research, B24/25 (1987).

2. H. van Rinsvelt, S. Bauman, J.W. Nelson and J.W. Winchester, eds., Particle Induced X-Ray Emission and its Analytical Applications, Nuclear Instruments and Methods in Physics Research, B22, No. 1-3 (1987).

3. Duggan J.L. and Morgan I.L., Eds., Application of Accelerators in Research and Industry '88, Nuclear Instruments and Methods in Physics Research B40/41 (1989).

4. Mallory, M.L., Proc. 11th Int. Conf. on Cyclotrons and their Applications, Ionics, Tokyo, (1987) 558.

5. V. Valkovic, Analysis of Biological Material for Trace Elements Using X-Ray Spectroscopy, CRC Press Inc., Boca Raton, Florida, 1980.

6. J. Hoste and C. Vandecasteele, Nucl. Instr. and Meth. in Phys. Res. B40/41 (1989) 1182-1188.

7. V. Valkovic, Analysis of Biological Material for Trace Elements using X-ray spectroscopy CRC Press, Boca Raton, Florida, 1980.

8. J.A. Cookson, Nucl. Instr. and Meth. B30 (1988) 324.

9. F. Watt, G.W. Grime and C.C. Perry, Nucl. Instr. and Meth. B30 (1988) 331.

10. K. Themner, Per Spanne, K.W. Jones, Nucl. Instr. Meth. In Phys. Res. B49 (1990) 52-59.

11. Y. Llabador, D. Bertault, J. C. Gouillard and Ph. Moreto, Nucl. Instr. Meth. in Phys. Res. B49 (1990) 435-440.

12. I. Orlic, F. van Langevelde and R.D. Vis, Nucl. Instr. Meth. in Phys. Res. B49 (1990) 74-77.

13. N.E.G. Lövestam and E. Swietlicki, Nucl. Instr. Meth. Phys. Res. B43 (1989) 104-111.

14. G.J.F. Legge and A. Saint, Nucl. Instr. Meth. in Phys. Res. B49 (1990) 418-424.

15. G.V. Iyengar and K. Kasperek, J. Radioanal. Chem. 39 (1977) 301.

16. N.P.O. Larsson, U.A.S. Tapper, K. Sturesson, R. Odselius, A. Brun, Nucl. Instr. Meth. in Phys. Res. B49 (1990) 472-475.

17. E. Rokita, T. Cichocki, S. Divoux, B. Gonsior, M. Höfert, L. Jarczyk, A. Strzalkowski and M. Sych, Nucl. Instr. Meth. in Phys. Res. B49 (1990) 441-445.

18. G.P. Diakun, L. Fairall, A. Klug, Nature 324 (1986) 698.

19. J.C. Philipps, K.J. Baldwin, W.F. Lehnert, A.D. Le Grand, C.T. Prewitt, Nuclear Instruments and Methods in Physics Research A246 (1986) 182.

20. R. Spal, R.C. Dobbyn, H.E. Burdette, G.G. Long, W.J. Boettinger and M. Kuriyama, Nucl. Instr. Meth. 221, 463 (1984).

21. E. Johansson and W. Lindh, Nucl. Instr. and Meth., B3 (1984) 637.

22. E. Johansson, U. Lindh, H. Johansson, C. Sundström, Nucl. Instr. and Meth. in Phys. Res., B22 (1987) 179.

23. K. Maeda, Y. Sasa, H. Kusuyama, K. Yoshida, M. Uda, Nucl. Instr. Meth. in Phys. Res. B49 (1990) 228-230.

24. P. Pakarinen, A.-K. Ekholm and J. Pollón, Nucl. Instr. Meth. in Phys. Res. B49 (1990) 241-244.

25. H.R. Shao, Q.X. Liu, G. Enzi, G. Moschini, G. Sergi, Nucl. Instr. Meth. in Phys. Res. B49 (1990) 238-240.

26. S.N. Nwosu and H.J. Fischbeck, Nucl. Instr. Meth. in Phys. Res. B40/41 (1989) 833-837.

27. F. Watt and G.W. Grime, Nucl. Instr. and Meth. in Phys. Res. B30 (1988) 356.

28. D. Heck, A. Ochs, A. Klempnow, K.P. Maier, C. Kratt, Nucl. Instr. and Meth. in Phys. Res., B22 (1987) 196.

29. T. Cichocki, D. Heck, L. Jarczyk, E. Rokita, A. Strzalkovski, M.°Sych, Nucl. Instr. and Meth. in Phys. Res., B22 (1987) 210.

30. U. Lindh, Medical Applications of the Nuclear Microprobe, Nucl. Instr. and Meth. in Phys. Res., B30 (1988) 404.

31. T. Pinheiro, J.M. Halpern, R. Fernandes, S. Ramos, S. Silva, H. Duflon, J. Cafmeyer, W. Maenhaut, Nucl. Instr. Meth. In Phys. Res. B49 (1990) 191-194.

32. D.J.T. Vaux, G.W. Grime and F. Watt, Biol. Trace Elem. Res. 13 (1987) 115.

33. J.M. Candy, J. Klinowski, R.H. Perry, E.K. Perry, A. Fairbairn, A.E. Oakely, T.A. Carpenter, J.R. Atack, G. Blessed and J.A. Edwardson, Lancet 15 (1986) 354.

34. S. Duckett and P. Galle, J. Neuropath. Exp. Neurol. 39 (1980) 350.

35. T. Nikaido, J. Austin, L. Trueb and R. Rinehart, Arch. Neurol. 27 (1972) 549.

36. F. Watt, G.W. Grime, G.M. Gadd, J.M. Candy, A.E. Oakley and J.A. Edwardson, Accelerator based analytical techniques: Elemental mapping of medical and biological samples using the Oxford Scanning Proton Microprobe, Proc. 11th Int. Congr. on X-ray Optics and Micoranalysis, University of Western Ontario, 1986, eds. J.D. Brown and R.H. Packwood.

37. F. Watt, G.W. Grime, D.N. Jamieson and B. McDonald, Internal report, ref. OUNP-89-17.

38. V. Valkovic, Human Hair, Vol. I and II, CRC Press, Boca Raton, Florida 1988.

39. M. Jaksic, S. Fazinic, J. Krmpotic-Nemanic, M. Budnar, Z. Smit and V. Valkovic, Nucl. Instr. and Meth. in Phys. Res., B22 (1987) 193.

40. R.M. Wheller, R.P. Chaturvedi, J.S. Onello, V. Valkovic and J. Krmpotic, Trace Elements in Human Nasal Cavity Bones, Nucl. Instr. and Meth. in Phys. Res., B24/25 (1987) 658.

41. V. Valkovic, M. Jaksic, J. Krmpotic-Nemnic, Biological Trace Element Research, 12 (1987) 375.

42. J. Krmpotic-Nemanic, V. Valkovic, G. Nemanic and M. Jaksic, Verh. Anat. Ges. 79 (1985) 211.

43. T. Tadic, M. Jaksic, J. Babic-Zivko, J. Cicek, S. Fazinic and V. Valkovic, Nucl. Instr. and Meth. in Phys. Res. B49 (1990) 211-215.

44. G.H.J. Tros, D.M. Lyaruu and R.D. Vis, Nucl. Instr. and Meth. in Phys. Res. B49 (1990) 476-480.

45. U. Tapper, H. Vuorinen and H. Mussalo-Rauhamaa, Nucl. Instr. Meth. in Phys. Res. B49 (1990) 245-249.

46. R.C. Kurten et al., J. Trace and Microprobe Techniques, 3 (1985) 303.

47. G.W. Grime et al., Nucl. Instr. and Meth. in Phys. Res., B22 (1987) 109.

48. D. Heck et al., Nucl. Instr. and Meth. in Phys. Res., B22 (1987) 196.

49. C.J. Sparks, Jr., Synchrotron Radiation Research, eds., H. Winick and S. Doniach (Plenum Press, New York, 1980), p. 459.

50. D.N. Slatkin, A.L. Hanson, K.W. Jones, H.W. Kraner, J.B. Worren and G.C.°Finkel, Nucl. Instr. and Meth. in Phys. Res., 227 (1984) 378.

51. U. Lindh, Nucl. Instr. Meth. in Phys. Res. B49 (1990) 451-469.

52. V. Valkovic, X-Ray Emission Spectroscopy in Environmental Sciences, CRC Press, Boca Raton, Florida, 1989.

53. M. Nagj, J. Injuk, V. Valkovic, Nucl. Instr. and Meth. in Phys. Res., B22 (1987) 465.

54. G. Hall, Nucl. Instr. Meth. in Phys. Res. B49 (1990) 60-67.

55. K. Pernestal, H.-K. Li, B. Jonsson, Nucl. Instr. Meth. in Phys. Res. B49 (1990) 261-263.

56. A.A. Marzai, J.S.C. McKec, Y.H. Yeo, D. Gallop and J. Medved, Nucl. Instr. Meth. in Phys. Res. B49 (1990) B13-317.

57. J.R. Morales, M.I. Dinator, P. Cerda, Nucl. Instr. Meth. in Phys. Res. B40/41 (1989) 1196-1198.

58. M. Yukawa, M. Sumiya, T. Furuse, Contribution to Seventh Int. Symp. on Trace Elements in Man and Animals, Dubrovnik, May 20-25, 1990.

59. E. Ricci, Nucl. Instr. Meth. 94 (1979) 565.

60. B. Sundqvist et al., Int. J. Appl. Radiat. Isotopes 25 (1974) 277.

61. C.R. Geard, Nucl. Instr. Meth. in Phys. Res. B40/41 (1989) 1372-1375.

62. M. Suzuki, M. Watanabe, K. Suzuki, K. Nakoano, I. Kaneko, Radiat. Res. 120 (1989) 468-476.

63. T.K. Hei, K. Komatsu, E.J. Hall, M. Zaider, Carcinogenesis 9 (1988) 747-750.

64. A. Perris, P. Pialoglou, A.A. Katsanos, E.G. Sideris, Int. J. Radiation Biol. Relat. Stud. Phys. Chem. Med. 50 (1986) 1093.

65. T.K. Hei, D.J. Brenner, E.J. Hall, Carcinogenesis 9 (1988) 1233-1236.

66. J.M. Slater, D.W. Miller, J.O. Archambean, Int. J. Radiat. Oncol. 14 (1988) 761-775.

67. J.R.Alonso, Nucl. Instr. Meth. Phys. Res. B40/41 (1989) 1340-1344.

68. R.L. Martin, CONF-881151-41.

69. H.G. Blosser, R.J. Burleigh, G.F. Blosser, E.B. Jemison, US Patent 4,641,104/A/.

70. R.A. Gough, LBL-23413, CONF-861271-1.

71. D.K. Bowley, Lancet, 7 Feb. 1987, No. 8528, 318-319.

72. G. Gademann, M. Wannenmacher, G.H. Hartmann, W.J. Lorenz, G. Kraft, Strahlenther. Onkol. 166 (1990) 34-39.

73. J.I. Fabrikant, R.P. Levy, M.H. Phillips, K.A. Frankel, J.T. Lyman, Nucl. Instr. Meth. Phys. Res. B40/41 (1989) 1376-1384.

74. A. Chatterjee, J. Llacer, M. Collier, T. Renner, S. Henderson, W. Chu, Report LBL-22962, 1987.

NUCLEAR MICROSCOPY IN MEDICAL DIAGNOSIS AND THE IMPACT ON TRACE ELEMENT BIOLOGY

Ulf Lindh

Department of Radiation Sciences, Division of Physical Biology, Uppsala University, S-751 21 Uppsala, Sweden

ABSTRACT

Nuclear microscopy is defined and its use in medical research is discussed. Results from the application of nuclear microscopy in medicine and trace element biology by various groups are presented. Mostly two types of samples are investigated by nuclear microscopy in the biomedical research, viz. individual cells and tissue sections. Some suggestions and guesses about the future of nuclear microscopy are presented.

INTRODUCTION

Microscopy has been used for very long in biology and medicine and is the necessary basis for the advancement of histological and histopathological knowledge. Today, light microscopes and electron microscopes are indispensable instruments in medical diagnostics and biomedicine. In trace element biology, however, microscopy has not yet found wide acceptance.

Electron microscopes, or microprobes, were used already several years ago for elemental determination of biological samples. By using either a static or a scanning beam, the elemental contents of a spot or an area could be determined. This technique is named *X-ray microanalysis* because the characteristic X-rays emitted from an irradiated sample was used for the determination of elements. One serious drawback is that only major elements ($\geq$0.01% concentration dry weight) can be determined quantitatively. The advantage of X-ray microanalysis carried out in a scanning transmission electron microscope is the possibility of producing simultaneously an electron micrograph and elemental maps of the sample.

NUCLEAR MICROSCOPY

The use of a beam of nuclear projectiles such as protons to investigate the elemental structure of biological objects attracts increasing interest. There are unique advantages to be gained by using a nuclear probe, which justifies the complex technology required. For this use of a nuclear microprobe the name *nuclear microscopy* is proposed.

Several powerful analytical techniques utilising ions, mostly protons and alphas of 1-5 MeV, are gaining popularity in laboratories around the world. The best known and also most used of these accelerator-based techniques are particle-induced X-ray emission (PIXE), nuclear reaction analysis (NRA) and Rutherford Backscattering (RBS).

The same general idea as in electron microscopy has been adopted for its nuclear correspondence, i.e. a finely focussed beam of charged particles is scanned over the sample area of interest. In nuclear microscopy, the main technique is PIXE, being the direct correspondence of X-ray microanalysis. In addition, however, there are several other uses of interactions between the primary beam of nuclear projectiles and the atoms of the sample to be exploited.

Nuclear microscopy is the youngest of the techniques discussed above and will irrevocably be compared with, in the first place, electron microscopy. The analytical technique dictates the range of elements that can be determined and at what sensitivity. The spatial resolution is the second parameter of main interest. The spatial resolution quoted for X-ray microanalysis is of the order of 0.1 μm or slightly better with extremely thin sections. The use of these sections cuts off most of the pear-shaped excitation volume under the sample surface. Nuclear microscopy can be exercised at 0.3-0.5 μm. The detection limit of nuclear microscopy is of the order of 0.0001% or better and the range of elements that can be determined is hydrogen through uranium. Using window-less detectors, the elements lithium through uranium can be detected by X-ray microanalysis. These detectors are, however, more complex to handle and more expensive than those equipped with a beryllium window. The thinnest windows are 5-7 μm and effectively absorb X-rays from light elements.

The Uppsala Nuclear Microscope.

The predecessor of the new Slim-Up (Scanning light ion microprobe in Uppsala) was the Studsvik Núclear Microprobe (SNMP). The latter was developed and established as a cooperation between the division of physical biology, Uppsala University and Atomenergi AB, Studsvik. It became operational in 1975, thereby being the fourth nuclear microprobe in the world. It has since been continuously improved and its final spatial resolution was 2.9x2.9 μm^2 at 100 pA of proton current. The nuclear microscopy community has agreed upon using a particle intensity corresponding to 100 pA as standard for comparisons of spatial resolution.

The SNMP was tranferred from Studsvik to the tandem accelerator of the The Svedberg Laboratory in Uppsala a couple of years ago and was eventually dismantled during the spring of 1989. The SNMP was used mostly in trace element biology. The applications of SNMP, however, also cover occupational medicine, internal medicine, oncology, toxicology and environmental sciences. It has now been replaced with a completely new construction promising spatial resolutions better than 0.5x0.5 μm^2 at 150 pA.

Below is presented a rather superficial description of the Slim-Up. For a more detailed technical account, the reader is referred to Sunde et al. [1].

The Slim-Up is dedicated primarily for the nuclear microscopy of environmental, biological and medical specimens. Recent developments in lens design (Oxford Microbeams 1989) enables submicron elemental mapping to be carried out routinely. The major innovation in the design of the new lenses is that the magnetic circuit is a single piece of stress-relieved magnet iron profiled to micron accuracy. The novel design has eliminated higher order parasitic aberrations.

The beam of light ions, usually protons, is delivered by an EN-tandem accelerator fed by an off-axis direct extract duoplasmatron ion source. Two feedback controlled magnetic deflectors align the beam with the optical axis of the probe-forming lenses. Upstream, and before the lenses, a magnetic beam scanning unit of current controlled ferrite cores is installed. The scan area at the sample position comprises a two-dimensional, 2x2 mm^2, matrix of 65,536x65,536 pixels. Arbitrarily shaped areas of the scan matrix can be irradiated repeatedly for multiples of 10^{-14} C or 1.0 μs.

The target chamber is an octogonal construction containing the following parts: (i) a sample holder movable in two directions perpendicular to the beam, (ii) a microscope, for viewing from the back of the target, with a highest magnification of 600x, with the turret inside the chamber, (iii) in one of the turret positions a Faraday cup for monitoring of beam intensity, (iv) a 150x stereo microscope yielding an oblique view of the sample front side, (v) a retractable 30 mm^2 Si(Li)-detector (147 eV resolution), covered by a 7 μm beryllium window, mounted at 135° to the beam direction with a maximum solid angle of 980 msr and (vi) a channel electron multiplier plate to yield a secondary electron image of the object.

In the near future the target chamber will be equipped with particle detectors to detect forward and backscattered particles. A Ge(Li)-detector will be introduced to make possible NRA. Equipment for image analysis will also be integrated with the nuclear microscope. The objective of this is to facilitate recognition of the region of interest.

The beam control and data acquisition are performed by Intel-286-based personal computers equipped with dedicated expansion cards. Data are read simultaneously from 8 channels with 16-bit resolution into a buffer and further to a mass storage memory device. One of the channels is devoted to multiplexed signals from ≤8 detectors with ≤8192 channels resolution. The other seven channels can be used for the necessary sampling of accumulated charge, time, coordinates etc.

MEDICINE AND TRACE ELEMENT BIOLOGY

Modern medicine has developed to be very dependent on laboratory tests and the interpretation of the results. Although electron microscopy has been used for decades, it is used in medical diagnostics more or less exclusively in pathological investigations of suspected neoplasias. Interest in trace elements has grown significantly during the last decades. The determination of trace elements in humans has, however, mostly been restricted to whole blood and plasma or serum and as bulk concentrations.

Table I: Essential and inessential elements

Essential:	Ca, Cl, Co, Cu, Fe, I, K, Mg, Mn, Mo, Na, Ni, Se, Zn
Probably essential:	As, Cr, F, Si, Sn, V
Inessential:	All other elements?

A trace element is defined as an element being present in a biological sample at concentrations below 0.01% wet weight. Trace elements are classified as essential or inessential according to whether an organism can grow and complete its life cycle in their absence [2]. It is, however, technically impossible to remove every atom of an element from the diet and environment of an organism and extremely difficult to produce deficiency symptoms in humans or even small mammals in most cases. Only in one case it has been proven that selenium deficiency was the cause of disease, viz. Keshan disease which is a cardiomyopathy [3].

There is always an optimum range of concentration in the diet, below which deficiency symptoms occur, and above which symptoms of toxicity begin. Our current knowledge of elements essential to mammals is imperfect and somewhat controversial, as shown by Table I [2].

Researchers in medicine and trace element biology have mostly concentrated on essential trace elements, especially copper, selenium and zinc in recent years. Detailed knowledge is abundant for the concentration interval where toxicity occurs. This is explained by the fact that higher concentrations are easier to determine quantitatively.

There has also been much interest in those inessential elements which have been increasing in the environment as a result of mans activities. The elements attracting most of the interest have been cadmium, mercury and lead, but silver, arsenic, chromium, manganese, antimony, tin and vanadium also fall in this category.

The list of trace elements being under investigation will inevitably grow substantially parallelling the introduction of analytical methods and instruments capable of multielement detection at very low concentrations.

Other methods for trace element analysis than nuclear microscopy will not be covered here because it would be beyond the scope of this article. A lot of experience has been gained by using bulk analytical multielement methods such as neutron activation analysis.

Trace Element Biology

Looking more thoroughly at trace element biology, more precise definitions will be necessary. Trace elements, or micronutrients, are intimately involved in physiological function and dysfunction at all levels of organisation. The classification of trace elements as essential or inessential is not fully accepted by all workers. Depending on the action, trace elements have been named either *essential* or *toxic*. The feature of being essential, of course, reflects the state of knowledge at a certain point of time. On the other hand, is does not seem logical to establish a group of toxic trace elements because toxicity is inherent in all elements and depends mainly on the concentration. The essentiality has been defined some years ago [4].

Trace elements are often found as important cofactors in enzymes, the most prominent example being zinc. The importance of zinc as an essential trace element is evident from the fact that more than 200 enzymes are zinc dependent [5]. Trace elements are necessary for maintenance and regulation of compartmentation of cell function, gene regulation and regulation of membrane functions, although some of the elements, by definition, cannot be regarded as trace elements. The development of life has brought about elaborate, diverse and subtle control mechanisms to provide for trace element homeostasis. The trace elements thus exert profound influence on human health and disease states.

The homeostasis and interaction of trace elements with each other and with nonessential trace elements such as heavy metals in biological systems is very complex. In spite of the research efforts spended, detailed knowledge of trace element functions is rather scarce. A lot of information has been gained through the use of radioactive nuclides in cell cultures and in experimental animal models. With the advent of Positron Emission Tomography it has been possible to perform dynamic studies with short-lived radionuclides. The gross concentration of major and trace elements in human tissues is well characterised by various analytical methods [6]. The literature, however, lacks a good compilation of major and trace element concentrations in animal tissues. Through the use of X-ray microanalysis in the scanning electron microscope and the nuclear microprobe, data are being gathered of trace element distributions in different tissues and cells. Trace element importance for the normal biological function and disease processes has hitherto only scarcely been studied.

SINGLE CELLS IN MEDICINE AND TRACE ELEMENT BIOLOGY

Single cells or individual cells are to be understood as the nuclear microscopy of cells one by one. Most users, or developers, of nuclear microprobes have been, casually or regularly, in this field. The reason might not be medical or biological but calibration needs; red blood cells from healthy individuals are very homogeneous in size (diameter ca. 8 μm) and they can consequently be used to calibrate the beam spot size.

Blood cells are readily available through ordinary venepuncture and subsequent centrifugation. In an attempt to estimate the medical value of nuclear microscopy of individual cells, the Uppsala group carried out a series of investigations on isolated individual blood cells from healthy control individuals [7] and patients suffering from various diseases such as leuchaemias [8], inflammatory connective-tissue diseases [9], Down's syndrome [10], Cystic fibrosis [11] etc. Examples from some of these applications and those undertaken by other groups are presented below.

Inflammatory Connective-Tissue Diseases

The alterations of trace element metabolism in association with chronic inflammatory diseases has been a field under study for many decades [12]. During recent years, clinical trace element research has come into focus partly due to the development of antirheumatic therapies which involve the manipulation of certain elements, but also due to the discovery of the central role certain elements have in free radical generation and lysosomal degranulation. The Uppsala group reported recently conspicuous accumulation of calcium and magnesium in granulocytes and erythrocytes from patients with rheumatoid arthritis and seronegative arthritides [13]. These results induced the speculation about consequences of cellular accumulation in the inflammatory process. However, later it became apparent that strontium [14] as well as manganese [15] also appeared in increased amounts in granulocytes from patients with rheumatoid arthritis while zinc appeared in reduced amounts [16].

Ankylosing spondylitis. Because of the genetic background, the inflammatory disease ankylosing spondylitis, or Bechterew's disease, was studied thoroughly. Only individuals having the tissue antigen HLA-B27 catch the disease. Four groups were established, viz. healthy age- and sex-matched controls, patients, HLA-B27$^+$ healthy relatives and healthy HLA-B27$^-$ relatives. The HLA are histocompatibility antigens, the antigens on living cells that distinguish self from nonself. They are glucoproteins on the surface of all nucleated cells and are called *human leukocyte antigens* (HLA) because they were first identified on circulating human lymphocytes. The genes that code for histocompatibility antigens are located in the major histocompatibility complex on the short arm of chromosome 6. The genes are found in three different loci: A, B and C.

Three blood cell types were isolated and studied by nuclear microscopy. The elemental profiles comprised magnesium, calcium, manganese, iron, copper, zinc and strontium. Compared with the healthy controls the investigated first degree B27$^+$ relatives showed significant accumulation of cellular magnesium, calcium, manganese and iron. Compared with the patients, relatives and controls showed significantly lower values of magnesium, calcium, manganese and iron [17].

The multielemental approach to individual cell analysis results in a hugh data set and it does not seem to be adequate to compare the results element by element. Multivariate statistics is a necessary prerequisite for the interpretation of interactions between the major and trace elements. Principal component analysis [18] was used to show that there was a similar lack of correlation between manganese and zinc in the control and HLA-B27$^-$ relatives. Zinc and manganese seemed to be similarly correlated in the patient group and the group with HLA-B27$^+$ relatives. In the same way calcium and iron were similarly correlated in the patient group and in the group with HLA-B27$^+$ relatives. Calcium and zinc correlate quite well in the control and HLA-B27$^-$ relative groups.

The multivariate approach certainly adds hints as to the direction of further studies.

The Chronic Fatigue Syndrome (CFS)

The main advantage that can be extracted from our experience with individual cells is that skew distributions of trace elements can be unveiled. Such skewness may be undetected using bulk analytical approaches. The cost and time spent analysing individual blood cells can thus in some cases be justified. The example presented below is such a case.

Anything that affects our environment is eventually going to affect man. Our primary concern here is heavy metal exposure. Unfortunately, there are other sources of heavy metal exposure than environmental, in the ordinary sense, and occupational, *e.g.* dental amalgams.

The debate about the adverse effects of dental amalgams has been particularly animated. The benefits of using silver amalgams have been questionned since it was first introduced in modern times. Some claim that we are now fighting the third amalgam war.

Dental amalgams are referred to as being alloys of silver, tin, copper and mercury. The liquid mercury (ca. 50%) which is mixed with the alloy powder to make the amalgam is not stable in the prepared filling. It was discovered in Sweden in the early 80s that a lot of people suffered from a diffuse but handicapping syndrome, the cause of which was veiled in obscurity. A series of tentative investigations led a Swedish group to summarise the experiences on the *metal syndrome* [19]. This report was later succeded by another one suggesting that the disease or diffuse syndrome was caused by the release of heavy metals from dental amalgams [20].

Lindh and Tveit [21] and Brune et al. [22] performed investigations in the late 70s and the early 80s, the results of which pointed to the fact that the silver amalgams were not stable filling materials and that they constituted a serious hazard through the leakage of heavy metals, especially mercury, to the human body.

The routes of exposure comprise inhalation of mercury vapour, swallowing of fractured amalgam particles and ionic transport.

The obvious investigation to perform on patients suffering from the diffuse syndrome (metal syndrome) was analysis of blood plasma. Although the patient group displayed significantly higher plasma concentrations of mercury than the control group, it was not possible to identify individuals in the patient group only by their plasma concentrations of mercury [23]. Analysis of individual blood cells revealed a lot better discriminative power than plasma concentrations. More than 80% of the patients had measurable concentrations of mercury in one or a few blood cells whereas the controls did not display one single cell with measurable mercury concentration [24].

To demonstrate the potential of high-resolution nuclear microscopy, the beam size was decreased from 3x3 μm^2 to 1.5x1.5 μm^2. This meant that the proton intensity was reduced to 30 pA, a proton current that can not be used for routine analytical purposes. The scan was collapsed to 10.5x10.5 μm^2 centering one red cell and the data collected were used to produce the surface plots in Figure 1. The magnesium and zinc plots emanate from one red cell belonging to a healthy control individual and the strontium and mercury plots from one red cell of a metal syndrome patient. In all plots, the biconcave shape of the erythrocyte is evident. To extract such

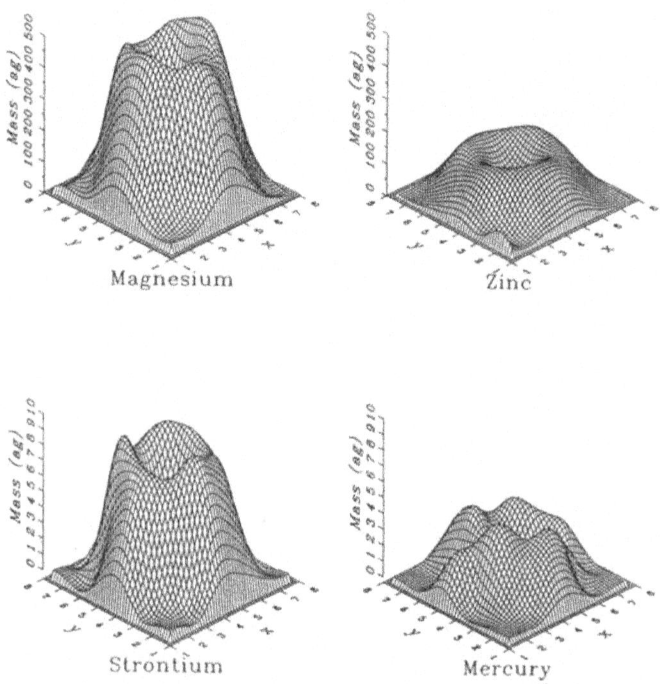

Figure 1: Surface plots of Mg and Zn from one single red blood cell from a healthy control individual and Sr and Hg from one red cell of a metal syndrome patient.

details, the probe has to be smaller than 2 μm. These detailed investigations are, however, too time-consuming and tedious to be used routinely.

The similarity of the symptoms of the patients suffering from the metal syndrome and those of patients with chronic mononucleosis, which is an Epstein-Barr virus associated disease [25] led us to perform a study with nuclear microscopy of individual blood cells from the latter group. The Epstein-Barr virus (EBV) is very well characterised. It is a lymphotropic virus causing infectious mononucleosis. Chronic illness has been associated with serological evidence of persistent active EBV-infection. The terms Chronic Mononucleosis and Chronic Fatigue Syndrome (CFS) have been used.

The results of invidual blood cell studies on the CFS patients revealed a striking similarity with those of the metal syndrome and it seems that the common denominator is the affection of the immune system by immunomodulating heavy metals. In fact, there are reasons to beleive that the two syndromes are the same.

The remarkably skew distribution of mercury among blood cells would probably have been undetected in bulk analyses because the small fraction of blood cells accumulating high amounts does not contribute significantly to the mean value.

Menkes' Disease

The Menkes' disease also called Menkes' steely hair syndrome or the "kinky hair syndrome", which is a sex-linked, inherited progressive fatal brain disease in infants was studied by Allen et al. [26]. They used the scanning proton microprobe to investigate trace elements, primarily copper, in human skin fibroblasts. Their very well designed experiment showed that Menkes' cells grown in a normal maintenance medium displayed an average intracellular copper level six times higher than normal fibroblasts.

Conclusions About Single Cell Analyses

However, even without an explanation of the changes, analyses of major and trace elements may be of significant value in research to elucidate the controversial border between health and disease, manifest or hidden disease in HLA-B27 positive individuals. Nuclear microscopy can also be of value for epidemiological studies concerning the propensity to inflammatory reactions in HLA-B27 positive individuals. The application of nuclear microscopy that has proved to be of direct diagnostic value is the one dealing with the chronic fatigue syndrome.

It seems unrealistic to use nuclear microscopy in routine clinical medicine because most investigations on biological objects are bound to be time-consuming. The uses in clinical research, though, seem to be unlimited.

NUCLEAR MICROSCOPY OF TISSUES

Most of the work performed with tissues has been concentrated upon animal models but also biopsy or autopsy material from humans has been used. The preparation technique is often cryosectioning of frozen tissue samples with subsequent air-drying or freeze-drying. To be able to locate the regions of inteterst, adjacent sections are stained according to ordinary histological practice.

Central Nervous Tissue

A fundamental study of the elemental distribution in human cerebellum was presented by the Oxford group [27]. They have also made use of the ratio of concentrations in elemental mapping to give chemical information. In the case of human cerebellum the concentration ratio of S:P produced an elemental map that corresponded very closely to an adjacent stained section [28].

In a collaboration with neuropathologists, the Uppsala group have presented data on lead uptake in the growing cerebellum of rats. The

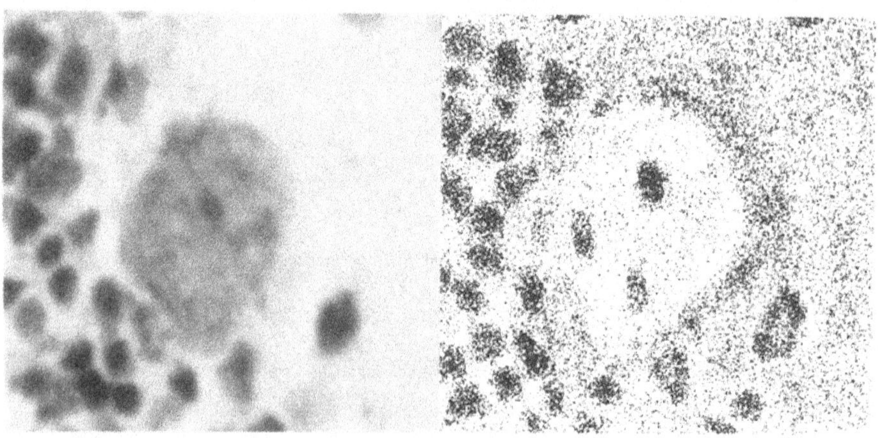

Figure 2: Light micrograph (left) of a part of rat cerebellum and a lead elemental map (right) of an adjacent unstained section.

incitament for the work has been the suspicion that children exposed to environmental lead might be mentally retarded. Rats were administered two different doses of lead, one of which gave growth affection and the other not. The rats were killed on day 15 after daily administration of lead and the cerebellum was prepared both for nuclear microscopy and histopathological analysis. The left part of Figure 2 shows a micrograph of part of the cerebellum from one of the high-dose rats. The leftmost dark part is the densely populated granular layer and the rightmost part is the sparsely populated molecular layer. The large ellipsoid structure in the middle is a Purkinje cell, the size of which is 30-50 μm. The right part of Figure 2 is a lead elemental map of the same part of cerebellum as in the micrograph.

The study revealed a higher incorporation of lead in the white matter of the cerebellum. There were also differences in the uptake in the different layers of gray matter as can be seen from Figure 2.

The Oxford group have also performed nuclear microscopy of Alzheimer's tissue. Alzheimer's disease is a progressive neurodegenerative disease characterised by the presence of large numbers of senile plaques and neurofibrillar tangles in the hippocampus and neocortex of the cerebrum. A combination of RBS and PIXE in the nuclear microscope revealed the concentration pattern of sulphur, phosporus, carbon, nitrogen, silicon, aluminium, titanium and calcium in a senile plaque [29]. Aluminium has been indicated in the aetiology of the disease. The most conspicuous finding of this study, however, was numerous micron-sized particles of titanium. An interesting hypothesis of the olfactory neuron as a pathway for small dust particles into the brain was also presented.

Nuclear Microscopy of Viscera

The liver is a very important organ for the detoxification of heavy metals. In an attempt to assess the distribution of an oral lead exposure nuclear microscopy was used to investigate mouse liver samples. The accumulation of lead in the tissue was demonstrated as were accompanying alterations of other elements [30].

Nuclear microscopy performed by Heck et al. [31] provided structural information about trace elements in liver samples from humans. Localised alterations of trace element concentrations wihtin liver lobules in alcohol- and haemochromatosis-induced cirrhosis were demonstrated. The role of copper in liver cirrhosis was also studied by the Karlsruhe group in an in vitro experiment [32].

Samples from primary biliary cirrhosis were studied with nuclear microscopy by the Oxford group. The copper to sulphur ratio in the elemental maps was consistent with the identification of the copper-rich areas as aggregates of metal-loaded metallothioneins [33]. Early primary haemochromatosis was also studied by this group. It was concluded that a combination of quantitation with examination of elemental distribution to distinguish parenchymal from reticulo-endothelial iron loading may enable the nuclear microscope to diagnose very early primary haemochromatosis on a liver biopsy from presymptomatic patients [34].

Using an experimental animal model, it was shown with nuclear microscopy on liver and kidney sections together with physiological data that selenium exerted a protecting effect against mercury toxicity [30]. In the same work it was concluded that lead has a protective effect against lead toxicity. A later study using nuclear microscopy showed that selenium also protects against cadmium toxicity [35].

Several years ago the lead distribution within individual osteons of human femur was investigated after occupational exposure. The nuclear microscopy was performed on autopsy samples [36].

Diabetes and Nuclar Microscopy of Pancreas

Pancreas is an important organ and one of its functions is to excrete insulin. The insulin secretion is performed by the β-cells in the islets of Langerhans. Certain trace elements are essential to the secretion of insulin. The impaired insulin secretion in diabetes mellitus might therefore arise from a defective β-cell handling of various elements that are directly engaged in the storage and/or release of the hormone. As part of a larger study, e.g. [37] and [38], the influence of starvation on trace element concentrations in β-cells was studied in a mouse model.

Two groups comprising nine mice each were either fed ad lib. or starved 48 hours. Freeze-dried pancreatic sections were then put into the nuclear microscope for elemental determination. Table II summarises the results.

Table II: Elemental concentrations in β-cells from mice fed ad lib. and mice starved 48 hours. The concentrations are given as median±interquartile range (n=9). The dimensions are μg/g dry weight.

Element	Fed ad lib.	Starved 48h
Na	580.4± 20.4	605.1± 28.1
Rb	26.2± 6.5	20.9± 11.5
Ca	1294±149	1475±315
Mg	709.4± 31.3	629.7± 57.8
Fe	237.8± 13.5	287.9± 43.4
Zn	322.4± 49.9	218.0± 39.6
Cu	7.1± 1.8	5.6± 2.2

Wilcoxon's rank sum test reveals that there are statistically significant differences in concentration for all the studied elements. The significance levels are <0.001 for all elements but for Ca and Cu where the levels are <0.05.

The conclusion of differences between the two groups is very easy to arrive at. It does not, however, give any idea about the handling of trace elements in the normal and starved states. We have used the multivariate technique principal component analysis to study possible covariations between the elements [24]. The analysis can be visualised through biplots for the first two principal components accounting for the largest variance or spread in the data. Figures 3 and 4 display such biplots for the two groups. The lines intersecting at the origin (0,0) represent the original variables. The length of each vector is proportional to its contribution to the princial components. The angle between any two is inversely proportional to the correlation between them. The two elements displaying a small change of correlation are zinc and rubidium whereas the correlation between calcium and iron is almost lost in the starved animals. The association between calcium and magnesium is less pronounced in the starved group.

Multivariate approaches are necessary if the multielement capacity of nuclear microscopy is to be adequately used. The biplots of principal

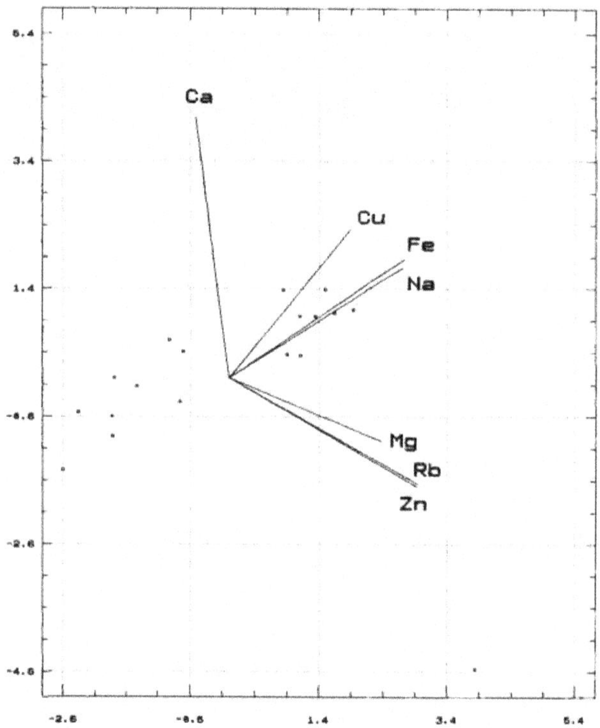

Figure 3: Biplot of two principal components for covariance of elements in mice fed ad lib.

components can be valuable in the design of new experiments in trace element biology.

Conclusions About Nuclear Microscopy of Tissues

Carefully prepared tissue sections from various organs of experimental animals will provide av very valuable tool in trace element biology when nuclear microscopy is applied. The use in animal models is probably limited only by the experimenter's imagation.

The medical use of tissue microscopy may be somewhat less common but it certainly has its place for the analysis of biopsies and autopsies.

The Future of Nuclear Microscopy

Nuclear microscopy will be incorporated into the experimental arsenal in many biological laboratories. The introduction most probably will be through collaboration with researchers at accelerator laboratories. The combination of nuclear microscopy with biochemical analytical methods will prove to be very versatile in providing new knowledge about trace element biology.

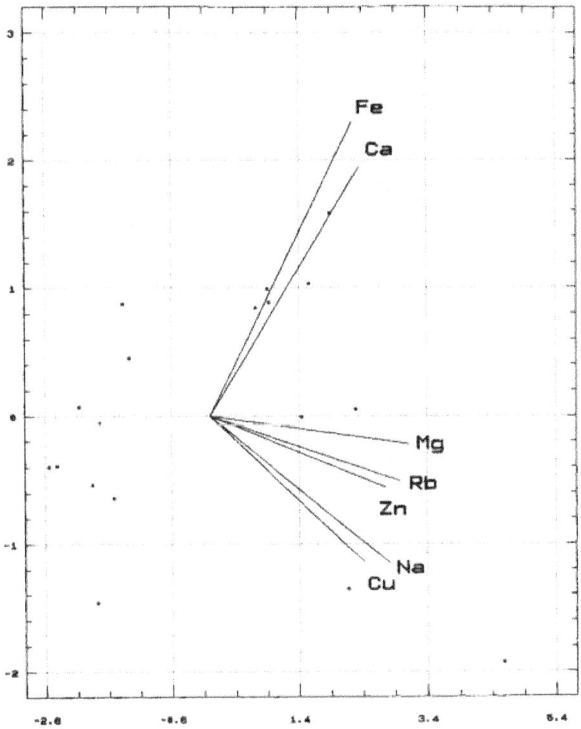

Figure 4: Biplot of two principal components for covariance of elements in mice starved 48 hours.

The resolution of nuclear microscopes will approach that of X-ray microanalysis in electron microscopes. Where we will meet the practical limit caused by radiation damages remains to be investigated. The trace element community will still have to await the first commercially available nuclear microscope.

REFERENCES

[1]. Sunde, T., Nyström, J. and Lindh, U., Nucl. Instr. and Meth., in press 1990.

[2]. Bowen, H.J.M., Environmental Chemistry of the Elements, Academic Press, London 1979, p. 129.

[3]. Keshan Disease Research Group of the Chinese Academy of Medical Sciences, Beijing, Chinese Med. J. 92, 477 (1979).

[4]. Underwood, E.J. and Mertz, W., Trace Elements in Human and Animal Nutrition, 5th ed., vol. 1, ed. W. Mertz, Academic Press, New York, 1986, p. 1.

[5]. Hambridge, K.M., Casey, C.E. and Krebs, N.F., Trace Elements in Human Nutrition, 5th ed., vol. 2, ed. W. Mertz, Academic Press, New York 1986, p. 1.

[6]. Iyengar, G.V., Kollmer, W.M. and Bowen, H.J.M., The Elemental Composition of Human Tissues and Body Fluids, Verlag Chemie, Weinheim 1978.

[7]. Lindh, U., Johansson, E. and Gille, L., Nucl. Instr. and Meth. B3, 631 (1984).

[8]. Lindh, U. and Johansson, E., Neurotoxicol. 4,3, 177 (1983).

[9]. Lindh, U., Johansson, E., Jameson, S. and Hällgren, R., Nutr. Res. Suppl. 1, 156 (1985).

[10]. Annerén, G., Lindh, U. and Johansson, E., Acta Paediatr. Scand. 74, 259 (1985).

[11]. Gustavsson, K.-H., Foucard, T., Mehary, G.-M., and Lindh, U., Acta Paediatr. Scand.

[12]. Sorensen, J.R.J, Inorg. Perspect. Biol. Med. 2, 1 (1978).

[13]. Hällgren, R., Svensson, K., Johansson, E. and Lindh, U., Arthritis Rheum. 28, 169 (1985).

[14]. Hällgren, R., Svensson, K., Johansson, E. and Lindh, U., J. Lab. Clin. Med. 104,, 893 (1984).

[15]. Hällgren, R., Svensson, K., Johansson, E. and Lindh, U., J. Rheumatol. 12, 876 (1986).

[16]. Svensson, K., Hällgren, R., Johansson, E. and Lindh, U., Inflammation 2, 189 (1985).

[17] Feltelius, N., Hällgren, R. and Lindh, U., J. Rheumatol. 15, 308 (1988).

[18]. Massart, D.L., Vandeginste, B.G.M., Deming, S.N., Michotte, Y. and Kaufman, L., Chemometrics: A Textbook, Elsevier, Amsterdam 1988.

[19]. Ahlrot-Westerlund, B., Carlmark, B., Grönquist, S.-O., Johansson, E., Lindh, U., Theorell, H. and de Vahl, K., Gustaf Werner Institute Report GWI-R 1/83, 1983, 52 pages (in Swedish).

[20]. Ahlrot-Westerlund, B., Carlmark, B., Grönquist, S.-O., Johansson, E., Lindh, U., Theorell, H. and de Vahl, K., Nutr. Res. Suppl. I, 156 (1985).

[21]. Lindh, U. and Tveit, A.-B., J. Radioanal. Chem. 59, 167 (1980).

[22]. Brune, D., Brunell, G. and Lindh, U., Nucl. Instr. and Meth. 197, 209 (1982).

[23]. Lindh, U., Grönquist, S.-O. and Lindvall, A., 11th Intl. Conference on X-Ray Optics and Microanalysis, eds. Brown, J.D. and Packwood, R.H. Faculty of Engineering Science, The University of Western Ontario, London 1987, p. 328.

[24]. Lindh, U., Nucl. Instr. and Meth. B30, 404 (1988).

[25]. Lindvall, A., Friman, G., Grönquist, S.-O., Linde, A. and Lindh, U., submitted for publication.

[26]. Allen, G.L., Camakaris, J. and Legge, G.J.F., Nucl. Instr. and Meth., in press.

[27]. Ilgren, E.B., Vaux, D.J.T., Watt, F., Grime, G.W. and Takacs, J., J. Neurosci. Meth. 12, 25 (1984).

[28]. Grime, G.W. and Watt, F., Nucl. Instr. and Meth. B50, 197 (1990).

[29]. Watt, F., Grime, G.W., Jamieson, D.N. and McDonald, B., Oxford University, Nuclear Physics Laboratory report OUNP-89-16 (1989).

[30]. Lindh, U., Nucl. Instr. and Meth. 197, 185 (1982).

[31]. Heck, D., Ochs, A., Kempnow, A, Maier, K.P. and Kratt, C., Nucl. Instr. and Meth. B22, 196 (1987).

[32]. Heck, D., Ochs, A., Buscher, H.P. and Thom, H., Nucl. Instr. and Meth. B30, 420 (1988).

[33]. Vaux, D.J.T., Watt, F., Grime, G.W. and Takacs, J., J. Clin. Path. 38, 653 (1985).

[34]. Vaux, D.J.T., Watt, F. and Grime, G.W., Principles and Applications of High-Energy Ion Microbeams, eds. F. Watt and G.W. Grime, Adam Hilger, Bristol 1987, p. 203.

[35]. Lindh, U., Nucl. Instr. and Meth. B49, 451 (1990).

[36]. Lindh, U., Anal. Chim. Acta 150, 233 (1983).

[37]. Juntti-Berggren, L., Lindh, U., Berggren, P.-O. and Frankel, B.J., Biosci. Reports 7, 33 (1987).

[38]. Juntti-Berggren, L., Lindh, U., Berggren, P.-O., Berglund, O. and Frankel, B.J., Biosci. Reports 10, 217 (1990).

STUDIES OF METABOLISM USING STABLE ISOTOPES AS TRACERS AND CHARGED PARTICLE ACTIVATION AS ANALYTICAL METHOD.

Niky MOLHO
Dipartimento di Fisica dell'Universita' and INFN Sezione di Milano
ITALY

INTRODUCTION

The importance of a detailed study of trace elements metabolism has considerably grown in the last years. This is due to the wide interest of the topic from different points of view, and to the increasing development of the measuring techniques.

It is worth noting that many oligoelements play a fundamental role in human life: everybody knows that calcium, potassium, iodine etc. are essential for man life, but many other elements like Zn, Se, Co, Cu etc. are of vital importance.

The essentiality of a trace element is mainly evaluated on the basis of the clinical effects shown in case of deficiency or more rarely of an excess state of that element. The correlation between a health disorder and an element imbalance is usually evaluated simply measuring the trace substance concentration in a particular tissue like blood or urine, and comparing the data with what is estimated as a "normal" value. Very seldom the metabolism of the element is well known so that very seldom the correction of the imbalance is well aimed.

A second reason that gives significance to metabolism studies is the problem of environmental pollution. It is very important to know that a certain level of a pollutant concentration in blood gives rise to clinical signals of health disorders, but much more important is to know the absorption details and the metabolic pathway of the pollutant in order to organize the medical intervention and to estimate the potential danger of an environment.

Another reason to study in detail the metabolism of particular trace elements is connected to the problem of nuclear accidents. The risk due to the fallout of radioactive products is substantially related to their biological mean life, to their absorption pathway and to their nuclear decay properties: it is immediate to evaluate that a particle emitter with long decay half life which is biologically accumulated in a specific organ is more dangerous than a gamma ray emitter diffused in the circulating blood with a very rapid biological clearance and a short nuclear half life.

The detailed knowlewdge of the metabolism is far from a conclusive form also for the most relevant oligoelements. Many specific topics must be deeply understood: biophysics, biochemistry, physiology, immunology are only part of the special knowlwdges that are necessary to give a complete answer to the problem, but as a first step a good global approximated model may be able to assess fundamental parameters from experimental data.

METABOLISM MODELS

The form and the details of a mathematical model of a metabolic system and the processes by which it is derived are principally determined by the purposes for which the model is required.

Usually one starts with a very rough and approximated model to simulate the gross structure of a process and afterwards, as the knowledge of the system goes into deeper details, the model is refined trying to take into account also secondary effects.

The fields of application of the models are mainly the investigations of physiological and biochemical mechanisms and the evaluations of quantitative parameters when a system has a known model structure.

Metabolic processes are very often associated with transfer of some material and in this context it is necessary to define biological compartments. A compartment is defined as the location or form in which a substance is found in a homogeneous form and from which it moves to another location or form. It is clear from definition that a compartment may be an organ like liver, a body fluid like urine or the pool of an element bound to a specific protein etc.

To show very briefly how a compartment model can be handled mathematically, the simple case of urea concentration in plasma will be considered. Urea is produced in liver as residue of metabolism and is filtered by kidneys. The simplest schematization of the process is the following:

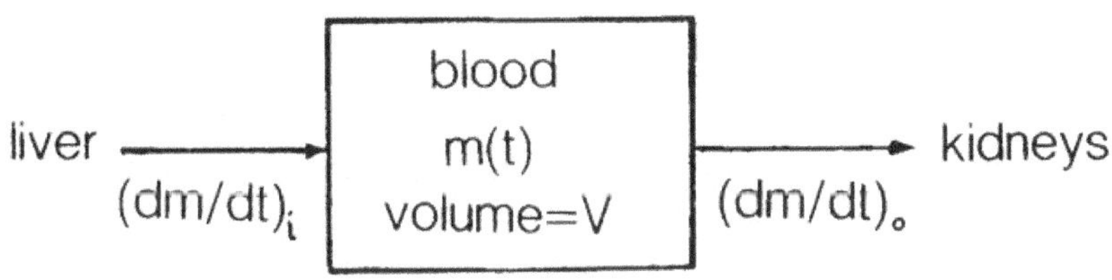

where dm/dt is the time rate of change of urea mass in input and output in the blood compartment.

The clearance in kidneys is represented by the simple equation:

$$(dm/dt)_o = -Kc(t) \qquad (1)$$

where K is the clearance constant and c=m/V the urea concentration in blood. The net mass balance is:

$$dm/dt = (dm/dt)_i + (dm/dt)_o \quad (2)$$

and substituting

$$V\, dc/dt + Kc = i(t) \quad (3)$$

where i(t) is the variable urea mass input per unit time in the blood compartment. Using Laplace transformations eq.(3) may be written in s space as:

$$V[sC(s)-c(0)] + KC(s) = I(s) \quad (4)$$

Assuming c(0)=0

$$C(s)[Vs+K] = I(s) \quad (5)$$

and

$$C(s)/I(s) = 1/(Vs+K) = G(s) \quad (6)$$

that is

$$C(s) = G(s)I(s) \quad (7)$$

Equation (7) shows that the concentration function C(s) is obtained applying the transfer function G(s) to the input function I(s). Coming back to time space one obtains from eq.(7) the convolution integral:

$$c(t) = \int_0^\infty g(t-\tau)i(\tau)d\tau \quad (8)$$

The importance of form (7) consists in the fact that the variable of interest C is related to the input variable I through a transfer function G; this allows to represent the

system as a block diagram

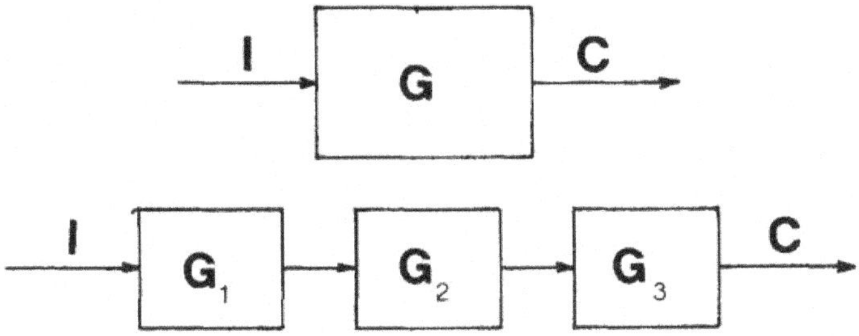

$$C = G_3 G_2 G_1 I$$

in which I and C are not real fluxes, but variables functionally connected. Complex systems may so be broken into a series of simple blocks, each one associated with a specific transfer function and the global transfer function may be expressed as a combination of the transfer functions of the single blocks.

TRACERS

To experimentally study the behaviour of a compartmental model of metabolism, the use of tracers is widely adopted.
As an example, to determine the clearance constant K of urea in blood, one must identify at least part of the urea pool and one must be able to measure the concentration of this part in blood samples withdrawn at different times.
To identify part of the urea pool a tracer must be used, that is some urea atoms must be marked in order that may be detected and recognized.
A tracer must fulfill the requirements to be administered in very small amounts in order not to perturb significantly the studied system, and to behave biologically as the natural substance (tracee).
A radioactive atom, if bound to the studied substance, is a

good tracer because of the ease of detection and identification by means of the emitted radiation: problems in wide use of radioactive tracers in humans arise because of the intrinsic danger of ionizing radiations especially for particular populations like pregnant women or children. Moreover sometimes it is difficult to have radioisotopes of an element with suitable decay characteristics.

To overcome these problems it is possible to use stable isotopes as tracers. If an isotopically enriched compound is introduced in a metabolic compartment and if the concentrations of the single isotopes can be measured, the concentration of the tracer introduced may be determined, also in presence of the intrinsic natural element, because of the isotopic ratio imbalance. This is particularly true in case of use of the rarest isotopes of the natural mixture.

To measure single isotope concentrations at trace level it is necessary to use techniques that can "sense" the nucleus and such techniques are mass spectroscopy, neutron activation and charged particle activation. Each technique has advantages and disadvantages and here only proton induced activation will be treated.

PROTON NUCLEAR ACTIVATION

The PNA method consists of the bombardment of the sample with a proton beam of appropriate energy, which induces predominantly (p,xn) reactions on the nuclei of the target. If the residual radioactive nuclei which originate from the isotopes under study as a result of the (p,xn) reaction have a sufficiently long mean life, the characteristic radiations can be detected off line and can identify the parent nuclei. This allows the quantitative analysis of the sample content. The particular (p,xn) reaction to be used is essentially chosen on the basis of the radioactive nuclei which can be obtained from the isotopes under study; in fact they must have convenient mean life and convenient decay characteristics. The mean life has to be sufficiently long to allow a significant reduction of the matrix background. Regarding the decay characteristics, no interference must exist among the most intense gamma-ray transitions of the nucleus considered and the other lines of the gamma-ray spectrum of the sample.

The absolute value of the amount m_x of the isotope contained in the sample, can be determined from the number I of gamma decays measured by the detector from a chosen transition of the corresponding radioactive nucleus. If t_i is the irradiation time, t_w the waiting interval between the irradiation and the collection of the spectrum, and t_m the time of spectrum collection, this number is:

$$I = \alpha\epsilon\sigma(\Omega/4\pi)N\Phi m_x\tau_x/A_x * [1-\exp(-t_i/\tau_x)] * [1-\exp(-t_m/\tau_x)] * \exp(-t_w/\tau_x) \qquad (9)$$

where A_x is the mass number of the isotope of interest, τ_x the mean life of the radioactive nucleus, and N the Avogadro's number. The other quantities are: α the percentage of the particular gamma transition for each decay of the parent nucleus, ϵ the efficiency of the detector for the particular gamma energy, Ω the solid angle subtended by the detector, σ the cross section for the (p,xn) reaction, and Φ the flux of incident protons.

As the last four parameters are not easily and precisely controllable, we use a method of measurement which does not require their determination. This method is based on the use of an internal reference element and on the comparison with a standard sample. Each sample is doped with a suitable known quantity of a reference element and a standard sample is prepared which contains also a known quantity of the isotope under study. After irradiation the ratio of the intensity of the gamma-ray transition, which identify the trace isotope, to the intensity of the gamma-ray transition, which identify the reference element, is then compared to the same ratio in the standard sample. In this way if t_i, t_w, and t_m, for the sample to be analyzed and for the standard sample are the same, the trace isotope content can be obtained directly. In any case, it is possible to normalize Eq.(9) to the same values of t_i, t_w, and t_m.

The sample preparation is a very delicate task in such kind of measurements because any contamination can completely modify the results. The ≈1 ml serum samples, doped with the reference element, are dryed, powderized and compressed in disk shape 9 mm in diameter and ≈0.8 mm thick.

The samples are then mounted in aluminium frames that are fixed along a circumference of a disc so that 30 samples may be irradiated in the same time because the proton beam is hitting the disk at a radius corresponding to the samples position.

The disk rotates at the speed of one turn per second and every sample is irradiated 20 ms every second.

The proton beam intensity is ≈8 μA, but the average beam current incident on each sample is ≈ 150 nA.

The thermal problem is very important because the power released by the beam into each sample is of the order of 0.5 W that in a very limited volume can considerably increase the temperature. To limit the sample temperature below ≈40 C a very effective cooling system has been used.

After irradiation and after a waiting time depending on the decay properties, a gamma ray spectrum is collected from each sample and is then analysed in order to determine the absolute value of the concentration in plasma of the studied element.

IRON METABOLISM

Disturbances of iron metabolism are frequently observed. The iron level in the human body under normal conditions is maintained constant over time through a balance process between the intestinal absorption and the iron losses. The total amount of iron in human body is ≈4 g, the most part (≈2.5 g) concentrated in red cells, the remainder in stores and tissues. Only a little fraction (≈3 mg) is present in plasma bound to a protein, the transferrin, but this fraction plays a fundamental role because any iron transfer from a compartment to another is performed through plasma. The situation may be described as a two cycle model:

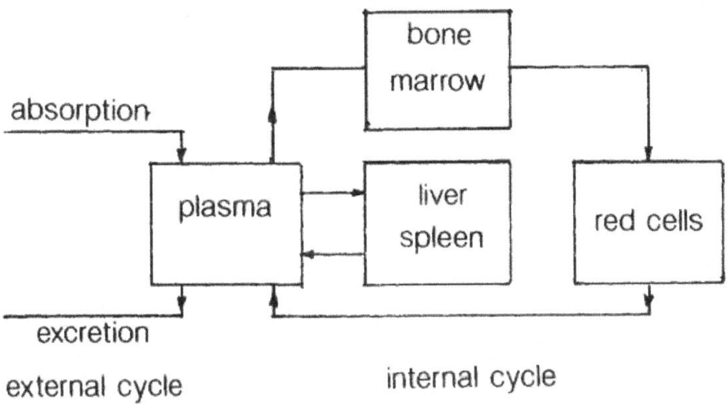

The external cycle takes into account the absorption and the excretion of iron, the internal cycle describes how the plasma iron goes to bone marrow and to stores.
To study iron absorption a simple model was used associated with the double tracer technique, in which ^{57}Fe was given orally and ^{58}Fe intravenously.

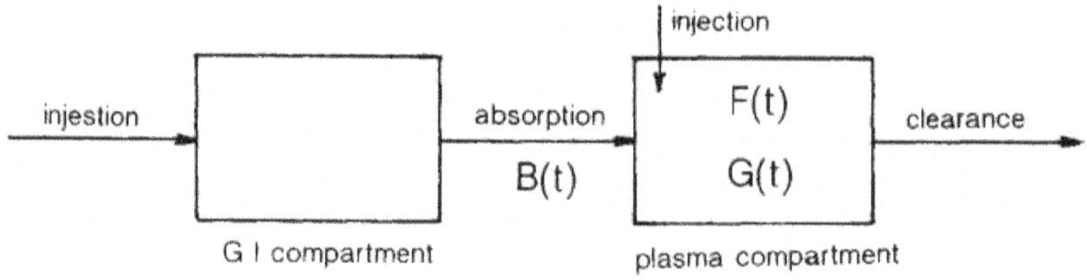

The time dependent ^{57}Fe and ^{58}Fe concentration in plasma [respectively called G(t) and F(t)] allowed to study the iron clearance and the iron intestinal absorption. In fact the nearly exponential decrease of F(t) allows to evaluate the clearance constant and the equation

$$G(t) = \int_0^t B(\tau)\, F(t-\tau)\, d\tau \qquad (10)$$

that expresses the mathematical model representing the behaviour of the fractional quantity G(t) of orally administered tracer as a function of the unknown entry rate B(t) and of the fractional quantity F(t) of the injected tracer, allows to determine the values of B(t) and so to estimate the total absorption fraction.

In order to optimize the conditions of measurement the proton energy corresponding to the maximum yield of the induced reactions was determined and possible interferences were analyzed.

A first test was performed on three rabbits and results obtained with stable isotopes and PNA were intraindividually compared with those obtained using radioactive tracers.

For the intravenous injection an iron citrate solution of enriched ^{58}Fe was used together with ^{55}Fe.

For the oral administration an iron ascorbate solution of enriched ^{57}Fe together with ^{59}Fe was given 40 min after injection, using an intragastric tube. Blood samples were drawn 5, 30, 60, 120, 180, 240, 360 and 480 min after oral dose.

The activities of ^{59}Fe and ^{55}Fe, and the ^{57}Fe and ^{58}Fe concentrations were measured for each sample.

The results are shown in figs.1 and 2.

The metabolic parameters obtained for three rabbits are the following:

Rabbit	clearance $(s^{-1})*10^{-4}$		intestinal absorption %	
	^{58}Fe	^{55}Fe	$^{57}Fe/^{58}Fe$	$^{59}Fe/^{55}Fe$
#1	1.2	1.4	3.3	3.2
#2	1.4	1.4	3.2	3.1
#3	1.7	1.5	2.7	3.2

The agreement between the results allowed to consider the PNA technique applied to stable isotopes as a reliable tool to measure metabolic parameters.

An investigation was then performed with only stable tracers in a female volunteer. The results are shown in figs. 3 and 4, the corresponding clearance is $6.9 \pm 0.8\ 10^{-3}\ min^{-1}$, well in the normal range, and the fractional intestinal absorption is 0.90 ± 0.08, significantly higher than the normal range 0.19-0.47. This high value is a signal of iron deficiency: in fact the subject investigated presents with latent iron deficiency due to a mild hypermenorrhea i.e. normal peripheral blood count, serum iron concentration and total iron binding capacity, but diminished serum ferritin concentration.

Considering that the use of stable isotopes allows to repeat the measurements on the same subjects in different conditions, a series of tests was performed to check the validity of the choice of grape juice as vehicle for iron supplementation.

In some areas of Germany there are diffused diet deficiencies on iron intake and to correct the imbalance an iron compound is usually supplied with grape juice as vehicle. The fact that tannins seem to act as depleting agents in iron absorption raise some doubts on the choice of the vehicle.

On the same female patient volunteer three more series of measurements were performed assuming as vehicle a grape juice, a german red wine produced from the same grape, and an italian red wine. The results obtained for intestinal absorption were respectively 0.15 ± 0.04, 0.12 ± 0.03, 0.10 ± 0.03 to be compared to the 0.90 ± 0.08 value obtained without special vehicles. The absorption depletion associated with grape juice assumption is significant and together with the results obtained with wines, that also contain tannins, seems to indicate an inhibition effect of tannins on iron intestinal absorption. These data, also if very preliminary, seem to indicate that grape juice is not the best choice as vehicle for iron supplementation.

MOLYBDENUM METABOLISM

Molybdenum is an essential trace element, for which few metabolic data may be found in literature and whose radioisotopes are included in nuclear fallout.

Molybdenum has apparently a large tolerance range in humans,also if excessive dietary Mo has been implicated in the induction of osteoporosis and joint abnormalities due to interference with Ca and P metabolism.

Following the same procedures already described for iron metabolism studies a measurement of Mo plasma clearance and Mo intestinal absorption has been performed.

The ^{95}Mo and ^{96}Mo isotopes have been chosen as stable tracers and ^{51}V as reference element.

The yield of production of ^{95}Tc and ^{96}Tc from natural Mo as a function of proton energy has been measured (fig. 5) and clearly shows that ^{95}Tc, also at relatively low energies, is likewise produced by (p,2n) reaction on ^{96}Mo. To separate the two contributions a specific preliminary irradiation was done of a sample containing a known amount of enriched ^{96}Mo: it was so possible to determine the ratio of the peak area corresponding to ^{95}Tc produced by ^{96}Mo to the peak area corresponding to ^{96}Tc. This ratio was later on used to correct the ^{95}Tc peak area in the sample spectra.

In fig. 6 a typical spectrum is shown.

Two male rabbits were given respectively 2.74 mg and 2.90 mg of ^{96}Mo using an intragastric tube. After 7 and 3 minutes respectively, 75 μg and 72 μg of ^{95}Mo were administered intravenously.

Seven blood samples within 360 minutes after the oral administration were drawn from each rabbit.

The figs. 7 and 8 show for one rabbit respectively the concentrations in plasma samples of the intravenously given tracer ^{95}Mo and of the orally given tracer ^{96}Mo as a function of time after oral administration.

To the other rabbit a known dose of radioactive ^{99}Mo was given together with ^{95}Mo. In fig. 9 the activity measurements are compared with the concentration measurements and show a good agreement.

From these data, by means of the same model utilized for iron absorption, fractional intestinal absorptions of 3.4 % and 3.3 % within 360 minutes were determined in the two rabbits.

Very recently the ethical committee approved the oral administration of Mo to humans. As a very preliminary result in fig. 10 the absorption curve of ^{96}Mo for a male patient volunteer is shown.

CONCLUSIONS

I have shown here how the application of simple models and the use of proton nuclear activation can give quantitative answers in the measurement of metabolic parameters.
Due to the high costs the PNA technique cannot be considered of routine use, but for scientific research it is a valuable tool that in many cases cannot be replaced by other instruments.

ACKNOWLEDGMENTS

This work could not have been done without the continuous collaboration of M.C. Cantone, D. De Bartolo, G. Gambarini, L. Pirola from Milan University and C. Hansen, P. Roth, E. Werner from GSU Frankfurt/M.
I whish here express my hearty thanks to all of them.

REFERENCES

1 D.G. Cramp Ed.-Quantitative approaches to metabolism- John Wiley and Sons (1982)

2 P.W. Horton -Radionuclide techniques in clinical investigations- Adam Hilger Ltd.,Bristol (1982)

3 M.C. Cantone, N. Molho, L. Pirola - Clin. Phys. Physiol. Meas.(1982),3,67-77

4 M.C. Cantone, N. Molho, L. Pirola, Ch. Hansen, P. Roth, and E. Werner - Med. Phys. (1987) 14, 223-227

5 M.C. Cantone, N. Molho, L. Pirola, G. Gambarini, Ch. Hansen, P. Roth and E. Werner - Med. Phys. (1988) 15, 862-866

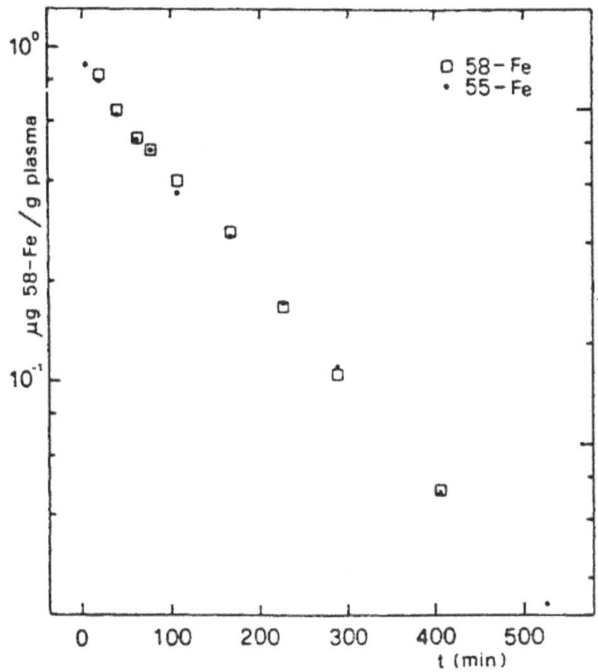

58-Fe
55-Fe
μg 58-Fe / g plasma
Bq 55-Fe / ml plasma
t (min)

fig 1

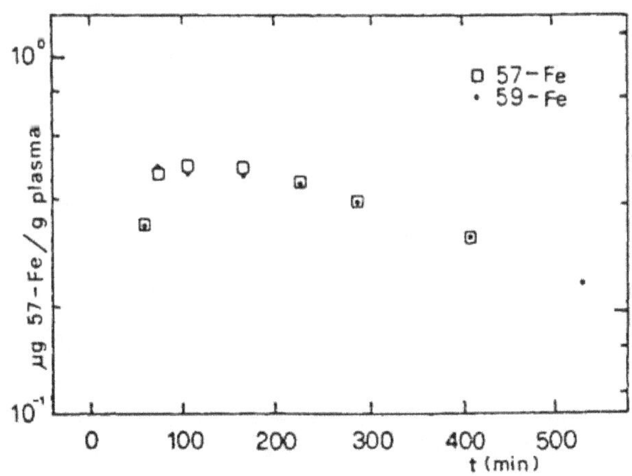

57-Fe
59-Fe
μg 57-Fe / g plasma
Bq 59-Fe / ml plasma
t (min)

fig 2

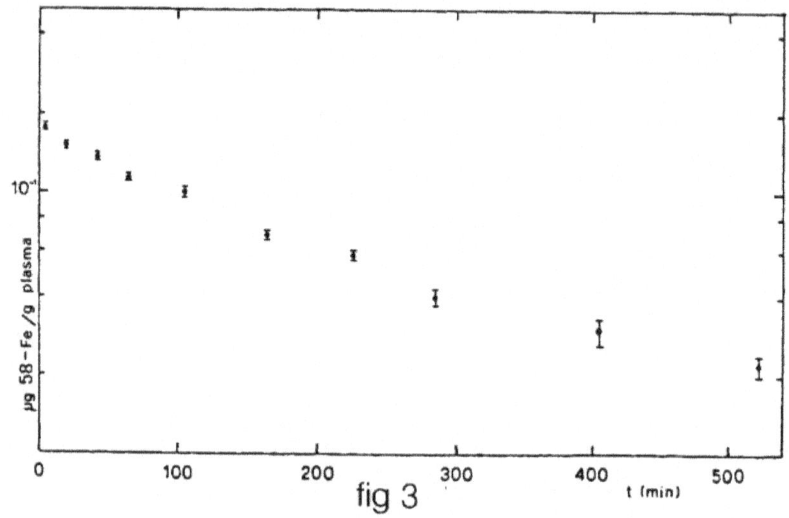

μg 58-Fe /g plasma
0
100
200
300
400
500
t (min)

fig 3

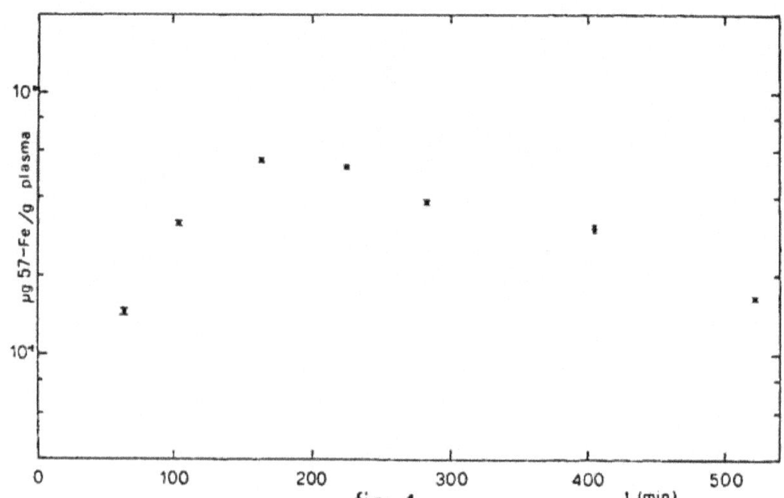

μg 57-Fe /g plasma
0
100
200
300
400
500
t (min)

fig 4

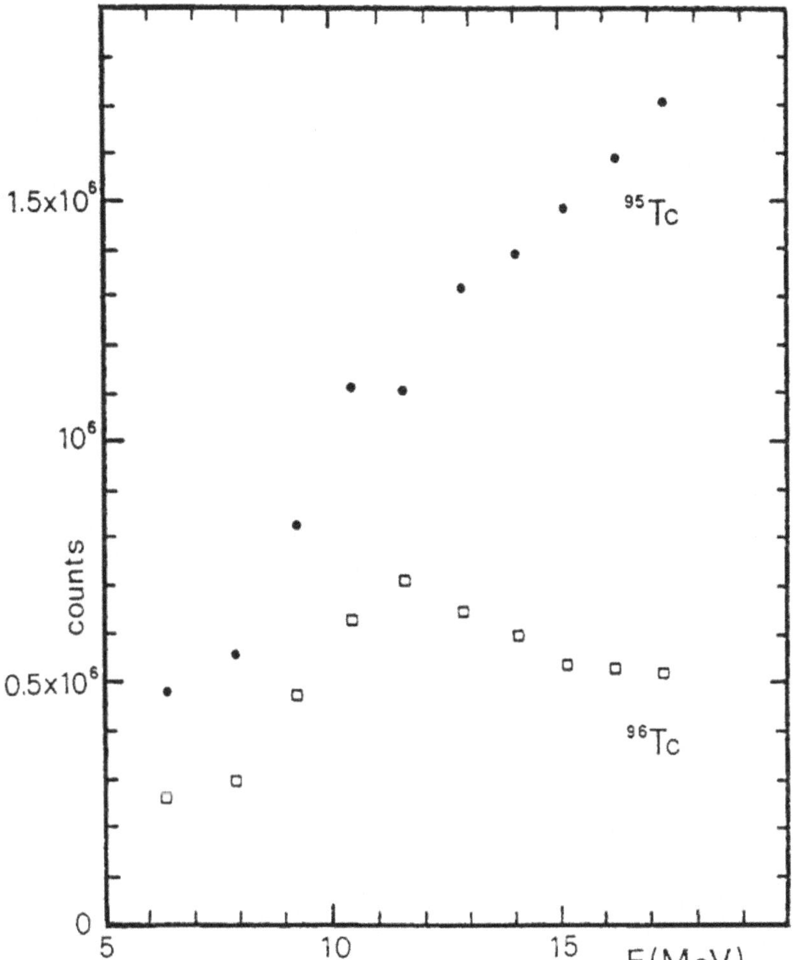

fig 5

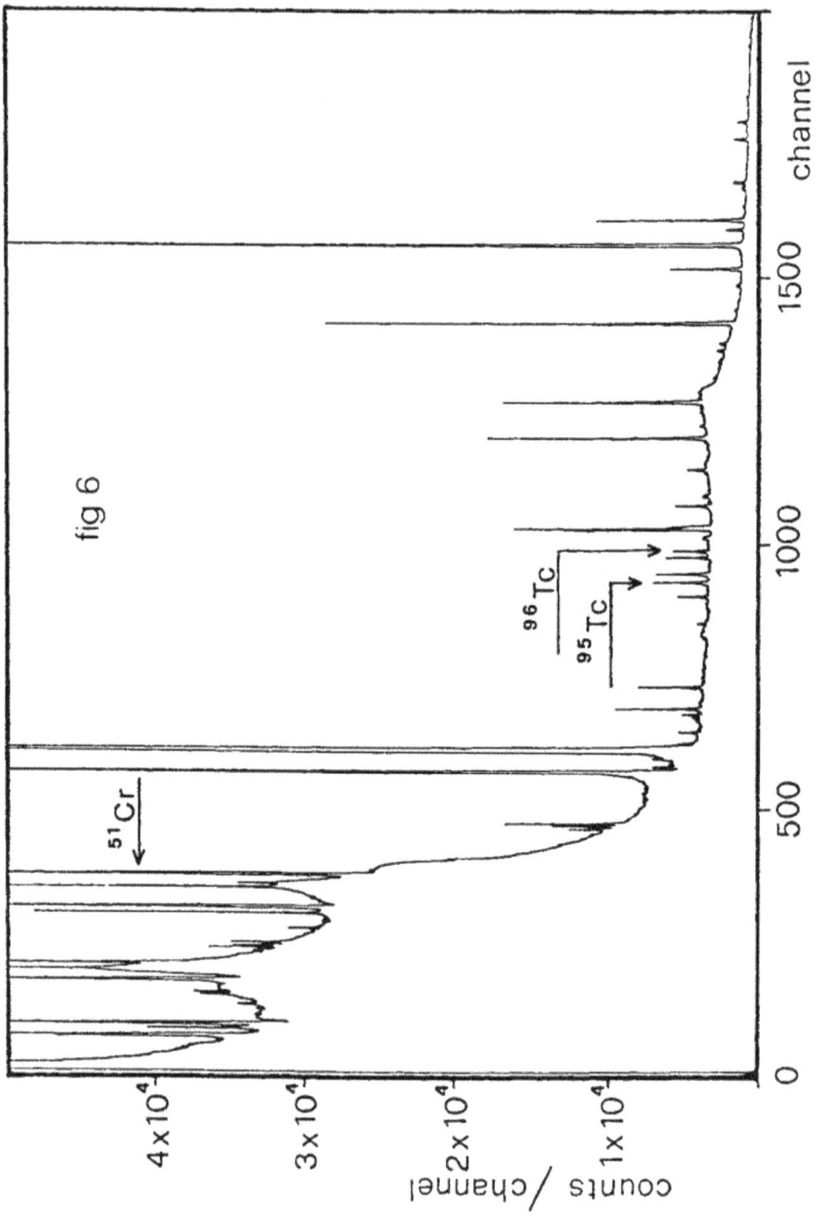
fig 6
51Cr
96Tc
95Tc
channel
0
500
1000
1500
counts / channel
1x10^4
2x10^4
3x10^4
4x10^4

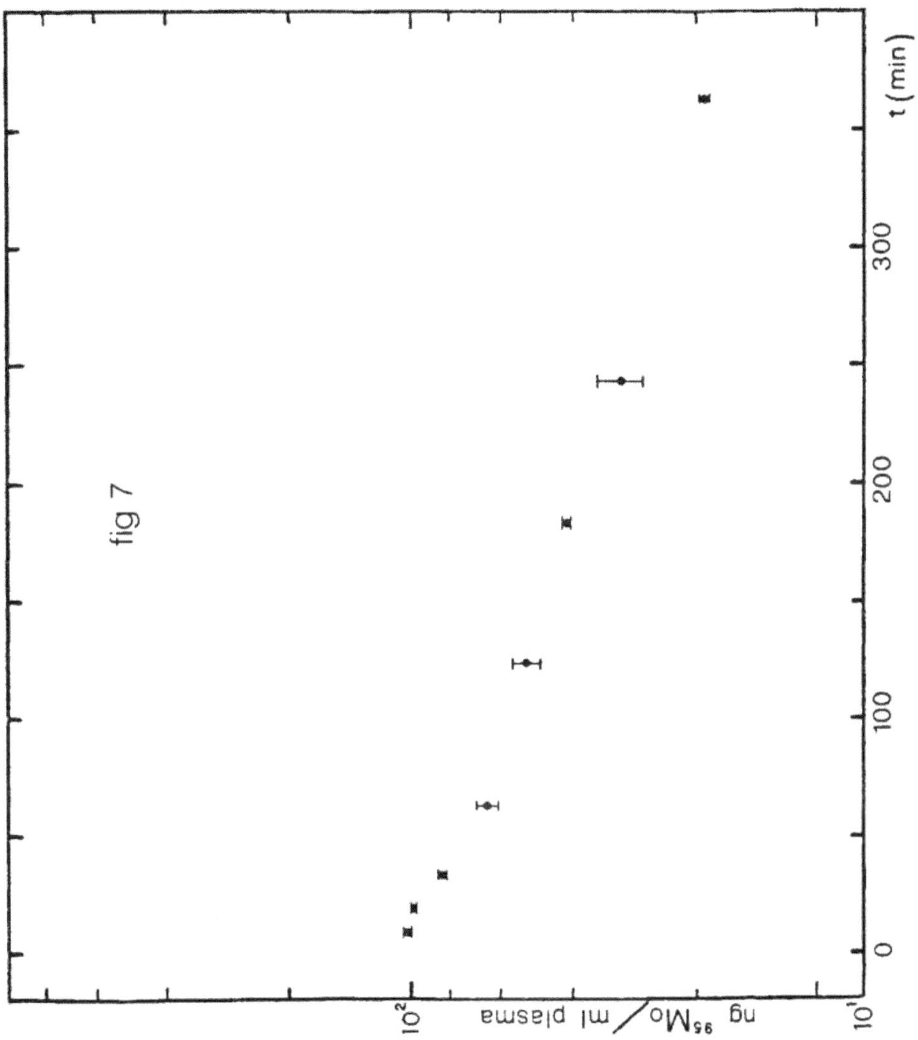
fig 7
t (min)
0
100
200
300
10^2
10^1
ng ^{95}Mo / ml plasma

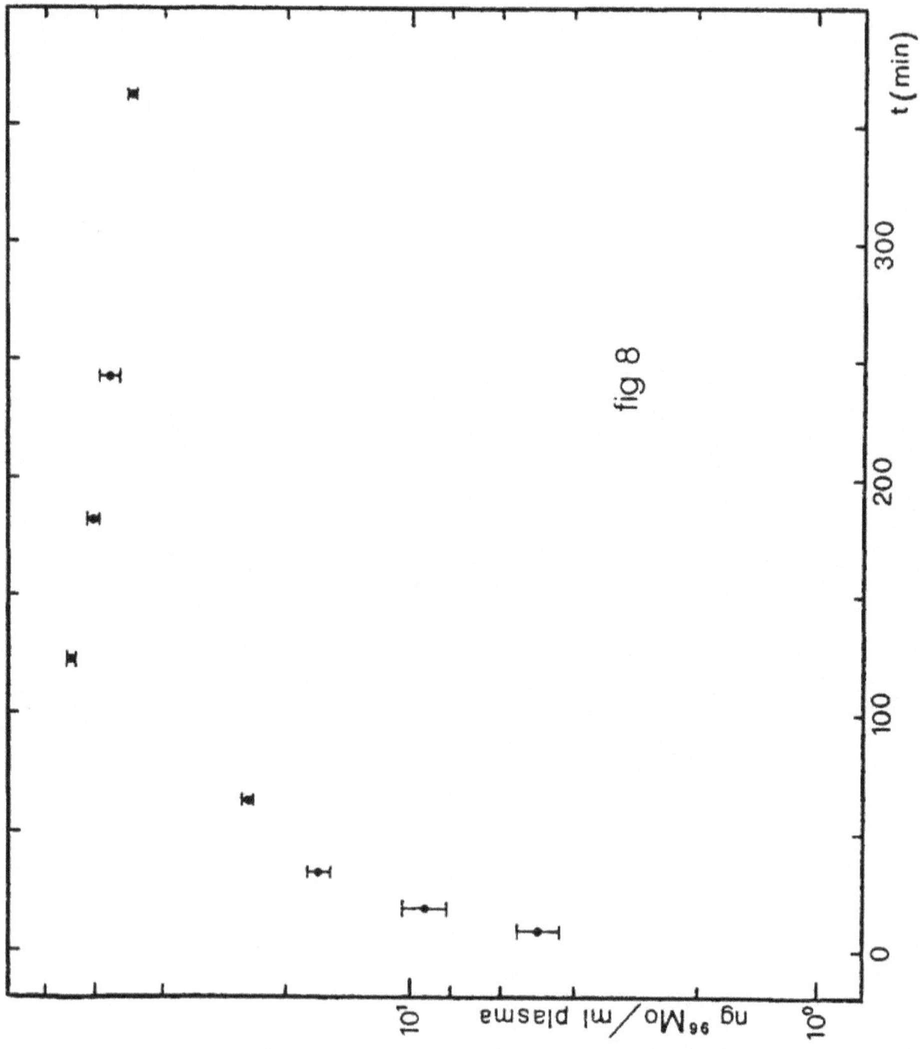

fig 8

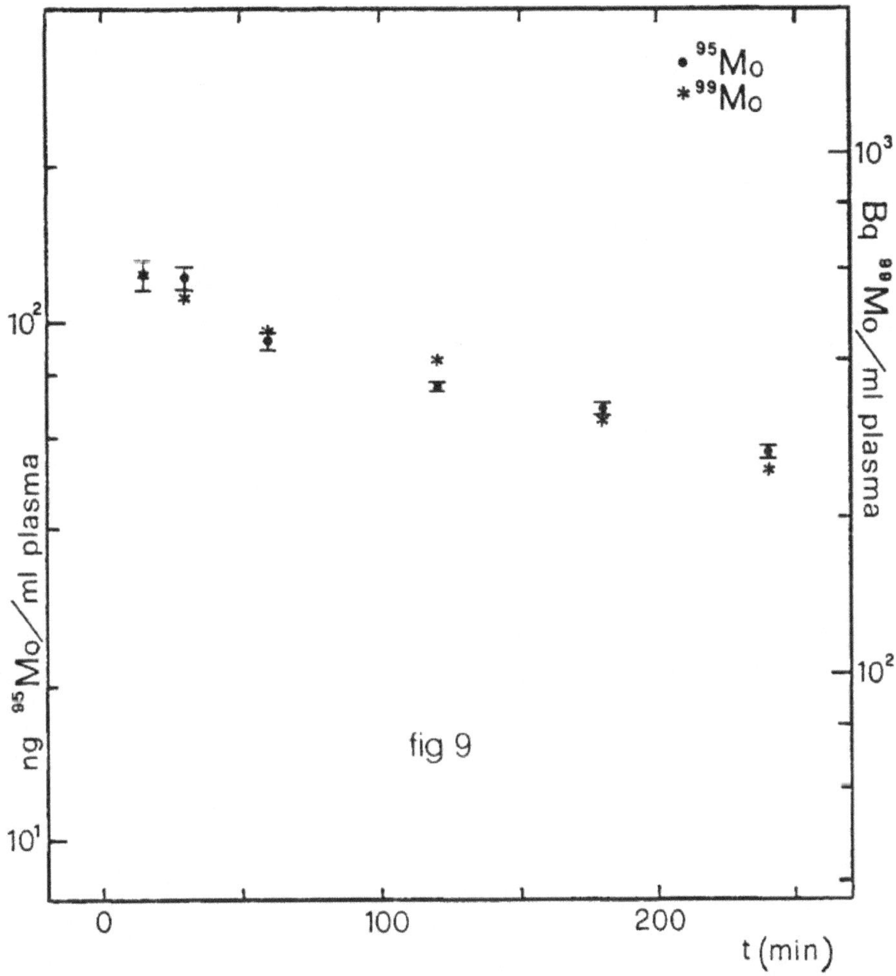
• ^{95}Mo
* ^{99}Mo
ng ^{95}Mo/ml plasma
Bq ^{99}Mo/ml plasma
10^{1}
10^{2}
10^{3}
0
100
200
t (min)
fig 9

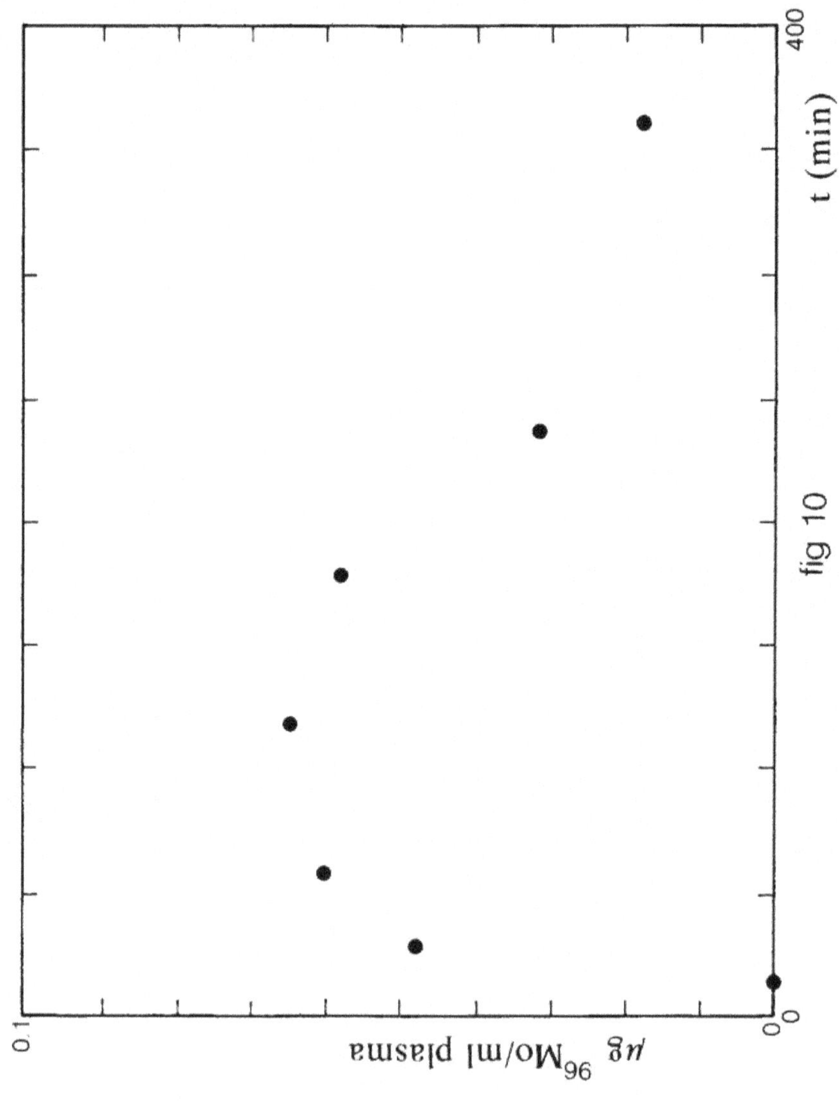

fig 10

THE ROLE OF PROTONS IN RADIOTHERAPY

D.E. Bonnett M.Sc., Ph.D.

M.R.C. Cyclotron Unit, Clatterbridge Hospital,
Bebington, Merseyside, L63 4JY, U.K.

ABSTRACT

This paper discusses the properties of proton beams and their potential advantage in radiotherapy, particularly in applications where the sparing of normal tissue or critical organs is of importance. The methods of production of clinically useful beams are discussed and reference is made to recent proposals for dedicated hospital based facilities. Finally some examples of treatments using both "low" energy protons (60-75 MeV) and "high" energy protons (160-250 MeV) are presented. It is concluded that protons will have a major role to play in cancer therapy.

INTRODUCTION

Modern radiotherapy techniques employing high energy photons and electrons have been very successful in treating many patients with localised tumours, however if the tumour is large or perhaps more importantly located close to a critical normal tissue it can be very difficult to give an optimal dose to the tumour without damaging adjacent organs. It has therefore been found necessary to continue to investigate ways of destroying the tumour whilst minimising the effects on normal tissue. In theory this can be achieved by either using a therapy regime that is more effective in killing tumour cells relative to healthy ones or by employing a type of radiation that provides a dose distribution more closely tailored to the target volume.

Figure 1 summarises the possible options in terms of the properties of beams available for radiation therapy. The various modalities are plotted in terms of the quality of the physical dose distribution versus Linear Energy Transfer (LET). From radiobiological experiments it can be shown that for high LET there is less radioresistance from hypoxic cells, less variation in radiation sensitivity with the mitotic cycle and no repair of radiation damage. (Wambersie et al. 1984[1]). All of these properties may bring some potential advantage in radiotherapy. On the other hand it is generally recognised that improving physical selectivity, that is the physical dose distribution, always brings a therapeutic benefit. It can be seen that the options range from neutrons which have a large potential biological advantage but relatively poor dose distribution, through to protons which have the best dose distribution but no inherent biological advantages. Ranged between the two are the heavy and light ions, with cobalt-60 and 20 MV x-rays representative of conventional therapy. It has been estimated that 10-20% of all patients would benefit from high LET radiation (Wambersie et al. 1984[1]) whilst a much higher percentage, perhaps 60-70% would benefit from improved physical selectivity. Protons give the best possibility of achieving this latter goal.

The object of this review is therefore to discuss the properties of protons, the methods of production of clinically useful beams and to give an introduction to their application in radiotherapy.

THE PROPERTIES OF PROTONS

Several authors have discussed the characteristics of proton beams for use in radiotherapy, for example Koehler and Preston (1972)[2], Raju (1980)[3] and Verhey and Munzenrider (1982)[4]. The properties of the proton beam can be summarised as follows:-

(1) A well defined range.
(2) Minimum scattering in tissue, hence small penumbra.
(3) The possibility of skin sparing, though in many instances this may depend on the use of multiple beams.
(4) The possibility of confining the high dose region to the target volume.

The potential value of protons compared to the more conventional radiations is illustrated in figure 2 which shows the energy deposition along the central axis of a high energy proton beam compared with more conventional radiations. The main energy deposition by protons in tissue takes the form of a Bragg curve. Protons undergo multiple scattering events with atomic electrons and the rate of energy loss is inversely proportional to the square of velocity of the proton and directly proportional to the charge. Thus the energy deposited increases as the proton slows down until its energy is dissipated. The range of the proton beam depends on the initial energy. The Bragg peak itself has a width of approximately a centimetre and therefore must be broadened in order to adequately treat a tumour volume. When the Bragg peak is spread over a given volume there some loss in skin sparing, that is the ratio of dose in the tumour to that at the entrance point is reduced (figure 2). It can also be seen in figure 2 that conventional radiations will always give exit doses beyond the target volume. A comparison of protons with high energy electron beams has also been given by Koehler and Preston (1972)[2]. In this comparison it is shown that the scattering with the proton beam is much less than for the electrons. The implications for treatments are that the well defined range and sharp penumbra of the proton beam could lead to substantial sparing of any critical organs located near or beyond the target volume. Modern radiotherapy however seldom uses single field treatments, rather multiple field techniques. Figure 3 compares isodoses for parallel opposed fields for Cobalt-60, 20 MV x-rays and 160 MeV protons (Koehler and Preston 1972[2]). The advantages of protons in terms of dose to normal tissue and the ability to confine the high dose region to the tumour volume are quite clear.

PRODUCTION OF PROTON BEAMS FOR CLINICAL USE

The possible use of protons in radiotherapy was first suggested by Wilson in 1946[5]. Until recently there has only been a limited investigation of this proposal. This is due to the energy of protons required to achieve sufficient penetration of tissue. Some proton ranges are given in table 1 (Janni 1982[6]). It can be seen that to achieve adequate penetration to treat deep seated tumours an energy of approximately 200 MeV is required. Until recently the only machines available to produce such beams have been physics research accelerators. At this present time two major developments are taking place, these are the conversion of older physics machines for use in radiotherapy, and the construction of dedicated proton therapy accelerators. The accelerators that could be used for proton beam production are the cyclotron, the synchrocyclotron, the synchrochroton and the linear accelerator. It is essential that any machine chosen for therapy applications should be simple to operate, reliable (incurring minimum down-time), easy to maintain and cost effective. Recently there have been several proposals for hospital based proton therapy machines of which four examples will now be considered.

The first is a fixed energy cyclotron capable of accelerating protons to 230 MeV proposed by IBA of Belgium and illustrated in figure 4. It uses novel high field conventional magnets (2.165 Tesla) and would be coupled to a treatment gantry (figure 5). This gantry has 360 degrees rotation and is 5m high (Jongen et al. 1990[7]). The accelerator would be capable of a high continuous output and energy selection would be achieved by the use of beam degraders. A further advantage of such a machine is that it is relatively easy to operate.

The second proposal is from the NSCL (National Superconducting Cyclotron Lab) at East Lansing (USA) for a clinical proton facility utilising a 250 MeV synchrocyclotron with superconducting magnets (figure 6). The use of such magnets greatly reduces the size of the accelerator and allows it to be mounted on the treatment gantry and could possibly act as counterbalance (Blosser 1987)[8], but does involve the added complication of helium cryogenics. The machine would be capable of producing a high intensity pulsed proton beam and would use beam degraders to achieve variation in energy.

The third is a treatment facility based on the use of a proton linear accelerator (linac) the concept of which has been proposed by AEA Technology (Oxford, UK). The facility would be based on a linac approximately 30m long with the beam extractable at four energies, i.e. 100, 150, 200 and 250 MeV. Further variations in energy will be achieved by the use of degraders. Beam output would be high and have a pulsed structure.

The final example of a therapy machine is based on a synchrotron and has been designed in conjunction with Fermilab and manufactured by SAIC. Such a machine was recently installed at the proton therapy facility at Loma Linda Medical Center in California and is described in detail by Slater et al. (1988)[9]. The accelerator has an energy range of 70 to 230 MeV and is 20m in diameter with a 2 MeV RFQ injector. Energy selection is restricted to discrete values with further variations achieved by beam degraders. There are three gantries and two horizontal fixed beam lines. The gantries are 11m in height and encompass a 3m diameter patient enclosure. They have a +/- 185° rotation at a maximum speed of 0.75 rev/min.

At present the optimum choice of accelerator is not obvious and may have to wait the installation and testing of the various options.

Clinical Beams

Having obtained a beam of the required energy it is then necessary to produce a clinically useful beam that will take advantage of the properties of the protons. Three stages are required to produce a clinically usable proton beam, namely range selection, beam profile flattening and beam modulation, and collimation. Most of these elements are the subject of design studies by several groups and an extensive review of the methods currently employed or under development have been given by Chu et al. (1989)[10]. Range selection is made by varying the proton energy from the accelerator or by introducing fixed thickness energy degraders in the beam line or a combination of both methods. For the production of a flat beam profile there are at present four possible approaches namely - scattering, beam wobbling, raster scanning and pixel or spot scanning and these are illustrated in figure 7. The scattering method was the first method to be used clinically (Koehler et al. 1977 [1], Gottschalk 1986 [2]) and is still the main method in use. The beam is scattered by either one or two foils with the possible use of stoppers or occluding rings. The narrow gaussian beam profile is broadened and excess particles in the centre of the beam removed by the central stopper or rings. The profile is then flattened by small angle scattering. The main disadvantage of this technique is the energy loss incurred by the use of the foils and the resultant lowering of machine efficiency due to beam loss. The use of a second scatterer may also give rise to a large effective spot size and hence a poor beam penumbra. Beam wobbling, is currently in use at Berkeley, California for the heavy ion beam used in radiotherapy. A rotating dipole magnet produces ring shaped fluxes of varying diameter (Chu et al. 1985[13]). The remaining methods are both currently under development (for example Blattmann et al. 1988[14]). For the raster scan method a pencil beam of protons is scanned across the required field area. This is repeated several times to reduce any errors. The pixel or spot scan method consists of dividing an irregular shaped area into small volume elements which are each irradiated individually in turn. The advantages of these last three methods are that they give a larger residual range for a given machine energy i.e. there are no energy losses, they give a sharper penumbra and larger field sizes are obtainable. The disadvantages are the greater safety requirements, the need for more beam monitoring and problems associated with patient movement which give rise to problems of under- and over-dosing within the target volume. Beam modulation is the process by which the Bragg peak is spread to either cover the whole tumour or slice of tumour depending on the technique in use. The principle is illustrated in figure 8. By varying the energy of the beam in small increments, in this case by using Perspex degraders, it is possible to move the position of the Bragg peak. If sufficient degraders are used and their contribution to the initial peak correctly calculated then the sum effect is as demonstrated in figure 8. Note that the sharpness of the fall off in range is preserved and there is a flat distribution across the volume in question but there is some loss of skin sparing. At present there are three main methods of beam modulation and collimation - fixed range modulation, variable range modulation and spot or pixel scanning which are illustrated in figure 9. Fixed modulation is the technique currently in use, the other two are again under development (Chu et al. 1989[10]). For fixed modulation the width of the Bragg peak is determined by the thickest part of the target volume in the direction of the beam. The field area is determined by a fixed irregular collimator and the distal surface can be shaped by the use of a tissue like compensator. Obviously there is still some unwanted dose to normal tissue. The other two methods attempt to deal with this problem. In the first case use is made of variable range modulation together with a variable collimator. The Bragg peak is spread to approximately 10mm and moved from slice to slice. At each position the collimator shape is adjusted to spare normal tissue. Compensators are still used and in this case the normal tissue will receive only the plateau dose. The final method, spot scanning, is a development of two-dimensional spot scanning as mentioned previously only in this case it is extended to three dimensions. This method may not require the use of beam collimators. These last two methods, variable modulation and spot scanning, do have the advantage of sparing more normal tissue but have the disadvantages of possible error due to patient movement and more stringent safety requirements. On the other hand the fixed modulation method is quite labour intensive and requires the manufacture of many patient specific devices, such as compensators, which can be quite complex, and irregular shaped collimators.

Clinical Use of Protons

There are currently some eight centres worldwide using protons of various energies (table 2). The centre with most experience being the Massachusetts General Hospital (MGH) working in collaboration with the

Harvard Cyclotron Laboratory. Several other centres are actively pursuing the construction of new facilities and these are given in table 3. Included in this list is the new centre at Loma Linda (California, USA) which is scheduled to begin treatments this year. A summary of some of the main sites of interest for proton treatment is given in table 4. This is a representative list and is not intended to be exhaustive. The sites are divided into low and high energy applications with the former consisting of one main treatment, that of ocular melanoma. The high and low energy applications will now be considered separately.

Treatment of Ocular Melanoma

Currently over 1,000 patients with ocular melanoma have been treated at Harvard in conjunction with the Massachusetts Ear and Eye (MEE) Infirmary, Boston and some 800 at PSI, Switzerland. Two other centres - Clatterbridge Hospital in Merseyside, UK and Uppsala in Sweden - have also recently started this work. Proton therapy represents a further development in the conservative management of melanoma occuring in the eye. Tumours have been treated by surgical excision, laser or radioactive plaques but there are many that are too large or too posterior for effective treatment by these means. Results of the proton treatment of tumours of all sizes have been reported in the literature (e.g. Seddon et al. 1985[16]), Seddon et al. 1986[17]), Gragoudas et al. 1987[18]), Munzenrider et al. 1989[19])). These results show that the rate of recurrence of a treated tumour is low, 2.3%, for dose schedules currently in use at MEE, and that patient survival was no worse than for cases where the eye had been enucleated. The survival at five years of the first 128 patients treated at MEE was 80±5%. It was also found that 64.8% of patients retained some useful visual function. This was dependent on the height, location of the tumour, i.e. the proximity to the optic disc or fovea, and retinal detachment involving the macula. Enucleation following proton therapy was only necessary in 5.7% of the first 1006 patients treated at Boston.

For each of these four main centres at present treating ocular melanoma there are many features of the treatment in common. These derive mainly from the method developed at the Massachusetts Ear and Eye (MEE) Infirmary in conjunction with the Harvard Cyclotron facility (Gragoudas et al. 1977[20], Gragoudas et al. 1980[21]). The first stage of the treatment is a surgical procedure. Three to five tantalum discs 2.5mm in diameter are sutured to the outer sclera of the eye in close proximity to the base of the tumour. The position of the discs relative to the tumour can be measured under transillumination. Further diagnostic information concerning tumour height and profile is provided by ultrasound scans. For both planning and treatment the patient is seated with their head movement restrained. The patient fixes their gaze on a small target light. Several pairs of axial and lateral x-ray films are taken simultaneously for different eye fixation positions. Using a three-dimensional planning programme developed in Boston (Goitein and Miller 1983[22]) it is then possible to reconstruct the eye and tumour position and select the best fixation position to avoid critical structures such as lens and optic nerve. Figure 10 shows such a reconstruction together with the outline of the collimator required for the treatment. The programme can be used to calculate isodoses for any given plane in the eye (figure 11). The treatment is either given as four fractions in four days with total dose of 60 Gy (Cobalt equivalent i.e. proton Gy X RBE 1.1) (Clatterbridge, Uppsala and PSI) or 70 Gy (Cobalt equivalent) in five fractions over 7-10 days (Harvard).

High Energy Proton Therapy

Much of the current interest in the use of high energy proton beams has been stimulated by the Harvard/MGH group. The results have been reviewed in the literature e.g. Larsson (1989)[23] and Austin-Seymour et al. (1990)[24]. Treatments include both proton only, and proton boost after a course of photon treatment. It is only possible to highlight some of the sites listed in table 4 as examples of what has been achieved with high energy protons. In each case reference is made to the results at Harvard/MGH.

Chordoma and chondrosarcoma are tumours which occur both at the base of the skull and the cervical spine and are very difficult to treat since they require both a high radiation dose, 68 Gy (1.9 Gy/fraction) and are often in very close proximity to normal structures such as the optic nerve, temporal lobes, spinal cord and brain stem. For a series of 68 patients 82% local control has been reported following proton therapy. This was for a median 34 month follow up and compares with an expected 33% local control with conventional techniques. Austin-Seymour (1990)[24] has also presented a study comparing treatment plans for protons and 10 MV x-rays for a tumour immediately adjacent to the brain stem and the left temporal lobe. In both plans the dose to the tumour is 70 Gy but the dose to the right optic nerve is 62 Gy for the x-ray treatment compared to 31 Gy for the protons. The right retina received 30 Gy with the x-rays but no dose with the proton treatment which also gave a much lower dose to the right temporal lobe. This result illustrates the ability of protons to spare critical normal tissue. In the same study an example of two treatment plans for a prostate tumour was also given. With the protons there was a much reduced dose to the rectum, bladder and both hip joints compared with the x-rays.

Soft tissue and bone sarcomas are examples of tumours that have been given proton boost to the photon dose applied to the treatment volume after surgery in order to limit the dose to adjacent normal tissue. In this case eye, spinal cord, gut, ovary, shoulder and hip joint. 86% local control has been reported but only a small group of patients were treated. The head and neck treatments were also given as a proton boost following photon therapy as were the prostatic cancers. In the latter case this enabled doses 15-20% above those conventionally employed to be given with no greater risk of bladder or rectal complications.

Proton beams have also been used in radiosurgical techniques treating such problems as pituitary adenoma (Kjellberg and Kliman 1980[25]) and arterio venous malformations (non-malignant growths in the brain) (Kjellberg et al. 1983[26]). The latter treatment was for patients whose lesions were considered inaccessible to surgical excision or embolization. It has been found that protection from death due to haemorrage develops gradually after one year.

CONCLUSION

The ability of protons to deliver a high therapeutic dose to a tumour volume whilst sparing critical tissues and preserving organ function has stimulated much interest in this technique. At present there are plans for several centres to begin such work so that there can be a critical evaluation of such treatments. This would be aided by the commencement of controlled clinical trials although there is already sufficient evidence to suggest that proton therapy is likely to be used as a treatment of choice for many applications and will have a major role to play in cancer therapy in the future.

ACKNOWLEDGEMENTS

I would like to thank many colleagues who have helped in the preparation of this review by the provision of information, material or in helpful discussions. In particular I would like to thank Michael Goitein, Janet Sisterson, Bernard Gottschalk, Andy Koehler, Charles Perret, Neil Griffiths and my colleagues at Clatterbridge, Andrzej Kacperek, Jane Kongerud, Martin Sheen and Douglas Errington.

REFERENCES

1) Wambersie,A., Barendsen,G.W. and Breteau,N., "Overview and Prospects of the Application of Fast Neutrons in Cancer Therapy", J. Eur. Radiother. 5, 248-264 (1984)

2) Koehler,A.M. and Preston,W.M., "Protons in Radiation Therapy", Radiology, 104, 191-195 (1972)

3) Raju,M.R., "Heavy Particle Radiotherapy", New York:Academic Press (1980)

4) Verhey,L.J. and Munzenrider,J.E., "Proton Beam Therapy", Ann. Rev. Biophys, Bioeng., 11, 331-357 (1982)

5) Wilson,R.R., "Radiological Use of Fast Protons", Radiology, 47, 487-491 (1946)

6) Janni,J.F., "Proton Range-Energy Tables, 1keV - 10 GeV", Atomic Data and Nuclear Data Tables, 27, 147-339 (1982)

7) Jongen,Y., Laisne and Lannoye,G., "Preliminary Design of a Reduced Cost Proton Therapy Facility Using a Compact, High Field Isochronous Cyclotron", IBA report YJ/AMV/3/5/90 (1990)

8) Blosser,H.G., "Superconducting Synchrocyclotrons for Proton Therapy", NSCL internal report (1987)

9) Slater,J.M., Miller,D.W. and Archambeau,J.O., "Development of a Hospital Based Proton Beam Treatment Center", Int. J. Radiation Oncology Biol. Phys., 14, 761-775 (1988)

10) Chu,W.T., Renner,T.R. and Ludewigt,B.A., "Dynamic Beam Delivery for Three-Dimensional Conformal Therapy": In "The Proceedings of the Eulima Workshop on the Potential Value of Light Ion Beam Therapy", Nice France 3-5 Nov. 1988, Chauvel,P. and Wambersie,A. eds. (1989).

11) Koehler,A.M., Scneider,R.J. and Sisterson,J.M., "Flattening of Proton Dose Distributions for Large Field Radiotherapy", Medical Physics, 4, 297-301, (1977)

12) Gottschalk,B., "Double Scattering System with Optimum Dose Uniformity in Proton Radiotherapy", Harvard Cyclotron Laboratory technical note PTA 8/1/86 (1986)

13) Chu,W.T., Curtis,S.B., Lancer,J., Renner,T.R. and Sorenson, "Wobbler Facility for Biomedical Experiments at the Bevelac.", IEEE Trans. on Nucl. Sci. NS-32:3321-3323 (1985)

14) Blattmann,H., Coray,A., Pedroni,E. and Greiner,R., "Spot Scanning for 250 MeV Protons", PSI Life Sciences Newsletter, 18-20 (1988)

15) "Particles" Newsletter No. 5 January 1990, Sisterson,J. Editor, Harvard Cyclotron Laboratory, 44 Oxford St., Cambridge, MA 02138, USA

16) Seddon,J.M., Gragoudas,E.S., Albert,D.M., Hsieh,C., Polivogianis,L. and Friedenberg,G.R., "Comparison of Survival Rates with Patients with Uveal Melanoma after Treatment with Proton Beam Irradiation or Enucleation", American Journal of Ophthalmology, 99, 282-290 (1985)

17) Seddon,J.M., Gragoudas,E.S., Polivogianis,L., Hseih,C., Egan,K.M., Goitein,M., Verhey,L., Muzenrider,J., Austin-Seymour,M., Urie,M. and Koehler,A., "Visual Outcome after Proton Beam Irradiations of Uveal Melanomas", Ophthalmology, 93, 666-674 (1986)

18) Gragoudas,E.S., Seddon,J.M., Egan,K., Glynn,R., Munzenrider,J., Austin-Seymour,M., Goitein,M., Verhey,L., Urie,M. and Koehler,A., "Long Term Results of Proton Beam Irradiated Uveal Melanomas", Ophthalmology, 94, 349-353 (1987)

19) Munzenrider,J.E., Verhey,L., Gragoudas,E.S., Seddon,J.M., Urie,M., Gentry,R., Birnbaum,S., Ruotolo,D., Crowell,C., McManus,P., Finn,S., Sisterson,J., Johnson,K., Egan,K., Lento,D. and Bassin,P., "Conservative Treatment of Uveal Melanoma:Local Recurrence after Proton Beam Therapy", Int. J. Radiation Oncology Biol. Phys., 17, 493-498 (1989)

20) Gragoudas,E.S., Goitein,M., Koehler,A.M., Verhey,L., Tepper,J., Suit,H.D., Brockhurst,R. and Constable,I.J., "Proton Irradiation of Small Choroidal Malignant Melanomas", American Journal of Ophthalmology, 83, 665-673 (1977)

21) Gragoudas,E.S., Goitein,M., Verhey,L., Munzenrider,J., Suit,H.D. and Koehler,A., "Proton Beam Irradiation:An Alternative to Enucleation for Intraocular Melanomas", Ophthalmology, 87, 571-581 (1980)

22) Goitein,M. and Miller,T., "Planning Proton Therapy of the Eye", Medical Physics, 10, 275-283 (1983)

23) Larsson,B., "Proton Therapy:Review of the Clinical Results": In "The Proceedings of the Eulima Workshop on the Potential Value of Light Ion Beam Therapy", Nice France 3-5 Nov. 1988, Chauvel,P. and Wambersie,A. eds. (1989)

24) Austin-Seymour,M., Urie,M., Munzenrider,J., Willet,C., Goitein,M., Verhey,L., Gentry,R., McNulty,P., Koehler,A. and Suit,H., "Considerations in Fractionated Proton Radiation Therapy:Clinical Potential and Results", Radiotherapy aned Oncology, 17, 29-35 (1990)

25) Kjellberg,R.N. and Kliman,B., "Radiosurgery Therapy for Pituitary Adenoma": In "The Pituitary Adenoma" Post,K.D., Jacson,I.M.D. and Reichlin,S. eds., Plenum Publishing Corp., New York (1980)

26) Kjellberg,R.N., Hanamura,T., Davis,K.R., Lyons,S.L. and Adams,R., "Bragg-Peak Proton Therapy for Arteriovenous Malformations of the Brain", The New England Journal of Medicine, 309, 269-274 (1983)

Table 1

The Variation in Range of Protons with Energy

Energy (MeV)	Range in ICRU Muscle (cm)
50	2.2
100	7.7
150	15.8
200	26.0
250	38.0

Table 2

Centres Currently Using Proton Therapy *

Centre	Location	Total No. of Patients	Energy (MeV)
Harvard	USA	4841	160
Moscow	USSR	1993	200
Leningrad	USSR	508	1000
Chiba	Japan	65	90
Tskuba	Japan	158	250
PSI	Switzerland	719	72
Uppsala	Sweden	7	70
Clatterbridge	England	83	62

* Data taken from "Particles" No. 5 1990

Table 3

Proposed Proton Therapy Facilities *

Centre	Location	Energy (MeV)
Loma Linda (California)	USA	250
Orsay (Paris)	France	200
N.A.C.	South Africa	200
Nice	France	65
Harvard	USA	250
TRIUMF	Canada	200
Louvain	Belgium	90
Tsukuba	Japan	230

* Data taken from "Particles" No. 5 1990 [15)]

Table 4

Current Sites of Clinical Interest

"Low" Energy Protons (60-75 MeV)

Ocular melanoma

"High" Energy Protons (160-250 MeV)

Pituitary adenoma
Vascular disease (AVM)
Prostatic carcinoma
Chordoma
Chondrosarcoma
CNS tumours
Head and neck tumours
Sarcoma
Miscellaneous deep seated tumours

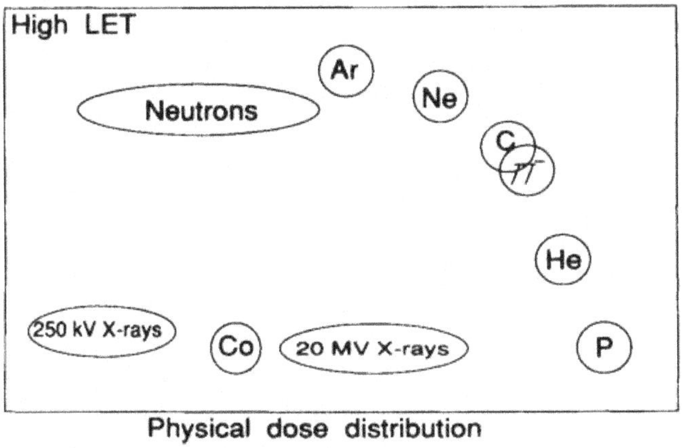

Figure 1: The properties of various therapy beams (redrawn from J.F. Fowler)

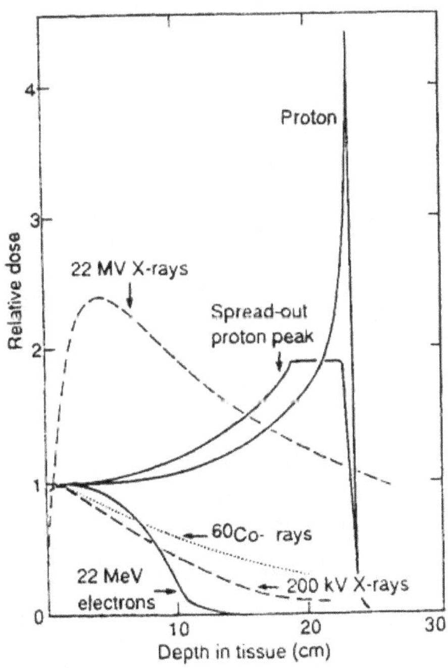

Figure 2: Depth dose distributions for various therapy beams compared with protons (adapted from the PARMS facility, University of Tsukuba, Japan)

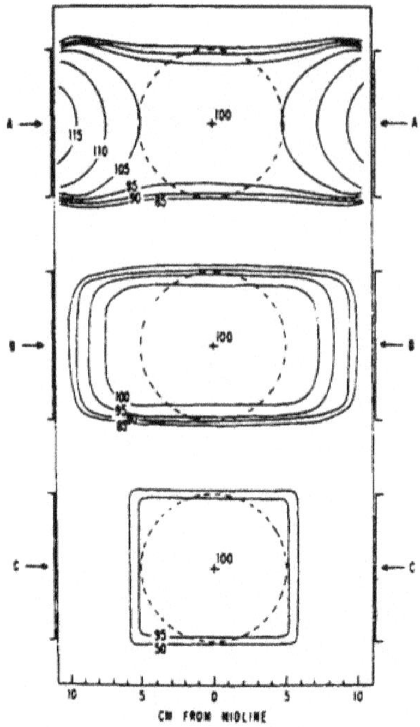

Figure 3: A comparison of isodose contours for parallel opposed fields for 3 clinical beams:- Co-60 gamma rays (top), 20 MeV X-rays (middle) and 160 MeV protons (bottom) (taken from Koehler and Preston 1972[2]).

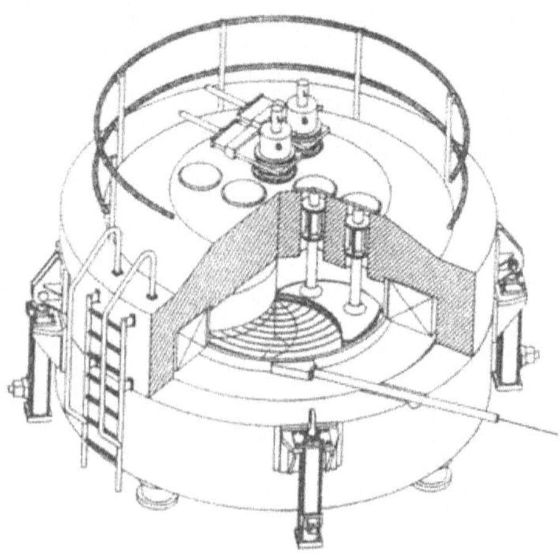

Figure 4: The IBA 230 MeV cyclotron for proton therapy (Yongen et al 1990[7])

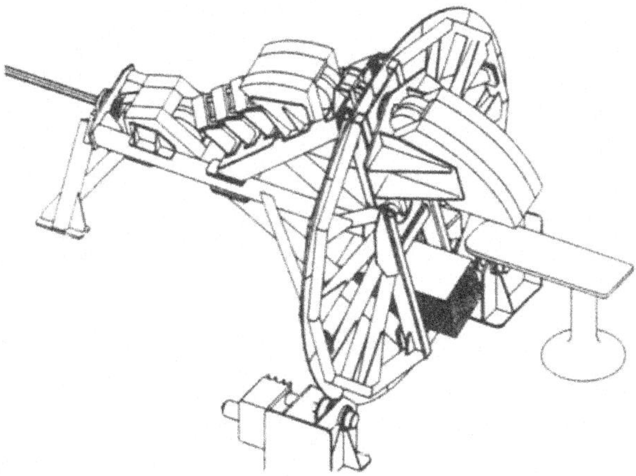

Figure 5: The treatment gantry for use with the IBA 230 MeV cyclotron (Yongen et al 1990)

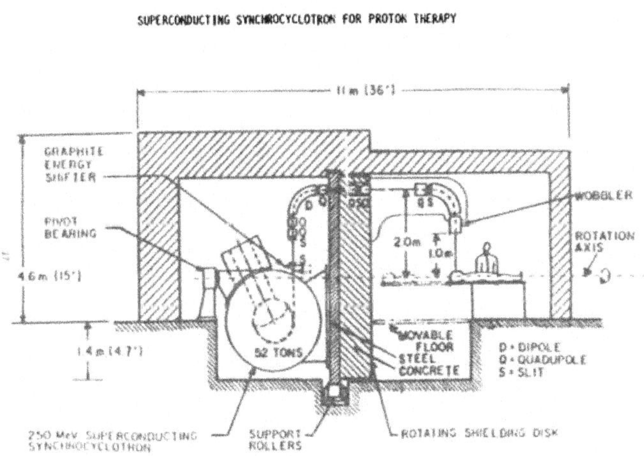

Figure 6: A proposal for a superconducting synchrocyclotron for proton therapy (Blosser 1987 [8])

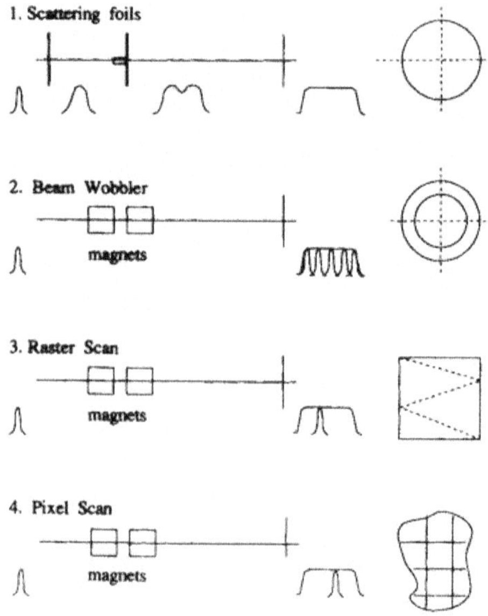

Figure 7: Four possible methods of field flattening (redrawn from Chu et al 1989[10])

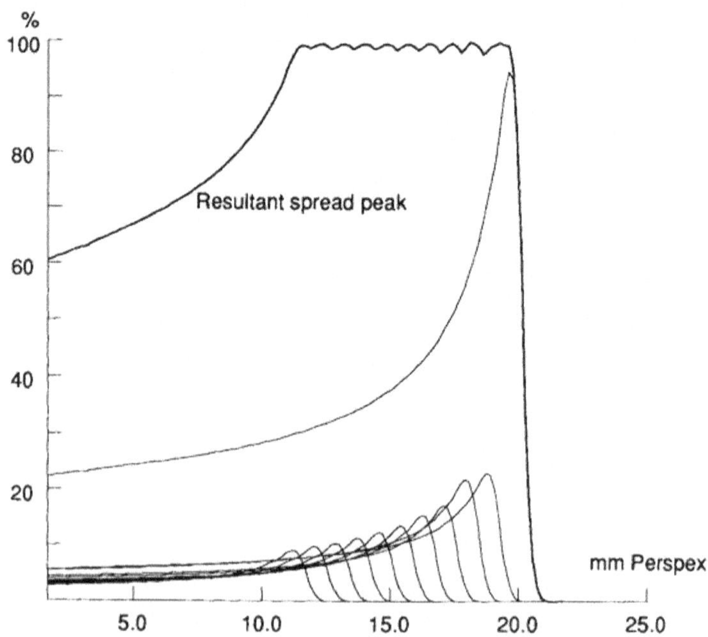

Figure 8: The process of proton beam modulation.
The final curve is the sum of the individual Bragg peaks.

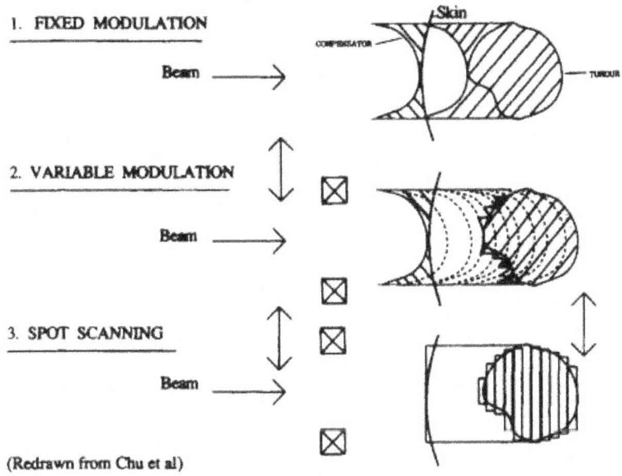

Figure 9: Three methods of irradiating a given tumour volume (redrawn from Chu et al 1989[10])

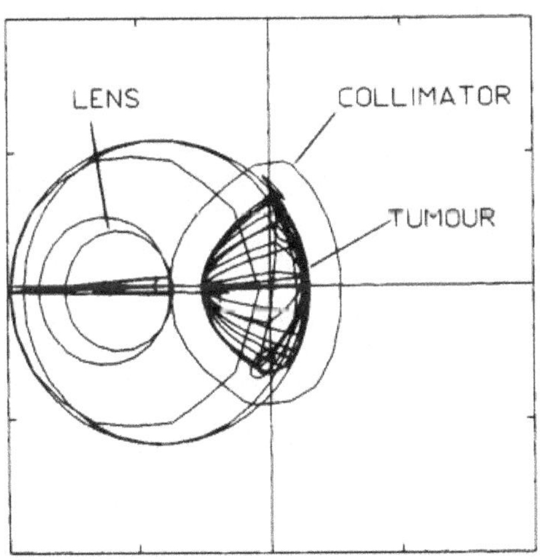

Figure 10: A beams eye view of a patient's eye showing tumour, lens and beam collimator

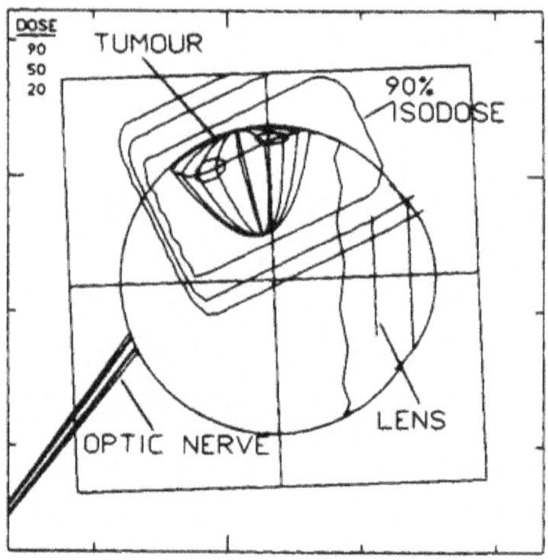

Figure 11: Isodose curves calculated for a plane through the eye, showing the sparing of the optic nerve and lens together with the uniform dose to the tumour

ACTIVITIES OF THE MEDICAL PHYSICS LABORATORY AT THE PHYSICS DEPT. OF MILAN UNIVERSITY, ITALY.

M.C.Cantone, D.deBartolo, G.Gambarini*, F.Groppi, N.Molho and L.Pirola

Dipartimento di Fisica dell'Università and I.N.F.N., Milano, Italy

* Istituto di Fisica del Politecnico and I.N.F.N., Milano, Italy

The activity of the group involving nuclear techniques is mainly devoted to two different topics. The prominent activity, that already achieved quite satisfactory results, concerns the determination of trace and ultratrace elements in biological samples. The second one regards the determination of the spectral structure of therapeutic X-ray beams.

1. TRACE AND ULTRATRACE ELEMENT DETERMINATION IN BIOLOGICAL SAMPLES.

Owing to the increasing knowledge of the fundamental role of trace elements in vital processes, in recent years the precise determination of trace element concentration in biological matrices has met more and more interest. Among the fields in which trace element analysis in biological samples is very important we can mention: a) the association of clinical conditions to abnormal observed levels of trace elements; b) the study of the metabolism of elements, to determine quantitative metabolic parameters; c) the study of the metabolism of dangerous elements in order to evaluate their absorption and accumulation.

For the determination of trace element concentrations in biological matrices several analytical techniques have been developed, utilizing various instruments and methods. These methods present different degrees of simplicity of operation, but also a different sensitivity and reliability.

Utilizing the Proton Nuclear Activation (PNA) we have developed a procedure which gives the possibility of determining trace element concentration with great sensitivity and reliability. This procedure, though expensive and time consuming, is generally useful for research studies and is also the only possible method in many specific situations. Moreover the method gives the possibility of determining the amount of single isotopes of the element.

Proton Nuclear Activation is obtained bombarding the sample to be analyzed with a proton beam of appropriate energy to induce (p,xn) reactions on the nuclei of the target. The intensities of the γ-rays emitted by deexcitation of the radioactive nuclei produced are proportional to the concentration of parent nuclei in the target.

The proton beam we have utilized for the irradiations up to two years ago was produced in the AVF cyclotron of the University of Milan; now we utilize the Scanditronix MC40 cyclotron of the laboratory of the EEC JRC in Ispra (Italy).

For the quantitative determination of trace elements we use the internal reference method, that is we make a comparison between the γ spectra of the sample and of a standard sample of the same matrix, both doped with a precisely known quantity of a suitable reference element; the standard is also doped with a precisely known amount of the element of interest.

With this methodology we carried out the investigation of the content of some elements in animal and human serum.

1.1 Sample preparation.

It is of great importance to avoid all possible contaminations, starting from blood withdrawal up to the γ spectrum collection. To this purpose one has to minimize chemical and mechanical manipulations.

Blood is drawn with teflon catheters. All the operations are carried out in a clean box to prevent sample contamination from the environment. Only non-metallic tools and laboratory-wares made of polyethylene, teflon or agate are used in all the steps of sample preparation. Particular attention is paid to avoid the loss of elements during handling, doping, drying and placing samples for irradiation.

1.2 Experimental.

Depending on the trace element to be measured, we determine the specific (p,xn) reaction that must be induced on the target nuclei to produce residual nuclei with suitable mean life-times and decay characteristics. The mean life values convenient for a good measurement must be a compromise between the intensity of the decay and the level of the background due to the very short living isotopes. For the choice of the reference element one has likewise to consider the mean life and the decay characteristics of the produced nucleus. Moreover it is necessary to control that there is no interference among the γ-ray to be analyzed and the γ-rays of the radioactive nuclei produced by all the possible reactions induced on the nuclei of the matrix by protons with the chosen energy.

For spectra collection the samples are positioned in front of a H.P.Ge detector, or of a Ge(Li) detector shielded, when necessary and useful, by an active NaI(Tl) anti-Compton suppressor.

For the precise determination of the best activation energy we make an experimental evaluation of the relative yield of the reaction induced on the isotope to be determined as a function of the incident proton energy. To this purpose we irradiate a stack of thin foils of the considered element, accurately controlled in homogeneity and thickness. The energy of the proton beam incident on each foil is calculated evaluating the energy loss in all the preceding foils. An example is shown in Fig.1 .

Some measurements are carried out in order to control the possible contaminations, the temperature effects, the response linearity and

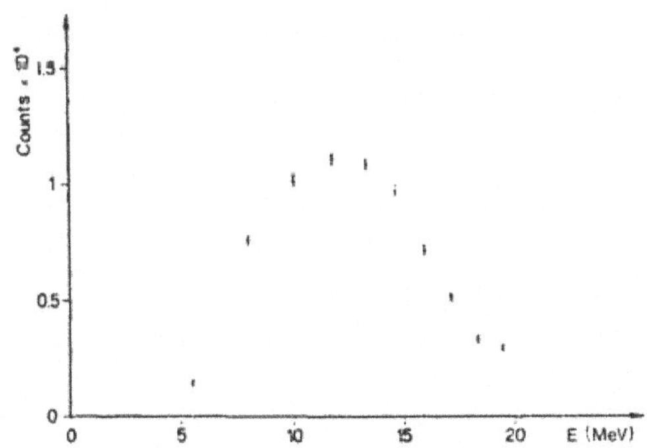

Fig.1 Relative yield of the reaction $^{66}Zn(p,n)^{66}Ga$.

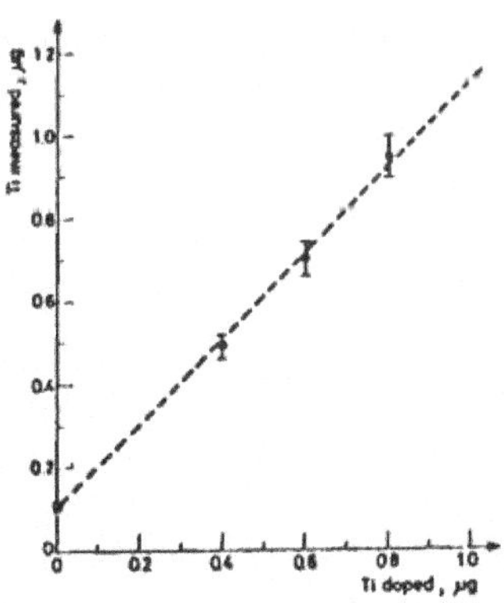

Fig.2 Linearity test.

reproducibility. The linearity test relative to Ti determination is shown as an example in Fig.2 . The quantity of Ti measured in a series of samples, prepared with the same serum and doped with different quantities of Ti, is shown as a function of doping amount.

To verify that no γ-line of radioisotopes produced from other intrinsic elements present in the matrix contribute to the γ-line chosen for the analysis, the intensity of this line is measured as a function of time, and its decrease is compared with that calculated from the mean life time of the corresponding radioactive nucleus (see Fig.3, where dashed line corresponds to the mean life time of the isotope).

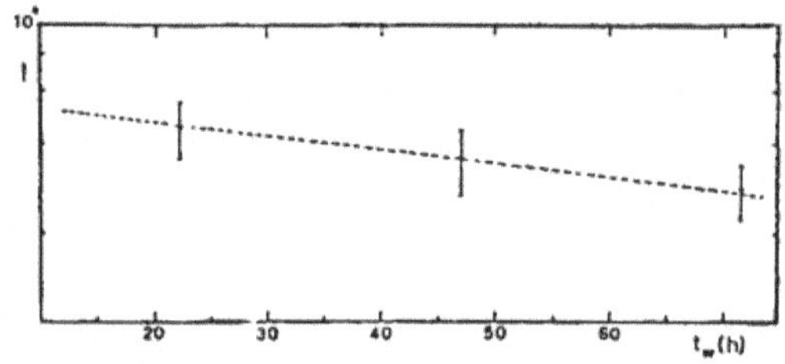

Fig.3 Intensity of the 245.35 keV line of the ^{111}In radioactive isotope as a function of time.

1.3 Some results.

As a first application of the technique we have determined the Zn content in 73 samples of serum from normal children and from children affected by growth failure and diabets. The mean values obtained are:

normal children	1.65 ± 0.40 μg/ml
children affected by diabets	1.32 ± 0.20 μg/ml
children with growth failure	1.74 ± 0.49 μg/ml

A simultaneous determination of Sr, Cu, Fe, Zn and Se content in serum from an individual has given the following results:

Sr	0.075 ± 0.009 μg/ml	Zn	1.45 ± 0.14 μg/ml
Cu	0.57 ± 0.06 μg/ml	Se	0.19 ± 0.01 μg/ml
Fe	1.69 ± 0.31 μg/ml		

and from the contemporary determination of Cd and Ti we obtained:

Cd	0.025 ± 0.005 μg/ml	Ti	0.19 ± 0.010 μg/ml

Chromium level determined in another individual was:

Cr 0.0170 ± 0.0019 μg/g

More recently utilizing this method we have widely studied the iron metabolism in rabbits and in humans. Now we are extending the studies to other elements, like molibdenum, tellurium and ruthenium.

2. STUDY OF THE SPECTRAL STRUCTURE OF A THERAPEUTIC X-RAY BEAM.

The knowledge of the spectral composition of the bremsstrahlung spectra produced by therapeutic accelerators is of great importance for a good evaluation of absorbed doses, but a direct measurement is nearly impossible due to the high fluxes involved (up to 10^9 fotons/s cm^2). We have developed an indirect method of measurement based on Compton scattering from a small target scanning the beam. In fact the diffused radiation has an intensity much lower than the incident one, and its energy range, by a convenient choice of the detection angle with respect to the primary beam direction, can be reduced to values suitable for conventional detectors. The geometry of the whole system moreover must be conveniently chosen, in order to have a biunivocal relation between energy and intensity of the primary beam and energy and intensity of the diffused one. We have made an evaluation of all the critical parameters in order to measure with an energy resolution of the order of $\simeq$1% a primary beam in an energy range from $\simeq$200 keV to $\simeq$10 MeV.

The detected spectrum is heavily deformed by the background produced in the detector itself, prevailingly due to spurious Compton events, and this prevents to reconstruct directly the primary continuous spectrum. We have therefore developed a convenient technique of deconvolution that allows to subtract this background and to obtain the true shape of the spectrum. This deconvolution procedure consists in subtracting from the detected spectrum the response function, suitably normalized, of the detector for the energy corresponding to every channel of the spectrum, iteratively from the last channel of the spectrum down to the first one. This implies the determination of the response functions for every energy channel of the spectrum. So we have set up a procedure to interpolate response functions between a limited number of measured ones. To this purpose we have obtained a set of response functions of a Ge(Li) detector by measuring the pulse height spectra from monochromatic γ-sources with energies between 165 keV and 910 keV at about 100 keV steps. For energies above 1 MeV the response functions were obtained by means of a Monte Carlo computer program.

To test the reliability of the methodology the experimental

spectrum from a ^{226}Ra source was deconvoluted. The result is shown in Fig.4 , were **a** is the measured ^{226}Ra spectrum, and **b** is the deconvoluted spectrum. The Compton background in the discrete spectrum has completely disappeared.

It is not possible to make the same test with a continuous spectrum, because the real spectrum is not known. In view of the experimental determination of a 10 MeV bremsstrahlung spectrum we have simulated the measurement in order to evaluate the feasibility and the critical parameters. Starting from a theoretical approximated calculation of a 10 MeV bremsstrahlung spectrum, the 45° Compton scattered spectrum from a small Pb target has been evaluated. Taking into account the intrinsic photoelectric efficiency of the detector we have determined the pure photoelectric spectrum resulting for 1 s measurement (Fig.5a); this spectrum is biunuvocally related to the primary beam in the sense that the intensity of every energy channel is directly related to the intensity of the primary beam in a definite energy interval. What is seen in the pulse height analyzer is the above mentioned spectrum distorted by the detector response functions by means of a convolution procedure (Fig.5b). One can deduce that in principle a 15 min measurement could be sufficient for a good experimental determination of the spectral structure of a 10 MeV bremsstrahlung spectrum.

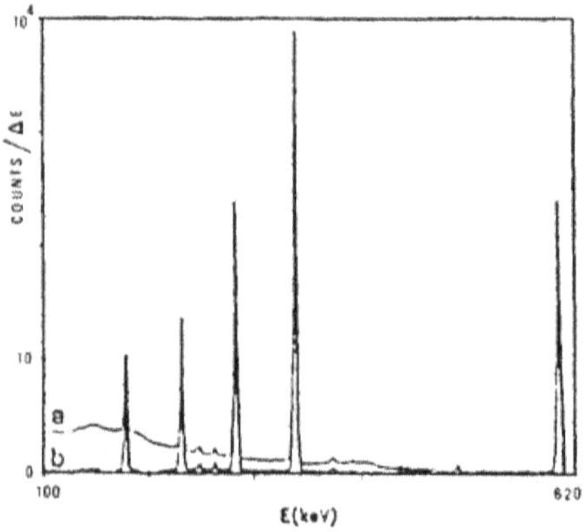

Fig. 4

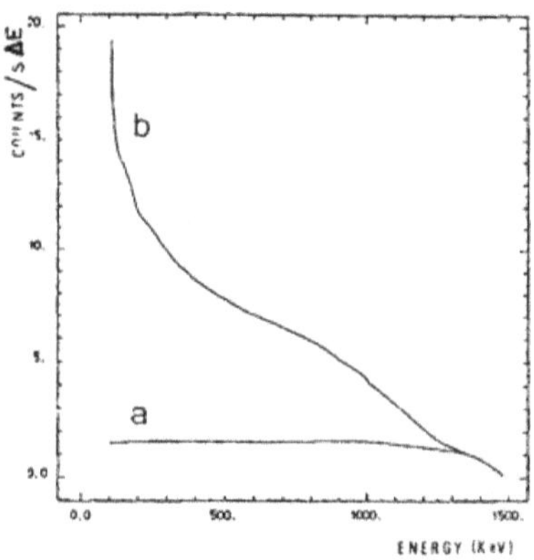

Fig. 5

MULTIELEMENT ANALYSIS OF HUMAN TOOTH ENAMEL AND DENTINE USING PROTON-INDUCED X-RAY EMISSION SPECTROSCOPY@

M.C. Buoso*, S. Fazinic*%, A.M.I. Haque*^, G. Moschini*#, A. Serna*&, A. Volpe+ and G.U. Caravello+

* Istituto Nazionale di Fisica Nucleare, Laboratori Nazionali di Legnaro (LNL), Legnaro (Padova), Italy

% Rudjer Boskovic Institute, Zagreb, Yugoslavia

\# Dipartimento di Fisica, Università di Padova, Padova, Italy

& Instituto de Fisica Corpuscular (IFIC), Valencia, Spain

\+ Istituto di Igiene, Università di Padova, Padova, Italy

ABSTRACT

Tooth samples, dating back to ancient age from Peru and Jesolo (Italy) up to the recent time, has been investigated with PIXE spectroscopy. Depth distribution of heavy elements (>Ca) from enamel to the tooth interior has been studied to apprehend the diffusion of these elements. Results are presented and discussed.

1. INTRODUCTION

The importance of teeth as dose indicators for the exposure of a person to foreign elements has enhanced the significance of their study through a variety of techniques. This in turn has motivated their diverse non-destructive studies by different authors in recent years[1-5]. In several cases, it was indicated that the accumulation of elements in dentine is related to their concentration in blood, from which they are fast segregated[3].

Tooth is chosen for the present study, since it contains several elements (Ca, P, Mg, Na, K, etc), which concentrate mainly during the period of its developement. Some of the cations could go far ahead to substitute the Ca^{2+} of the Hydroxyapatite and play a vital role for the chemical composition of the tooth itself. In this manner, the way is paved for the heavy elements to fix themselves permanently in the organic matrices and, hence, are detectable after a long elapse of time. With respect to bones, teeth are less fragile and, in general, are easily

---

@ Work partially supported by the Commission of European Communities [Contract N° CI 1-0331 - I(A)].

^ ICTP Fellow.

obtainable (for actual teeth). Moreover, they are fundamentally not subjected to further absorptions of elements from the environment at the posture. Naturally, it is rather difficult to obtain archeological tooth samples, since, for reasonable analysis with nuclear techniques, they must be sectioned.

The fact, the use of Proton induced X-ray emission spectroscopy (PIXE) has lowered the detection limit to the order of 1-10 ppm, has made it an exotic tool for the micronuclear analysis. Having the above importances of teeth in consideration, we have recently launched a detailed study of human tooth enamels and dentines, employing multielemental trace analysis method. The scope of our study is to evaluate if there exists a relation in the accumulation of heavy metals in the body in course of different epoches of human civilization. This would enable us to apprehend the eventual correlations existing between the concentrations of some heavy elements, measured through the techniques at our disposal (e.g. PIXE, neutron activation, etc.), and the historical and socio-cultural conditions prevailing in places of origin of these samples. For our study, we consider to measure the tooth samples ranging from the ancient/medieval time up to the present ages and originating from different parts of the globe.

2. EXPERIMENTAL PROCEDURE

Tooth samples are sectioned by diamond saw (Microtome, Leitz 1600), to subdivide the enamels and a number of sections for the dentine for each tooth. The sectionings of each tooth are done in succession, starting from the buccal or lingual plane to the direction of the pulp chamber, till it reaches the later. In order to apprehend the possible contaminations of the tooth, the saw is analyzed with the XRF technique. The sectioned teeth were washed in ultrasonic bath with distilled water. More than ten tooth samples are analyzed originating from ;

(i) Peru (600 B.C. - 900 A.D.)

(ii) Jesolo (near Venice), Italy (1100-1300 A.D.).

(iii) a recent age (Dental Patient).

In the present paper, we report the results of three tooth samples, one each representing each the above mentioned group.

For the internal calibration of the trace elements to be measured, the elements were introduced in known quantity in the matrices of calcium hydroxyapatite

[$Ca_{10}(PO_4)_6(OH)_2$, 39.9% Ca, weighted atomic number 14.4] as standards. The choice of combinations of elements in standards were done in such a way that the X-ray K_α and K_β peaks from different elements do not overlap.

In the actual experiment, the X-rays are induced by 1.8 MeV proton beam from the AN-2000 Van de Graaff Accelerator of LNL. The target holder was designed in such a way, so that it was possible to make horizontal movement of the tooth samples, in order to measure several points for each tooth-section. The beam was collimated by 3 graphite collimators and subsequently by a 1 mm Aluminium collimator on the target samples. For measuring the integrated proton current, the tooth samples were overbrushed with a thin layer of liquid graphite coating. The emitted X-rays were detected by a Si(Li) detector (30 mm^2), which was subsequently connected with a standard electronic chain including a pile-up rejector, with a resolution of 190 eV (FWHM at 5.9 keV). The data acquisition was obtained through an IBM-personal computer. The single spectra were subsequently passed to VAX-6310 computer system, for the off-line analysis with coventional programs. A typical spectrum, for instance of enamel, is shown in Fig. 1

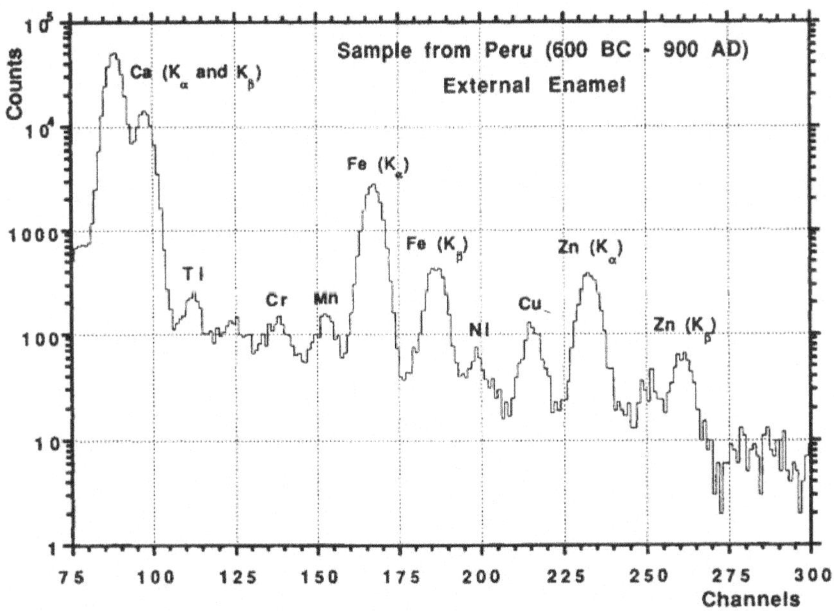

Fig. 1 Partial PIXE Spectrum for a Tooth Enamel sample from PERU

3. RESULTS AND DISCUSSION

The experimental results of the analysis for enamels and different sections for all the three tooth samples are shown in table 1. Fig 2 demonstrates the distribution of selected heavy elements extending from enamel to the different sections.

Table 1 a,b and c

Experimental results of the PIXE analysis of the three tooth samples.

a) Peru (600 BC-900 AD)

Samples (depth)	Ca %	Ti ppm % err	V ppm % err	Cr ppm % err	Mn ppm % err	Fe ppm % err	Co ppm % err	Ni ppm % err	Cu ppm % err	Zn ppm % err	Sr ppm % err	Pb ppm % err
Enamel	38.4	769 13		380 15	584 10	21646 1		374 18	1340 6	6672 2	2360 13	540 34
Dentine (1 mm)	28.2	1397 16		3584 5	1078 14	40347 1		2545 7	926 15	18236 2		
Dentine (1.4 mm)						219 6	33 41	420 4	51 33	289 7		
Dentine (1.8 mm)				137 17		1649 2		261 10	91 25	1275 3	285 38	

b) Jesolo (Venice), Italy, 1100-1300 AD

Samples (depth)	Ca %	Ti ppm %err	V ppm %err	Cr ppm %err	Mn ppm %err	Fe ppm %err	Co ppm %err	Ni ppm %err	Cu ppm %err	Zn ppm %err	Sr ppm %err	Pb ppm %err
Enamel	38.4	118 48	67 59	566 7	630 6	3700 2	451 10	607 7	314 12	1422 5	195 43	229 40
Dentine (1.4 mm)	28.2		182 80	465 47	260 42	4560 5		208 51	173 49	600 21	754 10	
Dentine (1.8 mm)			259 32	272 15	2471 5	3858 4		110 26	85 29	302 23	560 35	
Dentine (2.2 mm)				203 19	1393 5	3928 3		87 38	168 27	466 13	540 28	
Dentine (2.6 mm)			61 43	139 14	1816 2	3320 1	47 54	95 17	71 18	524 4	682 8	66 41

c) Actual tooth sample

Samples (depth)	Ca %	Ti ppm % err	V ppm % err	Cr ppm % err	Mn ppm % err	Fe ppm % err	Co ppm % err	Ni ppm % err	Cu ppm % err	Zn ppm % err	Sr ppm % err	Pb ppm % err
Enamel	38.4	247 39		150 33	126 39	3803 2.3		137 25	149 42	11189 2	2509 13	6389 7
Dentine (1.0 mm)	28.2					464 4		172 10	49 34	335 7	126 56	
Dentine (1.4 mm)						211 10		153 13		354 8		
Dentine (1.8 mm)						429 4		291 5	147 11	405 5	147 29	
Dentine (2.2 mm)				33 51		672 3		158 11	60 29	1010 3		

Fig. 2 Distribution of selected heavy elements in different sections of a single tooth for the three different samples

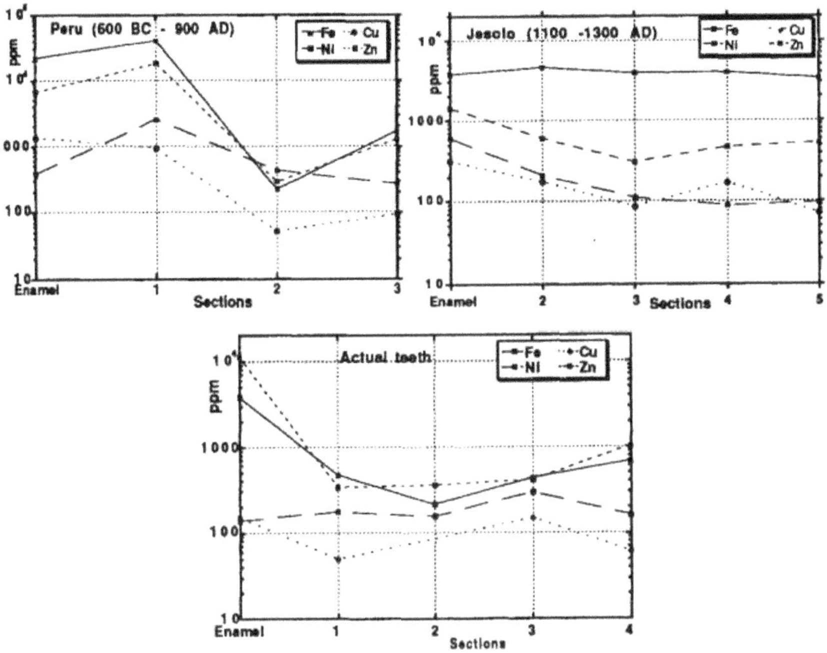

On the basis of table 1 and Fig. 2 the following hypothesis can be drawn ;

(i) The enamel (the first section) principally absorbs the heavy elements from the external environment (air, water, alimentation etc.), whereas for the dentines nearer to the pulp chamber, the absorption is attained through the blood. The most internal dentine-sections are hence less affected from the external environmental effects.

(ii) Comparing the above three distribution figures, one finds that there is a clear diffusion of heavy elements from the enamel towards the dentines. The effect is time dependent and more prominent when the teeth are not "in-vivo". In the samples from Peru, which are most ancient among the three groups, the concentrations are much elevated towards the internal dentine sections than the enamel itself. On the other hand, for the samples from Jesolo (less ancient), this diffusion characteristic is less apparent. For the tooth sample, originating from the actual age, the diffusion process towards the dentines is yet to initiate. Nevertheless the diffusion process is diverse for the particular elements concerned.

4. CONCLUSION

The tooth samples originating from different epoches of time, analyzed through the PIXE technique, indicate presence of diffusion of heavy elements, from the outer enamel towards the internal dentines. The effect is time dependent and rather varying for the elements in concern. Nevetheless, lot amount of samples, varying from different place of origin and different epoches of time, are still to be analyzed, in order to apprehend the characteristic of such diffusion processes and correlate the experimental data with the hitherto environmental effects.

5. REFERENCES

[1] Losee, F.L. and Ludwig, T.G., J. Dent. Res. 49 (1970) 1229.

[2] Needleman, H.L., Tuncay, O.C., and Shapiro, I.M., Nature 235 (1972) 111.

[3] Ahlberg, M. and Akselsson, R., Int. Journal of Appl. Radiat. and Isotopes, 27 (1976) 279.

[4] Tadic, T., Jaksic, M., Babic-Zivko, J., Cicek, J., Fazinic, S. and Valkovic, V. Nucl. Instr. and Meth.in Phys. Res. B49, 211 (1990).

[5] Cua, Florence T., Nucl. Inst. and Meth. in Phys. Res. B49, 205 (1990).

INVESTIGATION OF BIOLOGICAL NITROGEN FIXATION VIA NUCLEAR TECHNIQUES

C.E. Douka, *Institute of Biology*
and
A.C. Xenoulis, T. Paradellis and M. Pilakouta
Institute of Nuclear Physics

NCSR"Demokritos", 153 10 Aghia Paraskevi, Greece

ABSTRACT

The applicability of nuclear techniques in the study of biological nitrogen fixation is demonstrated. The degree of biological fixation achieved by a plant can be quantified with very good accuracy by means of nuclear-reaction γ-ray spectrometry. The application of the x-ray fluorescence technique demonstrates for first time that the elemental composition of plants is significantly affected by biological nitrogen fixation.

INTRODUCTION

Biological nitrogen fixation, which is an enzymatic reduction of atmospheric nitrogen to ammonia accessible to plants by soil bacteria of the genus *Rhizobium*, is achieved under suitable conditions through a symbiosis of rhizobia with leguminous plants. When such a symbiosis is succesfully established the plant yields increase while application of nitrogen fertilizers is not required since the plant can aquire the nitrogen needed for its growth directly from the unlimited supplies of the atmosphere.

The significance of biological nitrogen fixation to human welfare cannot be easily overemphasized. It is estimated, for instance, that about 80% of the available plant food protein is produced via that important process. Consequently, nitrogen fixation constitutes a subject of intense investigation worlwide, since its better exploitation, in addition to the obvious economic benefits, may contribute to the protection of the environment from the pollution caused by the headless use of fertilizers.

The purpose of the present study was to develop or adopt nuclear techniques required to facilitate the systematic study of biological nitrogen fixation.

THE INVESTIGATED PLANTS AND THEIR CONDITIONS OF GROWTH

The main plants used are *Medicago sativa* (alfalfa) and soybean. For the achievement of biological fixation the plants are inoculated with suitable rhizobia, of the genus *Rhizobium meliloti* and *Rhizobium Japonicum*, previously isolated in this laboratory from Greek soils. The growth of the plants took place on several media either in the laboratory in nitrogen-free agar medium[1] and soil-agar medium[2] or in field experiments.

MEASUREMENT OF INDUCED BIOLOGICAL NITROGEN FIXATION

A plant undergoing biological nitrogen fixation, can obtain nitrogen from the atmosphere, the growth medium and from a fertilizer which may be present. The estimation of the nitrogen obtained from the atmoshpere requires the application of an ^{15}N-labelled fertilizer as well as the measurement of the total nitrogen and the $^{14}N/^{15}N$ isotopic content on the plant. Prior to the present work the total nitrogen content was obtained by conventional chemical analysis while the isotopic ratio was measured with a mass separator.

We have developed a nuclear technique which allows the simultaneous measurement of both quantities with very good accuracy. Important advantages of the technique are that it requires very small amount of sample, a feature which permits to carry out relevant investigation on a single plant basis, and that it is non-destructive thus leaving the sample available for subsequent measurements.

The γ-rays emitted in the nuclear reactions $^{14}N(p,p'\gamma_1)^{14}N$ and $^{15}N(p,\alpha\gamma_1)^{12}C$ simultaneously induced by protons of 4.3 MeV on the corresponding isotopes of the sample are measured[3]. The applicability of the $^{14}N(p,p\gamma_1)$ reaction in the determination of absolute abundance of ^{14}N with a 2% uncertainty has been demonstrated with the aid of inorganic standards[4].

The application of the above technique, in the measurement of nitrogen content of plant material, however, provided initially inconsistent and overstimated results, which led eventually to the realization that the γ-ray yields from various isotopes during a charged-particle bombardment of a biological specimen increase, up to a saturation level, with increasing amount of total beam deposited on the target[5]. Significant alterations, as high as 60%, have been observed to be induced.

In order to eliminate the errors introduced by these yield increases a simple target-moving system was developed and tested[6]. Althought there is a limit in the total beam which may be safely deposited on the moving biological target, the duration of the safe bombardment is sufficiently prolonged by the motion to allow the accumulation of meaningful statistics.

An example of the information about biological nitrogen fixation, achieved under specific conditions, obtained from total and isotopic nitrogen content measurements is demonstrated in Fig. 1, where the extracted % nitrogen derived from fixation is shown. Two soybean varieties were tested in this field experiment. One soybean variety (indicated by N) is known to fix nitrogen, while the second (indicated by n) does not. The parameters examined were the effect of application of phosphorous and of the time of the harvest . The results associated with the n variety indicate zero nitrogen fixed in agreement with the characterization of the n soybean as non-fixing. With respect to the fixing soybean we note that the addition of phosphorous has no effect on nitrogen fixation. The application, therefore, of phosphorous fertilizer is deemed unnecessary. On the other hand, the time of the plant harvest is an important factor, since the amount of nitrogen fixed in the latter (third) harvest is significantly larger than that obtained at an earlier (second) harvest.

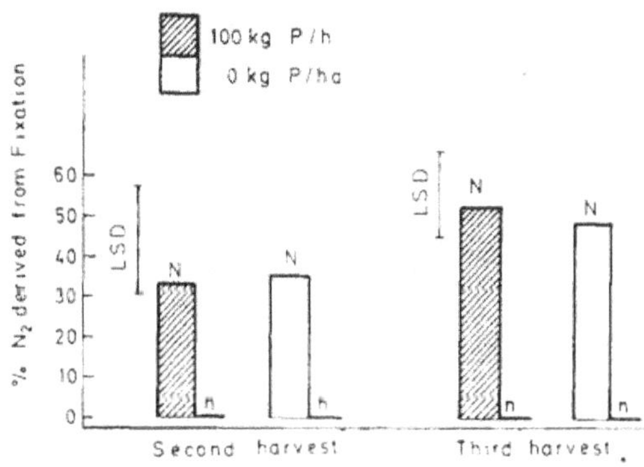

Fig. 1 ^{15}N - determined time course of percentage of N_2 derived from fixation at 0 and 100 kg P_2O_5/ha fertilizer application.

One can thus test the various factors affecting biological fixation and select the optimum conditions for maximum production.

EFFECT OF BIOLOGICAL NITROGEN FIXATION ON THE ELEMENTAL COMPOSITION OF PLANTS

Little attention has been attached prior to the present investigation to the efffects that biological nitrogen fixation may have on the elemental composition of plants, although identification of such effects may have important practical as well as methodological consequences.

The concentration of several elements in plants grown with and without inoculation with rhizobia under otherwise the same conditions was measured using the x-ray fluorescence technique and compared in order to determine whether and to what extent the elemental composition of plants is effected by biological nitrogen fixation.

Such a comparison is demonstrated in Table 1, associated with *Medicago sativa* plants grown with and without inoculation in a laboratory experiment. Significant differences in the elemental composition of these plants are observed.

To ensure that such concentration alterations are caused by nitrogen fixation independently of the nature of the growth medium and the conditions of growth, a great variety of experiments either in the field or in the laboratory have been already carried out.

The elements the concentrations of which have been identified to be definitely affected by nitrogen fixation may be summarized as follows:

i) The concentration of Mo, Mn, Sr, Cu and Zn significantly decreases

ii) The concentration of K and Rb increases

iii) The concentration of Ni and Ca is not affected.

TABLE 1. Effect of inoculation with effective Rhizobium melilotι strain on the elemetnal concentraιion of Medicago sativa plants grown on nitrogen-free agar slopes

Element	Elemental concentration in ppm or %	
	uminoculated	inoculated
Cl	≤1%	0.8 ± 0.3%
K	2.9 ± 0.5%	2.1 ± 0.2%
Ca	2.4 ± 0.3%	1.2 ± 0.2%
Fe	630 ± 70	320 ± 15
Mu	219 ± 10	113 ± 8
Zn	214 ± 10	84 ± 7
Co	5.1 ± 0.5	2.2 ± 0.2
Ni	9 ± 3	7 ± 1
Cu	28 ± 3	24 ± 2
Br	47 ± 4	50 ± 4
Rb	2.8 ± 0.2	4 ± 0.3
Sr	35 ± 2	22 ± 2
Mo	51±3	10 ± 1

The observed elemental composition alterations caused by biological nitrogen fixation have important practical as well as foundamental implications. From the practical point of view, the observed reductions may cancel toxic effects of certain elements. Concentrations of 20 ppm of Mo, for instance, in pastures are known to be toxic to ruminant animals. Apparently, the growth of pastures via biological nitrogen fixation aleviates that toxicity since in that case the concentration of Mo is reduced almost by an order of magnitude.

From a methodological point of view, furthermore, the observed significant effects of biological fixation on the concentration of Mo seem of particular interest. A large number of relevant experimental data suggest that a correlation may exist between the degree of induced nitrogen fixation and the Mo concentration of a nodulated plant. An explicit form of such a correlation between plant yield of inoculated Medicago sativa and Mo concentration has been identified. By a linear fit to the data of the correlation the following equation was derived

$$\text{plant yield} = 23.9 - 0.53 \times [\text{Mo}], \qquad (1)$$

where the plant yield is given in mg per plant and the concentration of Mo in ppm. The above result suggests that it may be possible to predict a plant yield from a rather simple measurement of the plant Mo concentration.

In conclusion the application of nuclear techniques can contribute substantially to a better understanding and utilization of the very important biological nitrogen fixation process.

REFERENCES

1. Jensen, H.L., "Nitrogen fixation in leguminous plants", Proc. Linean Soc. New South Wales **66**, 98-108 (1942).
2. Douka, C.E., "Unsterilized soil-agar technique for studying Rhizobium plant relationships", J. Appl. Bact. **46**, 615-622(1979).
3. Xenoulis, A.C. and Douka, C.E., "Determination of relative and absolute abundance of stable nitrogen isotopes by proton induced nuclear reactions", J. Radioan. Chem. **54**, 205-210(1979).
4. Xenoulis, A.C., Douka, C.E., Paradellis, T. and Katsanos, A., "Stopping power effects in the determination of ^{14}N by a resonance reaction", J. Radioan. Chem. **63**, 65-71(1981).
5. Xenoulis, A.C., Aravantinos, A.E. and Douka, C.E., "The stability of biological specimens during charged particle bombardment", J. Radioan. Chem. **77**, 207-222(1983).
6. Xenoulis, A.C., Trouposkiadis, F. and Douka, C.E., "Suppresion of charge-induced yield increases in PIGE analysis of biological specimens", J. Radioan. Nucl. Chem. **98**, 115-120(1986)

COMPARISON OF XRF AND PIXE METHODS FOR MEASUREMENT OF EXTRACELLULAR WATER IN HUMAN BODY@

H.R. Shao*=, Q.X. Liu*=, G. Sergi+, P. Perini+, G. Enzi+ and G. Moschini*#

* INFN, Laboratori Nazionali di Legnaro (LNL), Legnaro, Padova, Italy
= Institute of High Energy Physics, Beijing, China
\+ Dipartimento di Medicina Interna, Università di Padova, Padova, Italy
\# Dipartimento di Fisica, Università di Padova, Padova, Italy

ABSTRACT
Extracellular water is the compartment where the transport of gases, nutrients and catabolites of cell metabolism occurs. Oral load of sodium bromide is an accurate and non-invasive method for the evaluation of extracellular water. Bromine concentration in blood plasma was measured by XRF. PIXE was used as reference.

1. INTRODUCTION

Extracellular water (ECW) is the compartment where transport of gases, nutrients and catabolites of cell metabolism occurs. From the medical point of view, it is extremely important to evaluate ECW because of its variation with age[1,2,3], malnutritional status[4], heart[5] and kidney diseases, etc.

Normally, in the past, ECW had been measured using radioisotopes (^{82}Sr, ^{34}S, ^{24}Na) as tracers. Due to its invasivity, however, this technique cannot be used routinely. Non-radioactive bromide is also employed: bromide space measured by an oral load of Sodium bromide is an accurate and non-invasive method[3] for evaluation of ECW. Available spectrophotometric methods, however, require Br-concentration in plasma over 500 μg/ml and, therefore, this method cannot be applied to humans.

Low concentrations of bromine (μg/ml) in the biological fluids can be measured by three physical methods : Neutron Activation Analysis (NAA)[6], X-ray Emission induced by Protons (PIXE)[7] and Photons (XRF). However, NAA and PIXE are rather not suitable for routine analysis.

In the present study, we have tested the accuracy and precision of XRF, an inexpensive and simple method for determination of bromide in

@ Work partially supported by the Commission of European Communities [Contract N° CI 1-0331 - I(A)].

plasma, utilizing a small annular source of ^{241}Am. PIXE has been used as reference method.

2. PRINCIPLE AND METHOD

Bromine was absorbed by body and diluted by extracellular water. Extracellular water was calculated using the following formula[8]:

$$ECW = (Br\ dose\ /\ Br\ in\ plasma) \times 0.90 \times 0.95 \times 0.94$$

where 0.90 is the correction for nonextracellular distribution, 0.95 is the Donnan equilibrium factor, and 0.94 is the correction for percentage of water in the plasma.

Eleven healthy volunteers were orally supplemented with sodium bromide (30 mg/kg body weight). For each person 3 ml of heparinized blood was drawn before and at the intervals of 2, 3 and 4 hours after the drug administration. 1 ml of blood plasma was mixed with 0.5 mg yttrium and then freeze-dried. Yttrium was chosen as the internal standard since it does not naturally occur in blood. The weight of dried plasma was about 100 mg. The dried plasma was ground by the quartz mortar for the further homogenization. Then 50 mg of homogenized dried plasma powder was pressed into a pellet.

For calibration, 60, 120, 180 and 240 μg of bromine were added separately into normal blood plasma and then the samples were prepared in the identical way. Standard reference material of animal blood (IAEA A-13) was similarly prepared.

Bromine concentration in blood plasma were determined by the XRF method using an annular source of of ^{241}Am ($1.1*10^9$ Bq) with beryllium window. The Si(Li) detector was shielded and collimated by Cd. The distance between the source and the sample was about 2 cm.

Some samples were determined by PIXE as well. The same dried plasma powder was mixed with nuclear grade graphite and pressed into a pellet. The proton energy was 1.8 Mev. The beam current at the target was usually 7 - 10 nA. A 30 μm thick Al-film was placed in front of the detector to reduce the low energy radiations.

Both X-ray spectra data were analysed by X-ray spectral analysis program AXIL.

3. RESULTS AND DISCUSSION

Figure 1 shows XRF spectrum before and after drug administration. The calibration curve is shown in figure 2. It is a straight line. The bromine concentration in normal blood plasma was indicated at about 4.5±1.0 μg/ml. The bromine concentration in standard reference material animal blood was measured also. The result obtained is 23 ± 3 ppm. This is in good agreement with the certified value of 22 ppm.

Table 1 shows bromine concentrations after the drug administration and the extracellular water values. Bromine can be absorbed fast in the body. Equilibrium of bromine concentration in blood was established about 2 hours after the drug administration. Thereafter, the bromine concentration decreases very slowly. We choose the bromine concentration value obtained after 3 hours to calculate the value of extracellular water. The results indicate that no statistically significant difference could be found between slim and fat, and between men and women.

Table 2 shows a comparison between the XRF and the PIXE method. The results obtained by the XRF and the PIXE method are in a good agreement. However, for the PIXE method an accelerator is needed, which makes the application of this method more difficult. On the other hand, the XRF method uses a radioactive source which is an inexpensive, simple and reliable method for the routinary determination of bromine in plasma.

4 REFERENCES

[1] Cheek D.B., Wishart, J., MacLennam, A.H., Haslam,R. and Fitzgerald,A., "Hydration in the First 24 h of Postnatal Life in Normal Infants Born Vaginally or by CaesareanSection", Early Hum. Dev. 7, 323 (1982)

[2] Haschke, F., "Body Composition of Adolescent Males Part 1. Total Body Water in Normal Adolescent Males. Part 2. Body Composition of the male Reference Adolescent", Acta Paediatr. Scand. 307(suppl), 1 (1983)

[3] Cheek, D.B., "Extracellular Volume: Its Structure and measurement and the Influence of age and Disease", J. Pediatr. 58, 103 (1961)

[4] Shizgal, H.M., "The Effect of Malnutrition on Body Composition", Surg. Gynecol. Obstet. 152, 22 (1981)

[5] Cheek, D.B., "Human Growth: Body Composition, Cell Growth, Energy and Intelligence", (Lea and Febiger, Philadelphia, PA, 1968) pp. 150-164

[6] Jervis, R.E., Hancock, R.G.V., Hill, D.E. and Isles, K., "Biomedical and Health Studies with the New Canadian Slowpoke Reactor", J.Radioanal. Chem. 37, 463 (1977)

[7] Shao, H.R., Liu, Q.X., Enzi, G., Moschini, G. and Sergi, G. "Evaluation of the Extracellular Water in Human Body by Determination of Br Concentration in Blood Plasma", Nucl. Instru. and Meth. in Physics Research B49, 238 (1990)

[8] Bell, E.F., Ziegler, E.F. and Forbes, G.B., Letter to the Editor, Pediatr. Res. 18, 392 (1984)

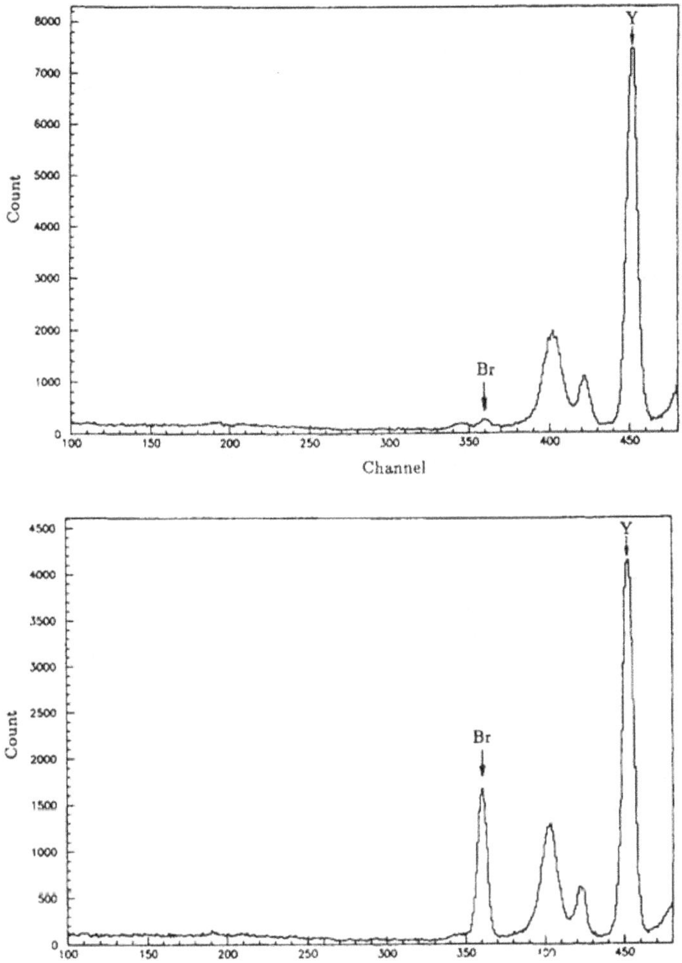

Fig.1, The XRF Spectrum of the blood plasma

a) before drug administration, measured time: 4000 sec.

b) 3 h after drug administration, measured time: 2400 sec.

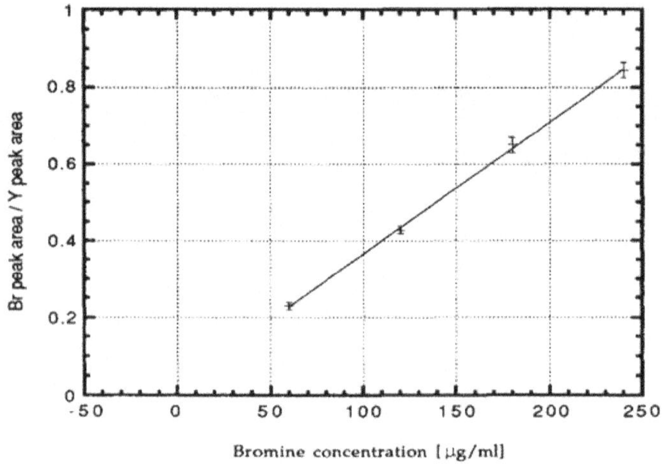

Fig. 2: Calibration curve of Bromine in plasma by XRF

Table 1. Bromine concentrations before and after drug administration, and the extracellular water values

Patient No.	Sex	Age	Weight [kg]	Height [cm]	H/W [kg/m]	Br a) [μg/ml]	Br b) [μg/ml]	ECW [%]
1	F	24	56.5	166	0.34	5.1	95±7	22±2
2	F	25	54.5	161	0.34	5.2	91±7	22±2
3	F	31	56.5	161	0.35	5.7	101±8	21±2
4	F	24	63.0	178	0.36	8.0	84±7	25±3
5	F	31	61.8	169	0.37	5.5	93±7	22±2
6	M	28	72.0	175	0.41	5.1	82±7	25±2
7	M	24	77.8	184	0.42	4.4	92±7	22±2
8	M	25	75.2	177	0.43	5.5	89±7	22±2

a) before drug administration *b) at 3 hours after drug administration*

Table 2. A comparison between the XRF and the PIXE method

Patient No.	Sex	Bromine concentration (Br) [μg/ml]		Extracellular water (ECW) [%]	
		PIXE	XRF	PIXE	XRF
		5.3	5.0		
9	F	87	86	23	23
		6.5	7.7		
10	M	87	86	23	23
		5.1	6.0		
11	M	92	90	21	22

DETECTION OF ELEMENTS AND TRACE ELEMENTS IN ENDOSCOPY BIOPSIES OF COLONIC MUCOSA IN NORMAL AND ULCERATIVE COLITIS PATIENTS

P.E. Lecis[+], M.C. Buoso[*], S.Fazinic[*%], S. Galassini[*@], M. Makarewitcz[°], G. Moschini[*#], R. Naccarato[+], R. Ogris[°], G.C. Sturniolo[+] and V. Valkovic[°]

* I.N.F.N., Laboratori Nazionali di Legnaro (LNL), Legnaro (Padova), Italy
% Rudjer Boskovic Institute, Zagreb, Yugoslavia
\# Dipartimento di Fisica, Università di Padova, Padova, Italy
@ Istituto di Chimica Biologica, Università di Verona,Verona, Italy
\+ Divisione di Gastroenterologia "R.Farini", Istituto Medicina Interna, Università di Padova, Padova, Italy.
° IAEA,Laboratories-Seibersdorf, Austria.

ABSTRACT

Many papers have reported a correlation between Trace Elements (T.E.) and colorectal cancer. Zinc (Zn), Selenium (Se) and Calcium (Ca) seem to be involved in the Colon Cancer development. In this work the sampling procedure, the analytical technique and the preliminary results of colon mucosal T.E. concentrations collected during endoscopic examination are presented. The samples were taken from patients with ulcerative colitis, a disease at high risk for colon cancer and from colon looking normal macroscopically and at histology.

1. INTRODUCTION

Colorectal cancer is one of the most diffused forms of cancer in man and the second cause of cancer related mortality in western countries[1, 2, 3)].

Carcinomas have different distributions in the various sites of the colon, which is normally divided into five segments (fig. 1):

1) caecum and ascending colon;
2) transverse colon;
3) descending colon;
4) sigmoid colon;
5) rectum.

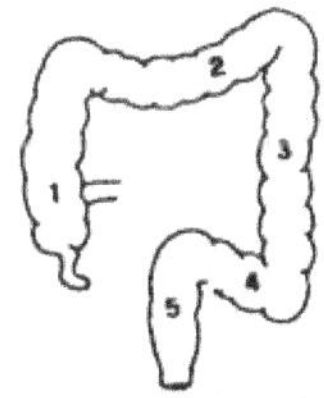

Fig.1 Colon segments

More than 60% of the carcinomas are localized in the left and distal parts of the colon (sigma and rectum)[2, 4)].

Epidemiological studies have clearly demonstrated that environmental exposures play an important role in the development of cancer,.suggesting that the occurrence of colorectal cancer could be closely related to dietary intake. The risk of colon cancer seems to increase with total fat, protein and calorie intake[5]. There is a strong evidence that this risk could depend on the bile acid metabolites (i.e. the deoxycholic and lithocholic acids) produced by the bacterial flora of the right colon, of people following a diet with high concentration of refined sugar and fatty acids[4,6].Several researchers tried to prevent colon cancer by modifying or supplementing diets with chemopreventive agents, like vitamins, trace elements and minerals.

It has been demonstrated, in the U.S.A. and Scandinavian Countries, that the colon cancer risk is inversely related to the concentrations of Ca and Vitamin D in the diet[7,8]. Experimental studies have demonstrated that Ca oral supplementation can reduce the altered cytoproliferative pattern in the rectal mucosa of rats, in the rectum of which were placed bile and fatty acids, as promoters of colonic cancerogenesis[9]. Ca supplements (1250 mg of $CaCO_3$ per day for eight weeks) to patients with familial colon cancer, have reversed to normal the hyperproliferative status of the colonic epithelial cells [10]. It is not known if the effect of Ca derive from the binding of bile and fatty acids into the colon or with a different intracellular mechanism [11].

Epidemiological reports have also underlined that a high risk for colon cancer in people living in particular areas is related to the low Se contents in both soil and diet[12]. Shamberg first noticed that cancer patients had significantly lower serum Se levels than non-cancer controls[13]. The Se chemiopreventive effect has been reported in rats treated with 1.2 dimethylhydrazine (DMH) as a carcinogenic agent [14].

Zinc is essential for more than seventy methalloenzymes, but can also be mutagenic[15]. Experimental Zn deficiency seems to be associated with suppression of colonocyte proliferation[16].

In the present work we will describe the methodological aspects, alongwith the preliminary experimental results, of the analytical techniques used for a multielemental analysis of the human colonic mucosa, in order to detect the concentrations of Ca, Se and Zn.

2. EXPERIMENTAL PROCEDURE

Intestinal mucosa specimens were collected during colonoscopy from patients with normal mucosa and from patients with ulcerative colitis, a chronic inflammatory bowel disease with a high colon cancer risk. During the examination, three biopsies from each colonic segment were collected, by means of steel pliers, immediately frozen at liquid nitrogen temperature and stored at -20°C. Later, the samples were freeze-dried (for 48 hours) before the acid digestion phase. Each sample has been made of three biopsies, in order to have enough quantity of organic material. Seven samples, obtained from three normal colon, were freeze-dried and then analysed by the Neutron Activation Analysis (NAA) technique, at I.A.E.A. Laboratories in Seibersdorf, (Wien), to detect Se and Zn concentrations

The results concerning the measurements of both Zn and Se are displayed in Table 1, where it may be evidenced the difference of the concentrations (for both the detected elements) between the right an left part of the colon.

Table 1 Zinc and Selenium Concentration in normal Population*

Patients N° Segment	Dry weight [mg]	Zn Conc. [mg/g]	Stand. Dev. %	MDL [mg/g]	Se Conc. [mg/g]	Stand. Dev. %	MDL [mg/g]
9 Right Colon	1.2	109	3	0.85	1.2	16	0.21
9 Left Colon	1.8	110	3	0.57	0.91	17	0.18
10 Right Colon	2.7	73	3	0.67	0.89	18	0.19
10 Left Colon	2.4	133	3	1.06	1.6	15	0.25
11 Right Colon	1.7	103	3	0.35	1.4	11	0.11
11 Right Colon	2.2	108	3	0.42	1.7	12	0.15
11 Left Colon	1.5	109	3	0.85	2.2	12	0.22
Bovine Liver		133	2	0.09	0.78	10	0.04

* Results obtained with NAA

Table 2. Concentration of Selenium in Colonic Mucosa Biopsies [ppm]

P.n.	PIXE						NAA	
	NORMAL			PATIENTS			NORMAL	
	Small Bowel	Right Colon	Left Colon	Small Bowel	Right Colon	Left Colon	Right Colon	Left Colon
1		< MDL	1.1					
2			4.8					
3		1.4	2.1					
4		< MDL						
5		< MDL						
6				0.9	0.5			
7					4.1	1.3		
8				1.1	0.8	2.4		
9							1.2	0.91
10							0.89	1.6
11							1.4	
11							1.7	2.2

Standard Bovine Liver : with PIXE 0.63 ppm
with NAA 0.78 ppm
certified value 0.71 ± 0.07 ppm

It should also be noted that the values obtained for Zn are in good agreement with those reported in literature[17]. The results obtained for the Se are grouped in Table 2, in an effort for a better comparison with the analogue results obtained by means of the PIXE technique.

To detect the Se concentrations, by applying the later technique, fourteen samples obtained from five normal people and three ulcerative colitis patients were prepared according to the wet ashing procedure, proposed by Maenhaut et al.[18] as the most suitable for biological materials. Each sample was placed into a pyrex glass flask, where 3 ml of concentrated 65% superpure nitric acid and 1 ml of 70% perchloric acid were added, and then heated until the volume of the solution was reduced to 1 ml. After stopping the heating, 4.5 ml of pure water were added. The final targets were prepared according to a chemical preconcentration technique, that uses Te both as a co-precipitant and as internal standard. This technique has been discussed elsewhere[19,20].

As exciting source a proton beam, with an energy of 1.8 MeV and a current intensity of about 40 - 50 nA from the AN-2000 Van de Graaff accelerator at Legnaro National Laboratories was used. A 30 mm^2 Si(Li) detector, connected with a standard electronic chain, including a pileup rejector, with an overall resolution of ~ 190 eV for the 5.9 KeV Mn-K_α line was used for the determination of the X-rays spectra. To reduce intense low energy radiations, an Aluminium absorber, with a thickness of 28 μm, was placed in front of the detector. To check the accuracy of the measurements a sample of bovine liver (NBS 1577) international standard was similarly prepared. The results thus obtained are shown in Table 2

To exclude the possible contamination from biopsies forceps, we examined the X-rays spectra of both the steel pliers obtained with a 1.1*10^9 Bq Fe^{55} source (Fig.2).

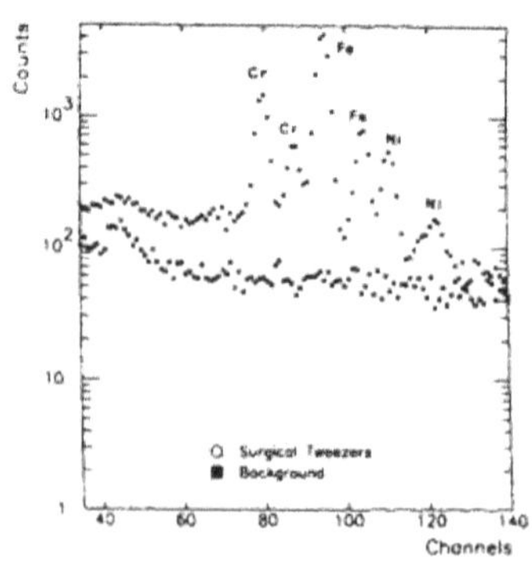

Fig.2 XRF Spectrum of a Surgical Tweezers.

To detect the Ca concentration, a normal colonic mucosa sample has been measured by means of the XRF technique employing a $4.4*10^8$ Bq Fe^{55} source. A freeze-dried sample, 3.3 mg in weight, was put onto a polypropylene foil and measured without any additional sample preparation. For the calibration we used a 20 μl drop from a 1 mg/ml Ca standard solution. Citrus leaves (NBS 1572) reference material was measured to check the accuracy of the measurements. Certified value 3.15 ± 0.10%, measured value 3.58 ± 0.55%. Results for Calcium and Potassium are reported in table 3

Table 3 Concentration of Calcium and Potassium in Colonic Mucosa*

Patient No. 4	Left Colon	
	Calcium	Potassium
concentration	640 ± 100 ppm	0.84 ± 0.08 %
M D L ppm]	60	180

3. CONCLUSIONS

The data about T.E. concentrations in the colonic mucosa are very scanty in literature and most of authors have analyzed the different T.E. concentrations in colon cancer and normal mucosa. The distribution of T.E. in cancer can be influenced by a lot of variables. The high cancer vascolarization can determine a T.E. imprisoning in the neoplastic tissue. The cancer can modify the T.E. balance because of a reduction of intake or an increasing in the excretion. It is impossible to know if the altered metabolism of T.E. is a cause or an effect of the disease. We analyzed persons with high cancer risk disease and not people with evident cancer, to understand the influence which T.E. can have on the cancer development. To do this, we need samples collected during endoscopy examination and not surgical samples. This represent a methodological problem because the sample size is about 3 mg weight, and not 200 mg like the surgical samples. We use amount of biological material in which the concentration of the elements is almost near to the Minimum Detection Limit (M.D.L.) for that method. The Se concentrations in our samples, detected by PIXE, is about 1 ppm. The same concentration was demonstrated with NAA. NAA and PIXE, methods have shown that there is a difference in Se concentration between right and left colon. It is higher in the left colon. The number of samples assayed are very limited and is impossible to draw final conclusions.

We have used a good method to study the Calcium concentration in the colonic biopsies without any digestion. Following calcium other T.E. could be detected. This was possible because the Ca concentration, we found in the colonic mucosa, is very high, i.e. about 600 ppm. PIXE is a good method to detect Se in small colonic mucosa and results obtained by NAA and PIXE are in good agreement. Nevertheless, our results need to be confirmed on a big scale. On the other hand, the XRF is an inexpensive, simple and reliable method to determine Ca in the colonic mucosa and to start a multielemental analysis in our small samples.

4. REFERENCES

[1] Winn R.J. and Levin B., Hem./Oncol. Clin.N.Am.3,1;(1989) 65-732
[2] Ponz de Leon M. et al., Cancer 60 (1987) 2848-2859
[3] Terracini B. and R. Puntoni, Fed. Med. XLIII 2 (1990) 106-115
[4] M. J. Hill, from "Diet and human carcinogenesis" J.V. Joossens et al. editors., Elsevier Science Publishers B.V.(1985) 149-163
[5] S. Graham et al., Am. J. Epidemiol.128 (1988) 490-503
[6] T. Narisawa et al., J. Natl. Cancer. Inst. 72 (1984) 1323
[7] Sorenson A., Slattery M.,Ford M., Nutri. Cancer. 11(3) (1988) 135-145
[8] Jensen OM, Maclennan R., Isr.J.Med.Sci. 15(4) (1979) 329-334
[9] Wargovich M.J. et al., Cancerogenesis 4 (1983) 1205-7
[10] Lipkin M, Newmark H.L., New Engl.J.Med. 313 (1985) 1381-1384
[11] Lipkin M. et al., Cancer Research 49 (1989) 248-254.
[12] Clark L.C., Fed. Proc. 44 (1985) 2584-4589
[13] Shamberg R.J. et al., JNCI 50 (1973) 867-870
[14] Medina M, Morrison D.G., Pathol. Immunopathol. Res. 7 (1988) 187-199
[15] Hsie A. et al., Trace Elements in health and disease, Raven press, New York (1979) 55-69
[16] Lawson M.J. et al., Biological trace elements research 18 (1988) 115-121
[17] Draskovic R.J., Bozanic M., J. Radian. Nucl. Chem. 116(2) (1987) 409-420
[18] Maenhaut W., De Reu L. and Vandenhaute J., Nucl. Instr. Meth. Phys. Res. B3 (1984) 135-140
[19] Liu Q.X., Shao H.R., Moschini G., J. Radianal. Nucl. Chem. ,Letters 144(1) (1990); 47-55
[20] Buso G.P. et al. Nucl. Instr. Meth. Phys. Res. B3 (1984) 177

INSTITUTE OSSERVATORIO GEOFISICO MODENA UNIVERSITY (ITALY) LABORATORY REPORT

TRACE AND ULTRATRACE ELEMENTS DETERMINED BY PIXE IN NATURAL WATERS AND SEDIMENTS TO STUDY POLLUTANT DISTRIBUTION AND THEIR TOXICOLOGICAL EFFECTS

R. Cecchi* and G. Ghermandi**
* Inst. Geofisico e Geodetico, Messina University (Italy)
** Inst. Osservatorio Geofisico, Modena University (Italy)

1. ABSTRACT

This report is a general description of systems, devices, and methods used by the authors for environmental research by means of PIXE technique.

2. INTRODUCTION

The Institute Osservatorio Geofisico has existed in Modena University (Italy) since 1826; its main activity is the observation and survey of geophysical parameters and the study of geophysical and ecological processes.

The PIXE research group of Inst. Osservatorio Geofisico began its work about fifteen years ago. The chief investigator of the group is Prof. R. Cecchi, his staff includes one researcher (Dr. G. Ghermandi) and one technician; post-graduates and students support the laboratory work and data handling.

Our group utilizes the PIXE (Particle Induced X-ray Emission) technique for research in the environmental field. At present, we are chiefly concerned with the study of trace and ultratrace metal transfer and distribution in aquatic systems, as it relates to biological aspects (metal uptake from water in marine fishes). In these typically interdisciplinary studies cooperation exists with other institutes for "in situ" measurements and sample operations. Our group provides the environmental sample preparation for

PIXE analysis. For this purpose a laboratory has been equipped, and a suitable methodology has been developed[1] for the preparation of thin targets for PIXE analysis from water and diluted sediment samples. The PIXE measurements are performed at the 2 MeV accelerator of the INFN Laboratory of Legnaro (Padua, Italy), using an especially designed irradiation chamber and experimental set-up. The PIXE spectral analysis and data handling are performed at Modena University, with suitable programs written by our group.

3. THE LABORATORY FOR TARGET PREPARATION

A separate laboratory is reserved only for the preparation of PIXE targets from environmental matrix samples; its environment is checked to avoid spurious contamination. This laboratory has suitable facilities for the pre-treatment of samples containing traces of pollutants to be analyzed. It has a) a safety cabinet class 100 (Gelaire BSB 6A) in which the total PIXE target preparation is performed, b) systems for liquid sample filtration, made of borosilicate glass, that are placed in the cabinet, c) a PVC glove-box containing the ultra pure reagents, standard solutions and reference materials used in the chemical sample treatment, in the calibration tests and the glassware washing operation, d) a plant for water purification (Millipore, producing water at 18 MΩcm resistivity) and a pure water distribution network that supplies the equipment for glassware washing and conditioning (Mielabor glass washer, ultrasonic bath) placed in this laboratory. The same network provides pure water also to another laboratory, used for the first step of glassware washing (by immersion in acid mixture).

The materials of laboratory vessels and tools are tested with great care.[2] All the glassware used are of borosilicate glass (DURAN). The glassware and the polyethylene and polypropelene accessories used in the sample treatment are cleaned and conditioned and finally kept in protected containers in the same laboratory. Blank targets are always prepared simultaneously with the environmental samples.

4. THE PIXE SET-UP AT LEGNARO LABORATORY

The targets are analyzed by means of the PIXE technique with the 2 MeV Van de Graaff accelerator of the INFN Laboratory of Legnaro (Padua, Italy). The experimental set-up that we use has been built with our cooperation by a research group of the Physics Department of Padua University, which also operates in the environmental field, applying the PIXE technique in aerosol analysis.

A proton beam of 1.8 MeV extracted from the accelerator is directed (at high vacuum, 10^{-6} Torr) to the irradiation chamber, where the targets are bombarded. At present, the sample holder in the chamber contains up to five samples, but it will be substituted early in the future with another one of larger capacity. The inner surfaces of the chamber are covered with carbon foil, in order to avoid other spurious signals. The beam is uniformly diffused by a 450 $\mu g/cm^2$ Ni foil and then collimated to obtain a homogeneous beam fraction with suitable form and size. A Faraday cup behind the chamber measures the charge collected on the target. The X-ray detector is placed at 135° to the beam direction and in front of the sample. Si(Li) detectors are used to count the characteristic X-rays and background radiation coming from the samples. Depending on the energies of the X-rays of interest an ORTEC Si(Li) detector with 6mm active diameter (0.0127mm Be window, resolution 160 eV at 5.9 keV energy) or another ORTEC Si(Li) with 10mm active diameter (0.245mm Be window, resolution 180 eV at 5.9 keV) may be used. Additional mylar absorbers may be placed in front of the detector. By means of suitable electronic modules (ORTEC preamplifier, amplifier, and pile-up rejector) the spectra are finally collected in a MCA and then stored in the on-line personal computer.

The parameters of a PIXE measurement for a thin sample are typically: beam spot larger than the target, current 20-30 nA, charge collected 20-30 μC, count rate kept lower than 1000 cps, a mylar absorber 156μ thick in front of the detector window, measurement time about 10-15 minutes.

5. METHODOLOGIC ASPECTS OF THE WORK

5.1 Target preparation procedure and calibrations

We have performed a chemical treatment to preconcentrate and extract by precipitation as carbonates seventeen elements (Ti, V, Cr, Mn, Fe, Co, Ni, Cu, Zn, Se, Y, Mo, Cd, Ag, Hg, Pb, Bi) from aqueous liquid samples. The thin precipitate layer, collected by filtration on a Nuclepore filter, is the target ready for PIXE measurement. This procedure allows the analysis of concentrations - with the internal standard method - up to ppb (ng/ml) levels starting from 15 ml sample volumes, and up to ppt (pg/ml) levels from 500 ml samples.[3] The method has been calibrated with PIXE measurements of single-element and multi-element targets, obtained from A.A.S. standard solutions and from reference materials. The results are satisfactory: main percentage errors less than 10% and detection limits from about 2 ng/cm^2 (Ni, Cu) to about 40 ng/cm^2 (Cd, Ag). The recovery efficiency of the procedure has also been evaluated for the studied metals.[4]

5.2 Data handling

We have improved a PIXE data handling system,[5] suitable for ultra trace element analysis of thin targets chemically prepared; it is written in FORTRAN 77 and runs in fully automatic or interactive mode on a VAX (6310 or 750) computer, with low CPU use. For each submitted spectrum, the program provides experimental background subtraction (from blank samples) and data reduction: the linear least squares fit of a function (including up to six gaussians plus an exponential background) "runs" along the spectra from low to high energies; the fit resolves interferences of very closed lines (like Fe K_β and Co K_α). The evaluation of emission intensities, the identification of the corresponding transitions and finally the calculation of the concentration values in the original samples are supported by the program data "library" of X-ray emission energies and experimental parameters (relative intensities of element shells, transmission absorber curves, efficiency coefficients and percentage errors from calibration measurements). The program also performs pile-up correction. At present we are perfecting the program with the evaluation of the "matrix effect" (X-ray absorption in the target itself, secondary X-ray fluorescence, energy loss of the proton beam in the target), that may occur in PIXE measurements with targets thicker than 1 mg/cm^2 (the assumed limit for "thin" target). This is not the case in surface water PIXE analysis, but when interstitial water or solubilized sediments are tested the target thickness may be increased up to 2 - 3 mg/cm^2; in spite of that, the correction factors evaluated for our samples are of the order of a few percent.

6. ENVIRONMENTAL APPLICATIONS

6.1 Venice Lagoon system study

The PIXE technique has been used for the investigations of the pollution process in the Venice Lagoon; the objective is to survey the quality of this environment and to define a mathematical model of pollutant distribution processes in a shallow water system.

The first step of this work was a general overview of the Lagoon conditions. It began about ten years ago in cooperation with the National Research Council Institute "Grandi Masse" of Venice (that supports monitoring and sampling operations in to the Lagoon). At eight of the main channels flowing in the Lagoon, field measurements and sample collections were carried out (monthly and also more frequently) for some years (1982-1984). The quantity of pollutants yearly discharged in the Lagoon from its drainage basin have thus been estimated.[6]

It is more difficult to evaluate the pollutant transfer from the Lagoon to the Adriatic Sea, because of the complexity of the Lagoon dynamics and the pollutant exchange among the environmental phases (water, particulate, biota, sediment) with possible accumulation and release of substances on a variable time scale. A test area has been selected in the Northern Venice Lagoon that is similar to the mainland - lagoon interface. A map of heavy metal distribution in waters and in surface sediments of this area has been performed.[7] The metal enrichment of the surface materials of the floor of the lagoon has been investigated with reference to the sources of pollution input (discharging channel mouth) and the sediment grain size.[8] The metal concentration has been measured with the PIXE technique in diluted sediment samples (non residual fraction) and in interstitial waters extracted by centrifugation.

Further research is needed in order to define the diffusion mechanism and the metal residence times in the environmental phases of this shallow water system. Since 1987, sampling is periodically performed with direct field measurements of chemical, physical, and hydrological parameters in order to create a large data bank that allows reliable statistical processing.

6.2 Toxic effects of exposure to Se in marine fishes

This research is a portion of a larger study about Se up-take and cellular interaction after Se exposure in marine fishes.[9-10] PIXE research groups of Italy (Padua University, Legnaro Laboratory, Naples University) and of Yugoslavia (R. Boskovic Inst., Zagreb) with great experience in analysis of biological matrices are involved in this work.

Our role consists of monitoring the Se level in the sea water of the reservoirs in which the studied fishes live (Northern Adriatic): the Se acute toxicity is observed by Na_2SeO_3 administered via direct up-take from doped sea water.

7. REFERENCES

1. Aprilesi, G., Cecchi, R., Ghermandi, G., Magnoni, G., and Santangelo, R., Nucl. Instr. Meth., 231 (B3), 172-176 (1984).

2. Cecchi, R., Ghermandi, G., Mittner, P., and Calvelli, G., Nucl. Instr. Meth., B15, 146-151 (1986).
3. Cecchi, R., Ghermandi, G., and Calvelli, G., Nucl. Instr. Meth., B22, 460-464 (1987).
4. Cecchi, R., Ghermandi, G., and Zonta R., Nucl. Instr. Meth. B49, 283-287 (1990).
5. Cecchi, R., Ferrari, M.G., Ghermandi, G., and Rossi, M., Nucl. Instr. Meth., B49, 146-151 (1990).
6. Bernardi, S., Costa F., Vazzoler, S., Vincenzi, Z., Cecchi, R., Ghermandi, G., and Magnoni, G., Proc. VII ICSEM/IOC/UNEP, 91-97 (1985).
7. Bernardi, S., Costa, F., Vazzoler, S., Zonta, R., Cecchi, R., and Ghermandi, G., Il Nuovo Cimento 11C, 5-6, 667-678 (1988).
8. Zonta, R., Costa, F., Ghermandi, G., Perin, G., and Vazzoler, S., Proc. III Int. Conf. Environmental Contamination - Venice, Sept. 26-29, 557-559 (1988).
9. Tallandini, L., Giacobini, F., Turchetto, M., Galassini, S., Lui, Q.X., Shao, H.R., Moschini, G., Moro, R., Gialanella, G., Ghermandi, G., Cecchi, R., Injuk, J., and Valkovic, V., Nucl. Instr. Meth. B40/41, 630-634 (1989).
10. Tallandini, L., Giacobini, F., Turchetto, M., Galassini, S., Lui, Q.X., Shao, H.R., Moschini, G., Moro, R., Gialanella, G., Ghermandi, G., Cecchi, R., and Valkovic, V., Proc. 2nd Int. Workshop on XRF and PIXE Application in Life Science, Capri (Italy) June 29-30, 1989, 297-303.

This work has partly been supported by the International Atomic Energy Agency, RC 4861/R1/EP.

Lateral and depth profiling in Al alloy implanted with nitrogen using $^{27}Al(p,\gamma)^{28}Si$ and $^{15}N(p,\alpha\gamma)^{12}C$ resonant nuclear reactions with proton microprobe.

S.Lucas, G.Terwagne and F. Bodart.

Laboratoire d'Analyses par Réactions Nucléaires (LARN), Institute for Studies in Interfaces Sciences (ISIS), Facultés Universitaires N-D de la Paix, 22 rue Muzet, B-5000 Namur (Belgium).

M. Benmallek.

Centre de Recherche de Pechiney, 38340 Voreppe (France).

ABSTRACT

Nuclear reactions have been widely used for several years for the characterization of various samples because they allow quantitative measurements of light elements. When combined with proton microprobe, they allow elemental mapping at different depth.

Combining the nuclear resonant reactions $^{27}Al(p,\gamma)^{28}Si$, $^{15}N(p,\alpha\gamma)^{12}C$ and nuclear microprobe, we have profiled the nitrogen and aluminium surface concentrations in an Al-Si alloy implanted with 50 keV nitrogen at a dose of $5x10^{17}$ $^{15}N^{+}/cm^{2}$.

Three-dimensional distributions of nitrogen and aluminium have been measured from the surface to the bulk by collecting an aluminium and nitrogen concentration map at different depths.

Differences in the nitrogen depth distribution and lateral maps are observed and seem to be correlated with the aluminium and silicon distribution into the alloy.

1) INTRODUCTION.

This paper is the first part of a work based on the enhancement of the tribological properties by ion implantation of a cast aluminium-alloy.

The aluminium-alloy 'AS12UN' is a material used mainly in foundry due to its special ability of being easily casted. Its elemental composition is described in Table 1.

Table 1: Registered composition (atomic %) of the additional elements of the aluminium-alloy AS12UN.

Si	Fe	Cu	Mn	Mg	Ni	Sn	Pb	Ti
12	<.7	1	<.1	1.15	1	<.1	<.1	<.1

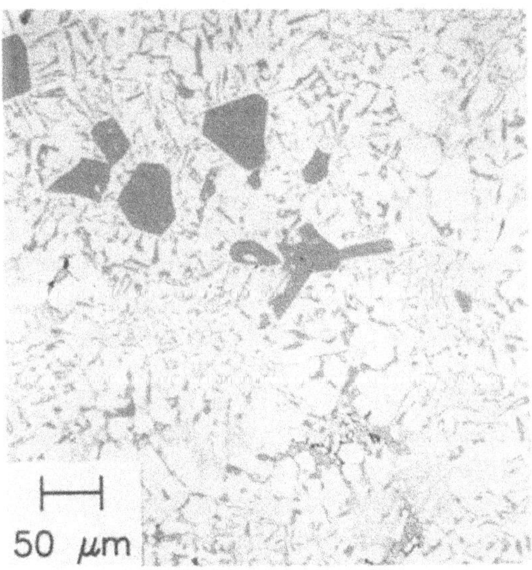

Figure 1:
Optical micrograph showing the different phases, silicon inclusions (dark grey), aluminium matrix (light grey) and eutectic phase (middle grey), in the aluminium-alloy 'AS12UN'.

Due to the high silicon concentration, this alloy has a very low dilatation coefficient (19.5×10^{-6} K^{-1}) which makes it suitable for use in automobile engines (eg pistons).The morphology of the alloy is constituted of mainly three phases which can be seen in the micrographs in Fig. 1. The darkest regions are the crystalline silicon grains included in the aluminium phase (light background).

The size of the silicon grain depends on the cooling rate of the molten alloy, and sizes as large as 150x200 μm on the surface and 500 μm in depth have been observed. Further, monocrystalline silicon phases can be found. Small elongated regions of intermediate contrast correspond to an Al-Si eutectic phase and possibly compounds like $NiAl_3$, AlCuNi or AlFeSi.

Unpublished pin-on-disk tribological measurements have shown that when implanted with nitrogen at 50 keV for doses ranging from $1x10^{17}$ N^+/cm^2 to $5x10^{17}$ N^+/cm^2, the wear of that alloy can be reduced by a factor of two. Since it is known that nitrogen implantation in pure aluminium[1)] and in pure silicon[2)] leads to the formation of the corresponding nitrides, the challenge is to investigate firstly what type of compounds the implantation process forms in the alloy and secondly the behaviour of the nitrogen after tribological experiments.

In this paper, we present the results of the first part. We show that by combining proton microprobe, nuclear and atomic techniques as Nuclear Reaction Analysis (NRA) and Proton Induced X-rays Emission (PIXE), it is possible to characterize in three dimensions (size and shape) the silicon inclusions in the aluminium alloy and further to determine precisely the differences in the nitrogen distributions implanted in the alloy.

2) EXPERIMENTAL PROCEDURE.

Prior the implantation, the surface of the sample was polished down to a grain size of 1 micron using different silicon carbide paper and diamond pastes. The sample was implanted with a rastered $^{15}N_2{}^+$ beam at 100 keV for a dose of $5x10^{17}$ N^+/cm^2. The beam current density was 10 $\mu A/cm^2$, and the temperature of the sample was less than 30°C throughout the implantation procedure. The use of ^{15}N for the implantation allows us to profile it using the narrow (Γ: 200 eV) nuclear resonant reaction $^{15}N(p,\alpha\gamma)^{12}C$ which occurs at 429 keV[3)]. The depth profile is obtained by varying the incident proton energy from 429 keV to 455 keV in steps of 2 keV and counting the prompt gammma-rays emitted by the ^{12}C residual nucleus (E_γ: 4.43 MeV). The profile is later deconvoluted and normalized with respect to a standard in order to get the true nitrogen concentration versus the depth[4)].

The PIXE method allows us to locate precisely the silicon inclusions, but due to strong absorption effects[5)] accurate quantitative and depth measurements are difficult. Further, proton-induced nuclear reactions on silicon have a very low cross-section leading us to choose the nuclear resonant reaction $^{27}Al(p,\gamma)^{28}Si$ which occurs at 992 keV[6]. So we profiled the aluminium around the silicon inclusion in order to get its shape. The depth profile is obtained in the same way as for the nitrogen: we vary the incident proton energy beam from 992 keV to 1007 keV steps by steps and we collect the number of the 1778 keV gamma-rays emitted from the residual nucleus (^{28}Si). at each steps. As gamma-rays detector, we used a 70 % efficiency germanium detector placed at 135 ° with respect to the incident beam direction.

All the experiments were achieved using the LARN microprobe experimental setup[7]. The following procedure was performed for both Al and N analysis. At each energy step the beam was focussed on the sample to a spot size of 15 μm or 5 μm diameter. While the sample is displaced in front of the beam, an elemental map (concentration versus position) is constructed from the emitted gamma-rays or x-rays. This procedure allowed us to build three dimensional concentration maps of aluminium, silicon, and the implanted nitrogen in the alloy.

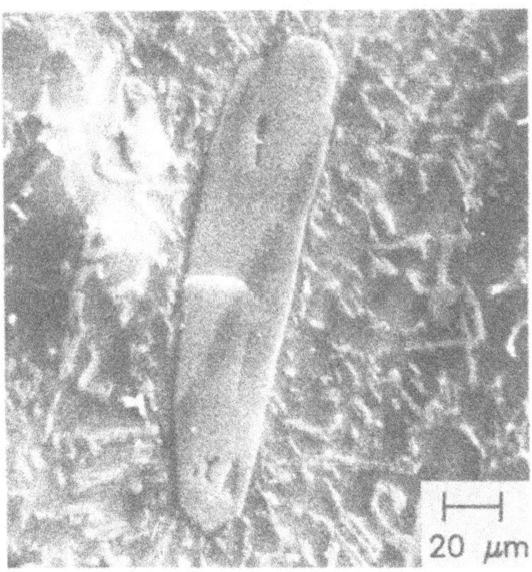

Figure 2:SEM micrograph showing the selected silicon grain included in the 'AS12UN' alloy.

3) RESULTS.

Figure 2 shows a SEM (Scanning Electron Microscopy) micrograph of a selected silicon grain. Around it can be seen the eutectic phase that was described earlier. Figure 3 shows a PIXE image of the same inclusion taken with a proton beam of 1 MeV focussed to 5 μ diameter , and a pixel resolution of the same size. The similarities between the two photos are easily seen. Nevertheless, the surface structure cannot be discriminated in the PIXE micrograph since the

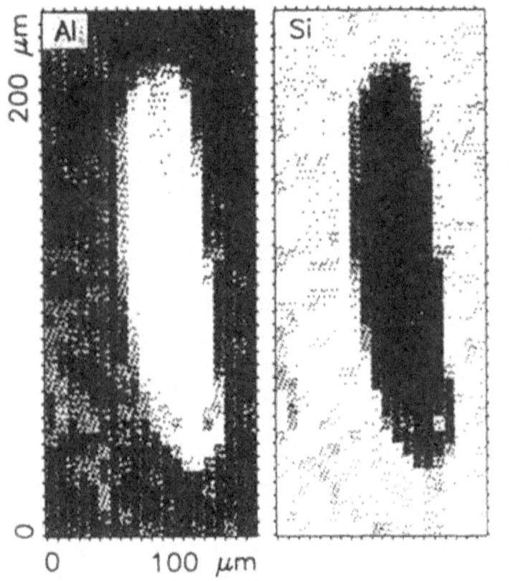

Figure 3:PIXE micro-beam map of aluminium and silicon distribution in the same grain shown in Fig. 2.

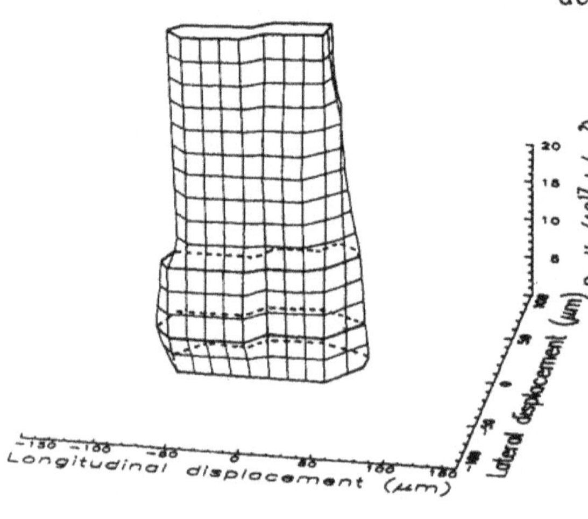

Figure 4:

Three dimensional representation of the silicon inclusion. This representation has been constructed from aluminium and silicon concentration maps determined by NRA at different depths

collected X-rays are coming from a depth which is greater than for the scanning electron microscopy technique. All our attention will be devoted to this silicon grain, and we will now present the results concerning its spatial distribution and the way nitrogen is distributed around it.

Since we know from the phase diagram that the silicon is nearly aluminium free (<2%), we constructed the silicon grain distribution from aluminium concentration maps taken at different depths.

We obtained successives maps with patterns containing very low aluminium signal.

These describe the cross-section of the grain at the considered depth. .The spatial distribution of that inclusion is shown in the figure 4. A linear interpolation has been made between the different planes in order to connect them. We observe that the body of the inclusion is not altered versus the

depth except that little protrusions appear on its sides. All the minor and major axes of the different planes have the same direction and the vertical axis of the inclusion is always perpendicular to the surface to the greatest depth analysed.
We have investigated the nitrogen depth distribution at different points in and out the inclusion. This is shown in the Figure 5 where the nitrogen distributions outside the grain (curve A, C), and in the middle of the grain (curve B) are presented.

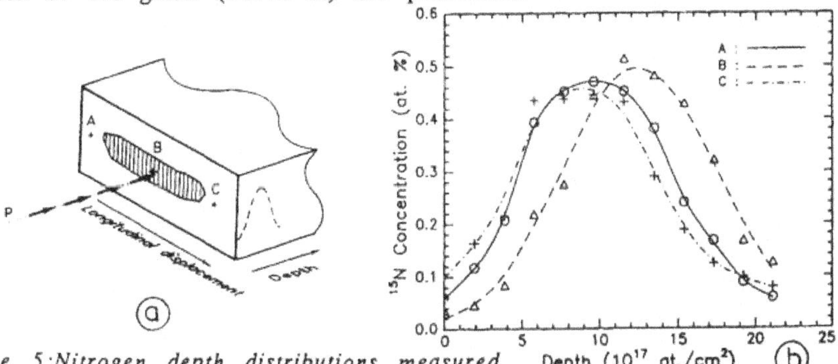

Figure 5:Nitrogen depth distributions measured at three impact positions shown in Fig. 5a where A and C are outside the Si inclusion and B in the middle. The nitrogen distributions curves shown in Fig. 5b correspond to these three positions.

The nitrogen distributions outside the grain (at position A, C) are similar to each other in shape and position, while the maximum of the distribution inside the grain (position C) is shifted to greater depth by more than $3x10^{17}$ at/cm^2. This is unexpected since the difference in the theoritical projected ranges of the nitrogen in pure aluminium and silicon is less then 4% [8)]
Table 2 presents the experimental mean projected range for nitrogen implanted at 50 keV in pure aluminium, silicon and in the aluminium and silicon phases in the alloy.

Table 2. Experimental mean projected ranges of $^{15}N^+$ implanted at 50 keV in pure aluminium, silicon, and in the aluminium and silicon phases of the alloy 'AS12UN'

R_p-pure Al 10^{17} at./cm^2	R_p-pure Si 10^{17} at./cm^2	R_p-Al(Alloy) 10^{17} at./cm^2	R_p-Si(alloy) 10^{17} at./cm^2
8.1	8.4	8.5	12.0

The expected range in the aluminium phase of the alloy is in good agreement with the one observed in pure aluminium. But we can see that a difference exists between the projected ranges of the nitrogen in the silicon inclusion and in the pure silicon. Since all depth were measured experimentally, the sputtering yield differences between Al and Si play no role in the observed phenomena. Diffusion during the irradiation is improbable since no broadening of the distribution is observed. A possible explanation could be channelling effects in monocrystalline silicon phase during the implantation process.

Figure 6 shows the nitrogen depth distributions at different points of a line crossing the inclusion along its long axis.

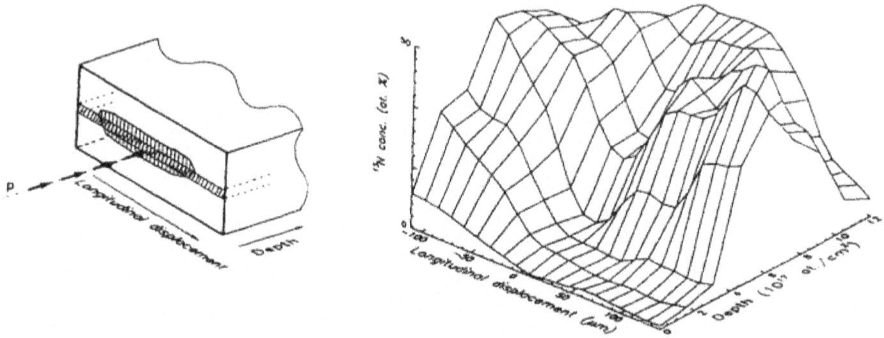

Figure 6:Nitrogen depth distributions measured along a line crossing the inclusion as its longer axis.

This Figure has been constructed from the nitrogen concentration maps taken at different depths. The origin (0,0,0) is the middle of the surface of the inclusion. We can see that the spatial nitrogen distribution in the middle of the silicon grain is shifted to greater depth compared to the distribution found outside the grain in the aluminium phase. This confirms the observations made from the three different points shown previously (Fig. 5). Futher, the width of the valley is lower than the width of the inclusion itself. This is not yet well understood.

In conclusion, we have shown that nuclear reaction analysis combined with a proton-microprobe are powerful tools for doing three dimensional measurements of the main constituents of the

'AS12UN' alloy. Further, the depth distributions of the nitrogen implanted in that alloy can also be investigated in the same way.

When implanted with nitrogen the mechanical behaviour of the alloy is modified. Among a lot of tribological tests, pin-on-disk experiments have shown a reduction in the wear rate of the alloy. At this point, we can conclude that the idea of measuring the nitrogen lateral and depth profiles along the wear tracks is feasible, while being able to distinguish the nitrogen distributions in the matrix from aluminium and silicon phases. These experiments are planned for the near future.

4) REFERENCES.

1)Lucas S., Terwagne G., Bodart F., Nucl. Inst. and Meth. B50, 401 (1990).

2)Hemment P.L.F., Peart R.F, et al Appl. Phys. Letts 46,952 (1985).

3)Terwagne G., Piette M., Bodart F. Nucl. Inst. and Meth. B19/20,145(1987).

4)Deconninck G., Van Oystayen B., Nucl. Inst. and Meth. 218, 165 (1983).

5)Mathot S., Demortier S., Nucl. Inst and Meth. B49, 505 (1990).

6)Deconninck G., Demortier G, Bodart F. "Prompt Gamma-ray Spectroscopy ...", Atomic Energy Review, N° 2, (1981).

7)Demortier G., "Principles and Applications of High Energy Ion Microbeams", (Watt F., Grime G.W. Ed.), Adam Hilger p. 333 (1987).

8)Ziegler J.F, Biersack J.P, Littmark U,"The Stopping and Range of Ions in Solids", (Pergamon press Inc,) N-Y, (1985).

QUANTITATIVE ANALYSIS OF ALUMINIUM IN BAUXITES USING PIGE

A. Savidou, X. Aslanoglou, A. Karydas, and T. Paradellis
Institute of Nuclear Physics
NCSR"Demokritos", 153 10 Aghia Paraskevi, Greece

ABSTRACT

Quantitative analysis of ^{27}Al in bauxites was performed utilizing 1.6 MeV protons. The 1779 KeV and 843 KeV γ-rays from $^{27}Al(p,\gamma)^{28}Si$ and $^{27}Al(p,p_1\gamma)\,^{27}Al$ respectively were detected. The stopping power for bauxites has been measured experimentaly for the two γ-rays and it is shown that they are the same for all bauxites. Using the experimental stopping powers, the Al content of bauxites is determined and found in good agreement with the analysis performed using chemical methods. The time for each measurement is about 20 min.

INTRODUCTION

The PIGE (proton induced gamma ray emission) technique is used particularly in low Z[1-6] elements analysis (from Li to P), where nuclear resonant reactions are intense due to the low Coulomb barrier.

The use of PIGE depends on the knowledge of the stopping power of the sample which is a function of all major elements in it.

Bauxites contain mainly Al_2O_3, Fe_2O_3, SiO_2, TiO_2, CaO and H_2O. Trace elements that are possible to exist, do not contribute to stopping power. For an accurate determination of the heavier elements, XRF measurements were performed using ^{109}Cd source (see table I). The XRF results were in good agreement with the chemical analysis.

TABLE I

X-ray results for different bauxites

% per weight	BAUX1	BAUX2	BAUX3	BAUX4	BAUX5	BAUX6
Fe_2O_3	21.85	20.4	20.3	19.5	18.9	19.4
CaO	1.26	0.71	0.91	1.37	1.56	0.97
TiO_2	2.5	2.4	2.2	2.5	2.5	2.6

The small variations of heavy elements between the XRF results and chemical analysis, or between different bauxites are not enough to cause a difference in the final stopping power. So bauxites can be considered to be composed of two matrices (see table II). The one is all the main Oxides and water the concentration of which is the nearly constant in all bauxites, and the other is Al_2O_3. The respective ratio of elements in these matrices are shown also in table II.

TABLE II

Percent by weight composition of the matrices

MATRIX I		MATRIX II	
Fe_2O_3, CaO SiO_2		Al_2O_3	
TiO_2, H_2O			
Fe	0.352	Al	0.529
Ti	0.040	O	0.471
Ca	0.023		
Si	0.033		
H	0.035		
0	0.516		

For a bauxite of known concentration in Aluminium, the stopping power can be estimated from tables using Bragg's rule:

$$S_{BAUX} = (1 - \alpha/100)\, S_I + (\alpha/100)\, S_{II}$$

where α% per weight is the Bauxite concentration in Al_2O_3 and S_I, S_{II} are the stopping powers for the two matrixes.

EXPERIMENTAL

The measurements were carried out employing the 5.5 MV Tandem Van de Graaff accelerator. For the detection of the γ-rays an intrinsic Ge detector was used having 1.9 KeV resolution at 1332 KeV.

The experimental set-up is shown in Fig. 1. The terminal section of the target holder tube was electrically isolated from the focusing and collimating parts and was connected to a current integrator. The terminal tube was lined with Ta to avoid interference from tube materials. The target was positioned at 45^0 with respect to the beam direction. In front of the target a secondary electron suppressor was installed for best current collection, operated with a 300V bias.

The detector was positioned at 90^0 with respect to the beam direction, 4cm from the target and it was shielded with 4cm lead collimator.

METHOD OF ANALYSIS AND DISCUSSION

Using the PIGE analysis in a sample which contains many elements, the proton energy must be chosen carefully, since interference may take place from different elements.

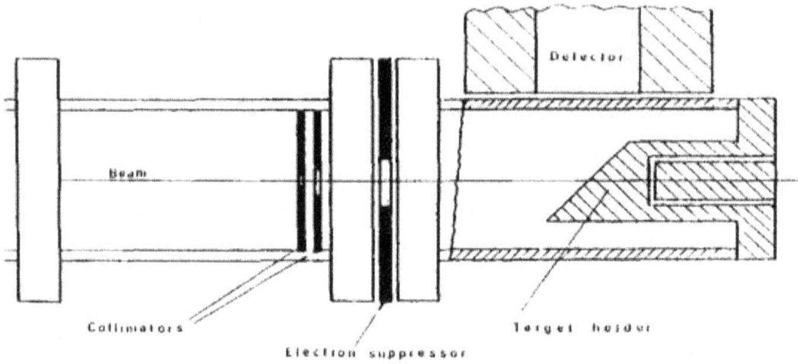

Figure 1. The target assembly used in the irradiation facility

The bombardment of the target with 1600 KeV protons ensures that the 1779 KeV γ-rays come only from $^{27}Al(p,\gamma)^{28}Si$. The 1779 KeV γ-rays can also be produced by inelastic scattering of protons on ^{28}Si when the proton energy is larger than 1842 KeV [5] as well as from the reaction $^{31}P(p,\alpha\gamma)^{28}Si$ provided that proton energy is larger than 1700 KeV [1].

The 843 KeV γ-rays is possible to be also produced from $^{26}Mg(p,\gamma)^{27}Al$, but the yield of this reaction is one order of magnitude less than the inelastic scattering on ^{27}Al. Furthermore Mg is a trace element in bauxites, contrary to Al which is of the dominant one.

The measurement of bauxites' stopping powers was carried out utilizing Fe-Al alloys and Bauxites of known concentration in Al. The proton energy was 1600 KeV. The 1779KeV and 843 KeV γ-rays from $^{27}Al(p,\gamma)^{28}Si$ and $^{27}Al(p,p_1\gamma)^{27}Al$ respectively were detected (see figure 2).

Bombarding at first the alloys with 1.6 MeV protons, the Ki coefficients can be determined for each γ-ray by fitting via the equation:

$$Y_i/Q = K_i \cdot C/S_i \qquad \begin{matrix} i = 1 \text{ for } 1779 \text{ KeV } \gamma\text{-rays} \\ i = 2 \text{ for } 843 \text{ KeV } \gamma\text{-rays} \end{matrix}$$

Y_i: Yield, number of net counts in the γ-ray peak

Q: Integrated charge

C: Concentration in Al

Si: Stopping power of the alloy calculated at the half of the Yield [5,7] using Northcliffe and Schilling tables.

Following that, using the Ki coefficients and bombarding bauxites of known concentration in Al, the stopping powers can be determined experimentaly for the two γ-rays via the relation:

$$S_i = K_i \cdot C / (Y_i/Q) \qquad \begin{matrix} i = 1 \text{ for } 1779 \text{ KeV} \\ i = 2 \text{ for } 843 \text{ KeV} \end{matrix}$$

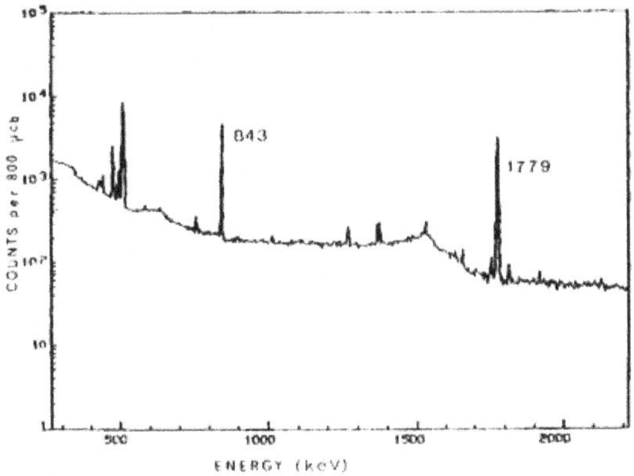

Figure 2. Gamma-ray spectrum produced by 1.6 MeV protons incident on a bauxite

The results are shown in tables III and IV.

For a given bauxite for which a complete chemical analysis was requested, the results using the presently determined experimental stopping powers are shown in table V.

TABLE III

Experimental stopping power of bauxites for the 843 keV γ-ray

S in keV/μg/cm²

File name	S(North. & Schil.)	S'exp	S"exp	Sexp
BAUX1	.155	.156	.154	. 155
BAUX2	.154	.151	.155	. 153
BAUX3	.154	.155	.152	.1535
BAUX4	.154	.152	.148	. 150
BAUX5	.155	.153	.155	. 154

S_1. exp = $(153.1 \pm 1.7)10^{-3}$

TABLE IV

Experimental stopping power of bauxites for the 1779 keV γ-ray

S in keV/μg/cm²

File name	S(North. & Schil.)	S'exp	S"exp	Sexp
BAUX1	.164	.164	.166	. 165
BAUX2	.164	.161	.167	. 164
BAUX3	.164	.162	.160	. 161
BAUX4	.164	.159	.158	.1585
BAUX5	.165	.163	.161	. 162

S_2. exp = $(162.1 \pm 2.3)10^{-3}$

TABLE V

Al determination of a specific bauxite sample

CHEM.ANAL.	PIGE 843 KeV	PIGE 1779 KeV	PIGE mean value
28.90%	29.04 ± 0.45%	28.62 ± 0.50%	28.83 ± 0.34%

CONCLUSIONS

As it is shown in tables III and IV the experimental stopping powers for the two γ-rays mentioned above are in accordance with the stopping powers estimated from Northcliffe and Schilling tables considering that bauxites are composed of the two matrixes.

Finally the PIGE quantitative analysis, is in good agreement with chemical analysis and is faster than chemical analysis (on the average 20 min for each measurement). In addition and simultaneously other trace elements (Li,B,F,Na,Mg) can also be determined, at the same bombarding energy and this is a subject of a current investigation.

REFERENCES

1. Demortier, G. and Bodart, F., "Quantitative analysis of phosphorus by (p, γ) reactions, "J. Radioanal. Chem., **12**, 209(1972).
2. Golicheff, I., Loeuillet, M., and Engelmann, C., "Analytical application of the direct observation of nuclear reactions induced by low-energy protons and leading to the emission of photons, which are measured," J. Radioanal. Chem., **12**, 233 (1972).
3. Asking, L., Swietlicki, E., and Garg, M.L., "PIGE analysis of sodium in thin aerosol samples," Nucl. Inst. Meth., **B22**, 368(1987).
4. Demortier, G. and Delsate, P., "Analysis of magnesium by prompt gamma rays induced by protons," Radiochem. radioanal. Lett., **21**, 219(1975).
5. Deconnick, G. and Demortier, G., "Quantitative analysis of aluminum by prompt nuclear reactions," J. Radioanal. Chem., **12**, 189(1972).
6. Hanson, A.L., Jones, K.W., and Corbin, D., "Determination of Si/Al ratios in zeolites by proton inelastic scattering," Nucl. Isnt. Meth. **B6**, 301(1985).
7. Ishii, K., Valladon, M., Sastri, C.S., and Debrun, J.L., "Accurate charged particle activation analysis: caclulation of the average energy by the average stopping power method," Nucl. Inst. Meth., **153**, 503(1978).

ON THE APPLICABILITY OF PIXE MICROANALYSIS AT 10^5 eV BOMBARDING ENERGIES

S. K. BISWAS+

I. BORBÉLY-KISS

U. M. EL-GHAWI++

E. KOLTAY

GY. SZABÓ

Institute of Nuclear Research, P.O.Box 51
Debrecen, H-4001, Hungary

Optimum detection parameters of PIXE can be achieved around 2 MeV bombarding energy due to the fact that Coulomb interaction between accelerated protons and atomic electrons of the inner shells gives maximum ionization cross sections for most elements in the proton energy range 2 - 10 MeV [1]. Therefore the routine application of PIXE is limited to laboratories with Van de Graaff or small cyclotron accelerators in the MeV range.

In earlier papers some attempts have been made to use PIXE on low energy accelerators - routinely used as neutron generators - with low beam energy of several hundred kilovolts [2],[3],[4]. Such machines are widely available, simple to service, less expensive and would offer an extension of research activity in a neutron physics laboratory toward microanalytical topics.

In these works the effect of low cross sections has been partly compensated with beam intensities increased to microampere range, i.e. targets have been subjected to thermal loads two-three orders of magnitude higher than its typical value in MeV bombarding energy range. Therefore, special measures were taken in cooling targets and the application was limited to fields where samples were able to withstand beams of 1 - 10 μA intensity.

+IAEA fellow, on leave from Atomic Energy Centre, P.O. Box 164, Dhaka-1000, Bangladesh.

++IAEA fellow, on leave from Tajura Nuclear Research Centre, P.O. Box 84464, Tripoli, Libya.

Some detailed investigations have been performed in order to clear up the analytical capabilities of sub-MeV PIXE [5],[6],[7]. The papers stress that the decrease of the maximum energy of secondary electrons to the range of a few hundred eV with low energy protons makes background of bremsstrahlung character practically vanishing, therefore, minimum detection limits for low atomic number elements are improved. Moreover, it is also stated that for the analysis of light elements the use of lower energy protons is preferable to the application of MeV energies.

In order to make a critical evaluation of the applicability of low beam energies the PIXE set-up [8] based on a 5 MV Van de Graaff accelerator and PIXE chamber capable to treat bombarding beam intensities below 100 nA as well as the evaluation routine [9] regularly used in conventional MeV energy PIXE in the present Institute were used for the determination of analytical capabilities in the energy range 2.5 - 0.25 MeV. Electron-optical properties of the accelerator set a lower limit of accelerator voltage at around 0.5 MV, therefore lower energy protons were obtained from H_2^+ and H_3^+ beams at double and triple terminal voltages, respectively. Full dissociation of the molecules was achieved by the application of a thin nickel scattering foil prior to the entrance to the PIXE analytical chamber. Due to an observed non-linearity of the beam intensity measurement appearing around 50 - 100 nA in the chamber the bombarding intensity was kept below this limit throughout the measurements.

An atomic ECPSSR model [9] has been used for calculating the cross section of K and L X-ray excitation as the function of bombarding energy, K X-ray cross section data are presented in Fig. 1 together with effective cross sections. The latter quantity is defined as the cross section integrated over the total proton range in thick samples corrected for self-absorption of X-rays and the energy dependence of X-ray detection efficiency. The above processes define the contribution of separate layers at different depth of the target to the total yield. Data on depth dependence calculated for different trace elements uniformly distributed in carbon and iron matrices are shown in Fig. 2 for different values of the initial beam energy as a parameter of the curve set. A comparison of measured and/or calculated X-ray intensities in characteristic lines with the background curve deduced from a χ^2 fitting procedure in spectrum evaluation minimum detection limits (MDL) have been obtained as the function of atomic number for separate bombarding energies. The results are illustrated in Fig. 3 presenting MDL-s for thick carbon matrix under the following experimental conditions: total collected charge 1 μC, solid angle 1.12 x 10^{-4} sr, detector resolution 210 eV for MnK_α.

On the basis of a preliminary evaluation the following conclusions can be drawn on the characteristic features of sub-MeV PIXE:

(i) on the level of 10^3 - 10^4 ppm detection limits conventional beam intensities can be used with realistic measuring times,

(ii) detection limits can be improved by nearly two orders of magnitude with respect to those shown in Fig. 3 when using beam intensities above the μA range. The practical applicability is than limited to cases with samples withstanding high beam loads,

(iii) the analytical information relates to the conditions at a shallow surface layer, the scanned depth is decreasing to less than 1 μm with decreasing bombarding energy. The lower range in depth profiling by PIXE [10],[11],[12] can be decreased by sub-MeV beam energies,

(iv) low beam rigidity at 10^5 eV energy region offers simple realisation of microbeam facilities with moderate lateral resolution, but still competitive with minimum detection limits available in Electron Probe Microanalysis (EPMA).

References

[1] Boro, R. and Cipolla, S., *Nucl. Instr. Meth.* 131 (1975) 343 .

[2] Bales, D. I., Landolt, R. R. and Toll, W. E., *Advan. X-ray Anal.* 18 (1975) 299.

[3] Musket, R. G. and Bauer, W., *J. Appl. Phys.* 43 (1972) 4786.

[4] Frank, W., Lenk, M. and Rossiger, V., *Isotopenpraxis* 20 (1984) 227.

[5] Moriya, Y., Ato, Y. and Miyagawa, S., *Nucl. Instr. Meth.* 150 (1978) 523.

[6] Lindner, G., *Nucl. Instr. and Meth.* *B3* (1984) 130.

[7] Shiokawa, T., Chu, T. C., Navarette, V. R., Kaji, H., Izawa, G., Ishii, K., Morita, S. and Tawara, H., *Nucl. Instr. Meth.* 142 (1977) 199.

[8] Borbély-Kiss, I., Koltay, E., László, S., Szabó, Gy. and Zolnai, L., *Nucl. Instr. Meth.* *B12* (1985) 496.

[9] Zolnai, L. and Szabó, Gy., *Nucl. Instr. Meth.* *B34* (1988) 118, 36 (1989) 88 and further references therein.

[10] Végh, J., Berényi, D., Koltay, E., Kiss, I., Seif El-Nasr, S. and Sarkadi, L., *Nucl. Instr. Meth.* 153 (1978) 553.

[11] Brissaud, I., Frontier, J. P. and Regnier, P., *Nucl. Instr. Meth.* *B12* (1985) 235.

[12] Jakšić, M., Vajić, M., Valković, V., *Nucl. Instr. Meth.* *B49* (1990) 111.

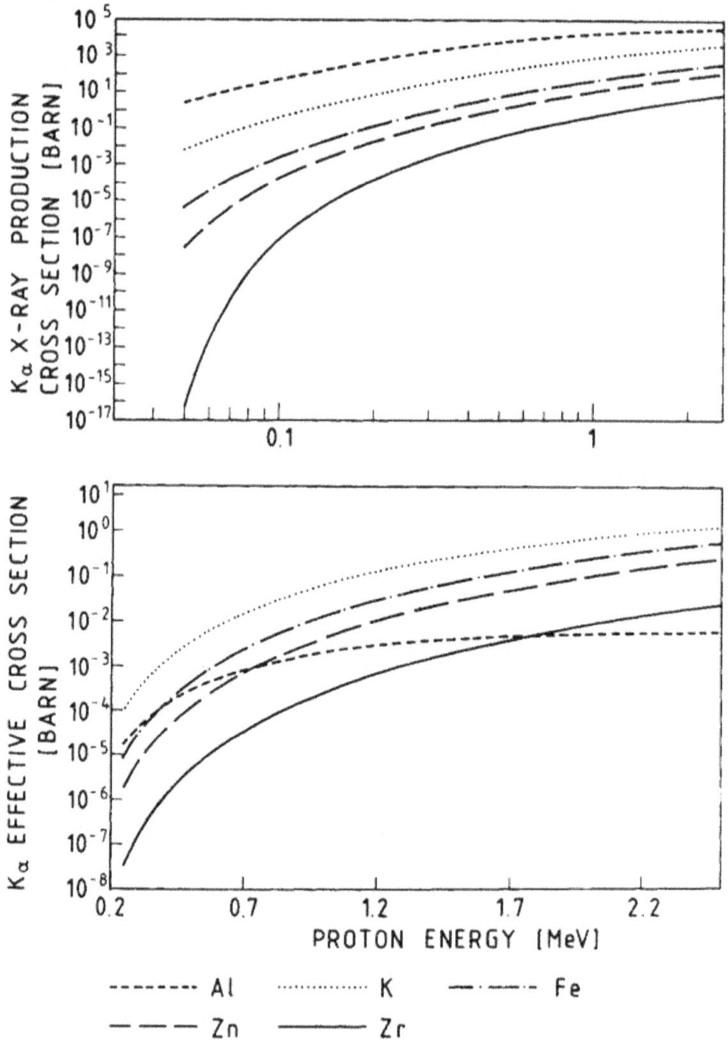

K_α X-RAY PRODUCTION CROSS SECTION [BARN]
10^5
10^3
10^1
10^{-1}
10^{-3}
10^{-5}
10^{-7}
10^{-9}
10^{-11}
10^{-13}
10^{-15}
10^{-17}
0.1
1
K_α EFFECTIVE CROSS SECTION [BARN]
10^1
10^0
10^{-1}
10^{-2}
10^{-3}
10^{-4}
10^{-5}
10^{-6}
10^{-7}
10^{-8}
0.2
0.7
1.2
1.7
2.2
PROTON ENERGY [MeV]
Al
K
Fe
Zn
Zr

Fig. 1.

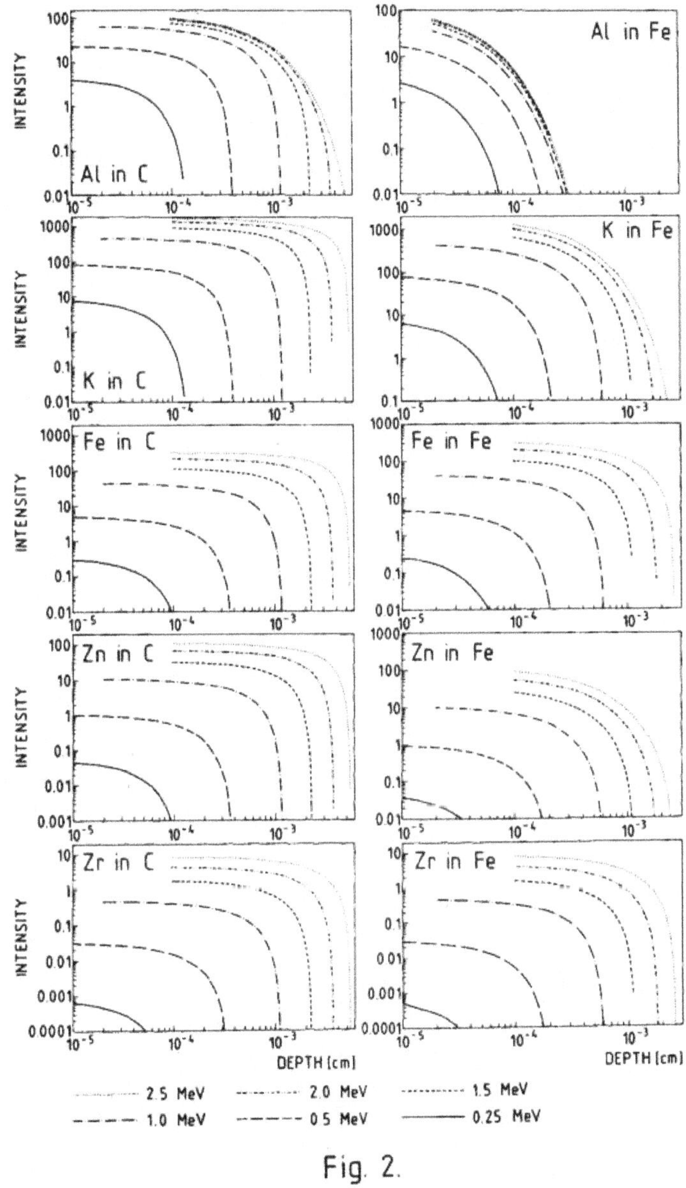
Al in C
Al in Fe
K in C
K in Fe
Fe in C
Fe in Fe
Zn in C
Zn in Fe
Zr in C
Zr in Fe
INTENSITY
DEPTH [cm]
2.5 MeV
2.0 MeV
1.5 MeV
1.0 MeV
0.5 MeV
0.25 MeV

Fig. 2.

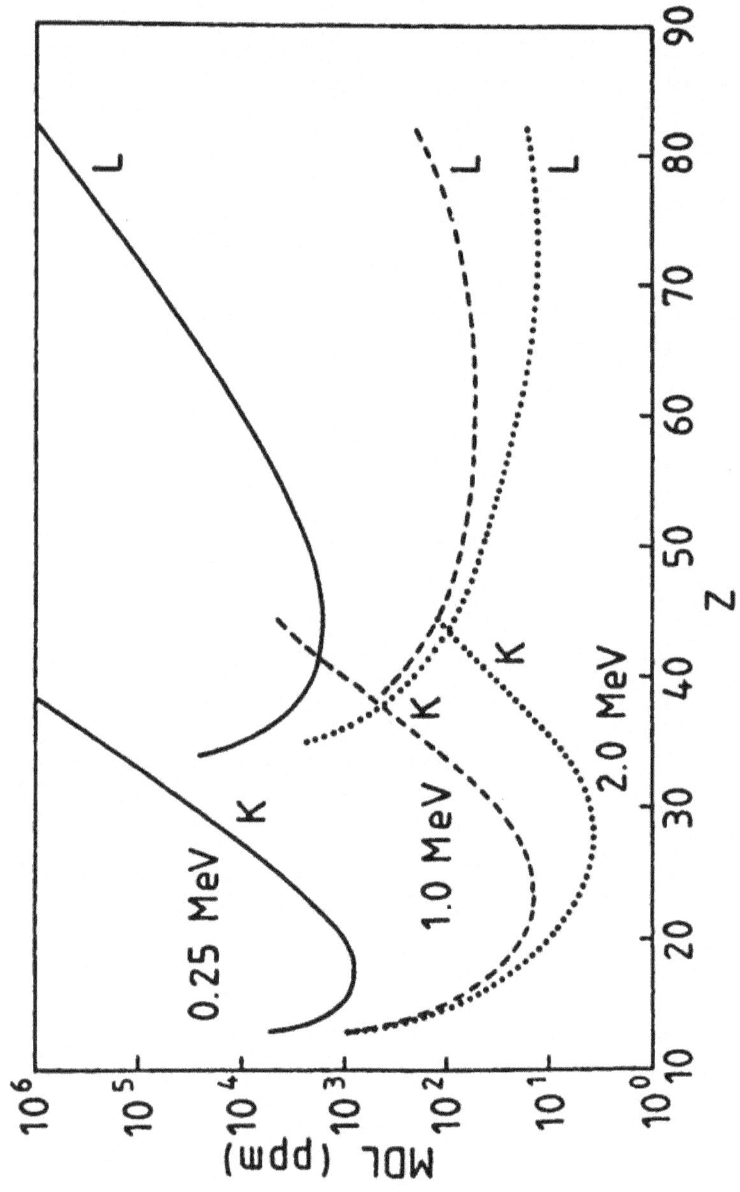
MDL (ppm)
Z
0.25 MeV
1.0 MeV
2.0 MeV
K
L
10
20
30
40
50
60
70
80
90
10^0
10^1
10^2
10^3
10^4
10^5
10^6

Fig. 3.

APPLICATION OF PIGE METHOD FOR THE DETERMINATION OF SODIUM AND BORON IN AEROSOL SAMPLES

S. K. Biswas* , Zs. Fülöp, Á. Z. Kiss, E. Koltay and E. Somorjai

Institute of Nuclear Research, Hungarian Academy of Sciences, P.O. Box 51, H-4001, Debrecen, Hungary

ABSTRACT

PIGE method has been used for the analysis of Na and B content in thin aerosol samples. A resonance free energy region around 2.64 MeV was chosen for all the measurements to obtain a reasonably stable yield. The gamma radiation was detected with a 105 cm^3 Ge(Li) detector. Calibration for concentration determination was made using evaporated targets. The minimum detection limit (MDL) for the method was found to be 150 ng/m^3 for Na at 439 keV γ-line and 460 ng/m^3 for B at 429 keV, respectively.

INTRODUCTION

In environmental research the detection of light elements in addition to PIXE analysis has special importance. PIGE method based on the measurement of prompt gamma rays emitted from $^{23}Na(p,p'\gamma)^{23}Na$, $^{10}B(p,\alpha\gamma)^{7}Be$ and $^{10}B(p,p'\gamma)^{10}B$ reactions[1] has been used for the analysis of Na and B content in thin aerosol samples[2],[3]. Na is important as a sea salt element in the atmospheric aerosols, while B is a tracer for coal burning emissions[4].

PIGE technique is based on the detection of prompt gamma rays emitted from (p,γ), $(p,p'\gamma)$ and $(p,\alpha\gamma)$ nuclear reactions. The quantitative analysis in this method is quite complicated, because, in general the cross section for such reactions is not a smooth function of the incident particle energies and geometrical factor, detector efficiency, etc. have to be taken into account as well. Using standards of similar thickness as the sample, most of the difficulties could be overcome. In that case, the ratio of the measured gamma yield N_γ from a particular element in the sample and

* *IAEA fellow, on leave from Atomic Energy Centre, P.O. Box 164, Dhaka-1000, Bangladesh.*

standard can be expressed as

$$\frac{N_\gamma}{N_{\gamma s}} = \frac{nt}{n_s t_s}, \tag{1}$$

where nt is the atomic density in $\mu g/cm^2$ and the subscript s stands for standards.

In this work we present some preliminary results for simultaneous determination of Na and B at our setup. For the calibration evaporated thin targets were used, where the target thickness was measured utilizing the resonance feature of the $^{23}Na + p$ process.

For Na there are two strong prompt γ–lines. The first line comes at 439.8 keV due to inelastic scattering of the incident protons on ^{23}Na and the other line is at 1634 keV followed by $^{23}Na(p,\alpha\gamma)^{20}Ne$. We used only the stronger 439.8 keV gamma line in the present experiment. For B there are also two lines suitable for analysis in our energy region. The first one is 429 keV γ-line originates from $^{10}B(p,\alpha\gamma)^{7}Be$ reaction and the second one at 718 keV from the $^{10}B(p,p'\gamma)^{10}B$ reaction. The broader but stronger 429 keV line was used for the present set of analysis.

The proton bombarding energy was selected at 2.64 MeV as it was suggested by Asking et al.[3]. It is also evident from the excitation curve that around that energy there is a valley between resonances with width greater than 30 keV, which is the estimated maximum proton energy loss within the actual aerosol samples. So, selecting this energy one can achieve a reasonably constant yield of gamma radiation even there is a small variation of sample thickness.

EXPERIMENTAL

Figure 1 depicts the general view of the PIGE setup used for the present experiment. The 5 MV Van-de-Graaff accelerator of the Institute provides the proton beam, which after analyzing and switching magnet enters into the beam line of the PIGE chamber. A tantalum collimator set was used before the chamber to define the beam and also to prevent the scattered protons from reaching the wall of the chamber as well as the sample holder. The inner wall of the chamber was also covered by thin layer of tantalum foil. A lead shielding of 4 cm thickness was used around the Ge(Li) detector to reduce laboratory background. Samples were mounted on target frame and placed into the chamber at an angle of 90° with the beam direction for irradiation. The beam spot on the target with such an arrangement was about 0.12 cm^2 wide.

The chamber was isolated, while the sample holder was electrically connected to the chamber and the beam current was measured on the chamber as a Faraday cup.

Standard samples were prepared by vacuum evaporation of analytical grade Na_2WO_4, $NaCl$ and $NaBO_2$ on 2.5 μm thick Mylar foil. The thickness of the samples were measured using a narrow $Na + p$ resonance at 2.763 MeV incident proton energy. The target thickness used for the present experiment were kept between 80 and 300 $\mu g/cm^2$, which correspond to minimum and maximum proton energy loss in the sample of 10 and 30 keV, respectively. These samples were used for the calibration of the system for quantitative analysis of Na and B in aerosol samples.

As a first attempt some aerosol samples were collected on nuclepore filters of 0.4 μm pore size with 11.6 cm^2 total exposed area. The sampling time were typically 8 hours with total pumped air volume around 4 m^3. Both the aerosol samples and the standards were mounted on the target holders and preserved in a vacuum desiccator until irradiation.

Gamma rays were detected with a 105 cm^3 Ge(Li) detector (resolution 3.0 keV at 1.33 MeV) placed at 90° with the beam direction. The target to detector distance was kept at 3.5 cm. The gamma ray signals from the detector preamplifier were fed to a PC based 4k channel Nucleus ADC card through an ORTEC 572 amplifier. The typical beam intensity used through out the experiment was around 15 nA, which was mainly determined by the moderate dead time (below 5%) of the data acquisition system. As a consequence no dead time correction was necessary. The spectra were evaluated by a PC version of the computer code developed by Székely[5].

RESULT AND DISCUSSION

The standard samples having different concentration of Na have been subjected to proton bombardment in the same geometrical arrangement as the actual aerosol samples. The incident proton energy was kept at 2.64 MeV as described earlier. The net area for the gamma peak of interest was plotted against the calculated concentrations of the element (Fig. 2). A linear correlation between the gamma ray yield and the concentration was obtained up to the thickness of 300 $\mu g/cm^2$ with correlation coefficient of 0.94. The error bar shown in the Fig. 2 represents the error in thickness measurements only.

The calibration for the boron content determination was made in the same way, the target thickness was measured on the Na resonance because of its presence in the $NaBO_2$.

The minimum detection limit (MDL) for both elements were calculated from the consideration of counting statistics and is defined here as the amount of the element which would yield a γ-ray intensity at least equal to 3σ of the background under the peak of interest in an interval of two times the FWHM of the peak. The MDL under the present experimental conditions was found to be 0.05 $\mu g/cm^2$ and 0.16 $\mu g/cm^2$ for sodium and boron, respectively, for $5\mu C$ of collected charge. Considering the typical collected air volume and exposed filter area these MDL-s correspond to 150 ng/m^3 and 460 ng/m^3.

In comparison with other results it turns out that our detection limit for Na is lower than the value reported by Askig et al.[3] by a factor of two and similar to Boni et al.[6]. For B the MDL is also lower with a factor of two than the value reported by Boni et el.[6] for a total collected charge of 60 μC.

To demonstrate the capability of the method some aerosol samples have been analysed using the present prescription and the results are shown in table 1. As for B content we could not find any detectable quantity in the present set of samples collected at rural area.

Table 1. Na concentrations in ng/m^3 measured in some aerosol samples.

Sample no.	1	2	3	4	5	6
conc.	935±66	339±37	880±47	1312±52	6922±115	263±15

The regular determination of Na and B content in aerosol samples require additional effort in experimental work, e.g., in collection and preparation of samples. Such program is in progress in the laboratory.

REFERENCES

[1] Kiss, Á. Z., Koltay, E., Nyakó, B., Somorjai, E., Anttila, A. and Räisänen, J., J. Radioanal Nucl. Chem. 89(1), 123 (1985).

[2] Boni, C., Caruso, E., Cereda, E., Braga Marcazzan, G. M. and Redaelli, P., Nucl. Instr. and Meth. *B*40/41, 620 (1989).

[3] Asking, L., Swietlicki, E. and Grag, M. L., Nucl. Instr and Meth. *B*22, 368 (1987).

[4] Fogg, T. R. and Rahn, K. A., Geophys. Res. Letters 1, 854 (1984).

[5] Székely, G., Comp. Phys. Commu. 34, 313 (1985).

[6] Boni, C., Caridi, A., Cereda, E. and Braga Marcazzan, G. M., Nucl. Instr. and Meth., *B*49, 106 (1990).

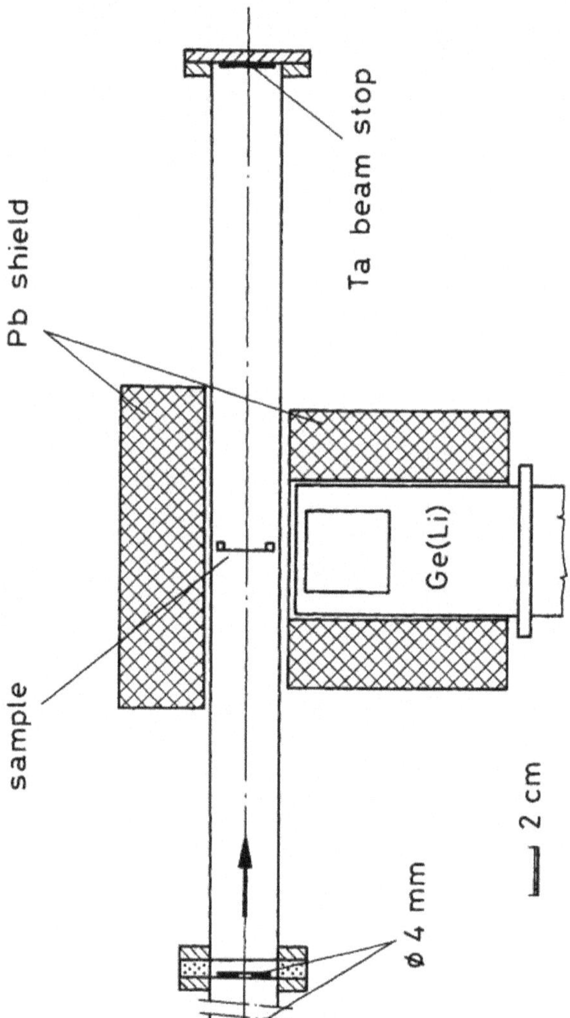

Fig. 1.

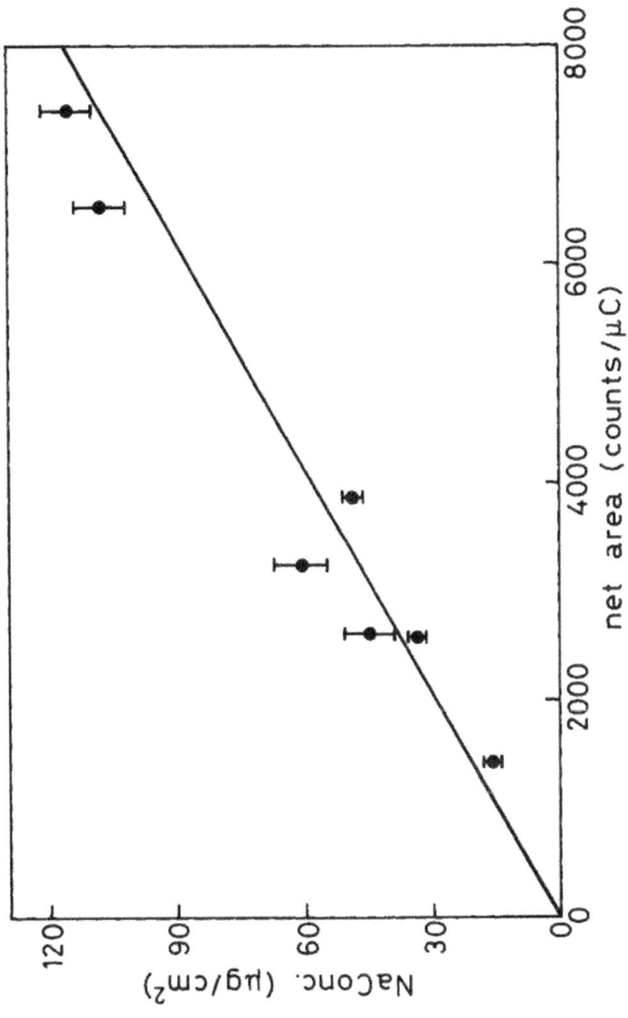

NaConc. (µg/cm²)
0
30
60
90
120
net area (counts/µC)
0
2000
4000
6000
8000

Fig. 2.

PROTON INDUCED MONOENERGETIC PHOTON BEAMS FOR ANALYTICAL PURPOSES

A.G. Karydas and T. Paradellis
Institute of Nuclear Physics
NCSR"Demokritos", 153 10 Aghia Paraskevi, Greece

ABSTRACT

Protons were utilized to produce copper and titanium characteristic x-rays. The two biological samples SRM 1571 and SRM 1577 were analyzed with the copper x-ray beam and two geological samples with the titanium one. The method was found to be accurate for the determination of the major and minor constituents. In addition for medium weight production targets the sensitivity of the method was found to be comparable with that obtained with synchrotron radiation.

INTRODUCTION

A serious problem in PIXE or XRF analysis is the detection of trace elements which have an atomic number close to that of the dominant one in the matrix. One solution to overcome this problem is the use of exciting x-rays which coincide with that of the dominant element in the matrix. In this work we use a method which is a combination of PIXE and XRF. The protons hit a pure thick target and produce the characteristic x-rays of this target. This secondary radiation is an intense monoenergetic photon beam which can in turn irradiate the sample to be analyzed by XRF. Of course, this photon beam can excite only those elements whose absorption edge is at lower energy than the energy of the beam.

This selectivity in excitation can offer the possibility of detecting small amounts of Co in the presense of high Ni concentrations, low detection limits for Cr in steel[1)] and for Fe in a copper matrix [2)]. In addition to the above, it must be noticed that the selective excitation is important not only because excitation of elements which can cause interferences is avoided, but also because the elements of interest are excited with greater cross sections than other modes of excitation. It has been shown[3)] that using monoenergetic photon beams of variable energy, 1-1000 more fluorescent events can be obtained in comparison with electrons or protons excitation.

As the sensitivity is concerned, if the photon beam has energy 1% greater than the absorption edge of a certain element, calculations have shown[4)] that one ppb of this element can be detected in carbon matrix if large photon fluences (10^{16} photons) and good energy detection resolution (5 eV) is available.

Proton energies 2 - 2.5 MeV for medium targets can provide enough high photon fluxes. For heavier targets (Mo, Ag) higher proton energies are required. In this situation the choice of the proton energy must be done carefully, because a limiting factor for the sensitivity of the system is the background due to Compton scattering within the detector and its surrounding materials of γ-rays or neutrons originating from the production target.

In this work two biological standard reference materials were analysed with copper x-ray beam to test the precision of the quantitative analysis, as well as the sensitivity of the method. In addition the possibility of detecting light elements (Al, Mg) in geological samples employing Ti as production target was investigated.

EXPERIMENTAL ARRANGEMENT

The experimental arrangement (fig. 1) is an improved version of a previous one [5]. The irradiation of the sample and the detection of the fluorescent x-rays are done now in vacum. A closed cooling system is used again to protect the target from excessive heating. For the absorption of the low energy bremsstrahlung inherent in PIXE, Copper (20 mg/cm^2) and Kapton foils (17 mg/cm^2) were used for Cu and Ti excitation respectively.

Protons of energy 2.15 MeV were used to excite copper characteristic x-rays and protons of energy 2.95 MeV for the titanium case. For the detection of the fluorescent x-rays a 30mm^2 x 3mm Si (Li) detector was used with resolution of 190 eV at 5.9 KeV.

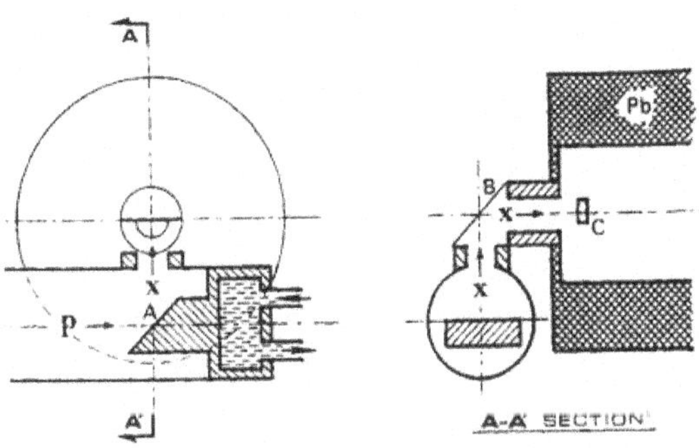

Fig 1 Experimental set up. A: production target, B: sample, C: Si(Li) detector

The calibration of the arrangement was made using thin and thick pure targets, while the concentration of elements has been determined by using fundamental parameter analysis[5] and experimental correction factors[6] for the absorption of the exciting and fluorescent radiation in the sample.

RESULTS

The results of the analysis of the Orchard leaves and Bovine liver are given in figs 2 and 3.

	This work	NBS
Cl	667 ± 72	(700)
K(%)	1.46 ± 0.07	1.47 ± 0.03
Ca(%)	1.6 ± 0.16	2.09 ± 0.03
Ti	21 ± 3	-
Mn	100 ± 3	91 ± 4
Fe	279 ± 20	300 ± 20

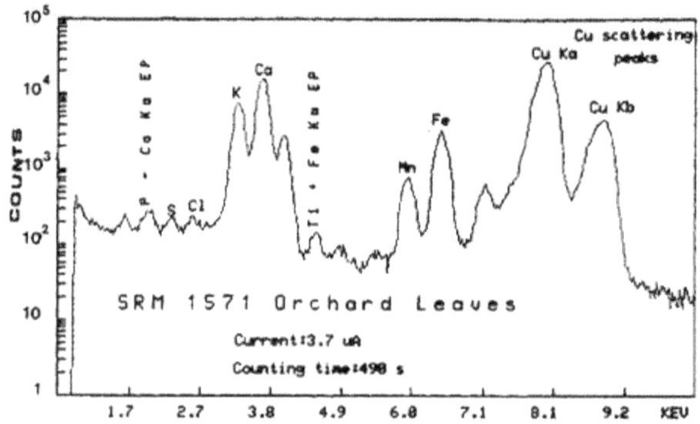

Fig. 2 Concentrations in ppm for SRM 1571 Orchard Leaves. The exciting source is Cu x-rays.

	This work	NBS
P	8583 ± 875	(8000)
S	6177 ± 626	-
Cl	2493 ± 253	2600
K	9646 ± 480	9700 ± 600
Ca	87 ± 9	124 ± 6
Mn	9.7 ± 0.4	10.3 ± 1
Fe	267 ± 6	268 ± 8

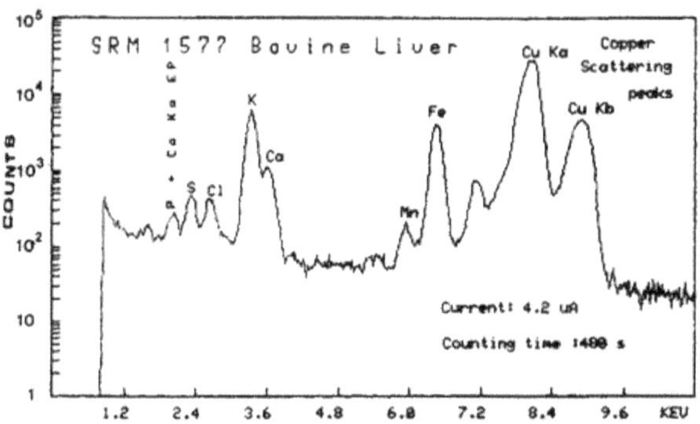

Fig. 3 Concentrations in ppm for SRM 1577 Bovine Liver. The exciting source is Cu x-rays.

Using the criterion $N_i=3.29\sqrt{N_b}$ for the minimum fluorescence counts for the element i, where N_b are the counts of background under the peak, the experimentaly determined MDL for the two standards are shown in fig. 4.

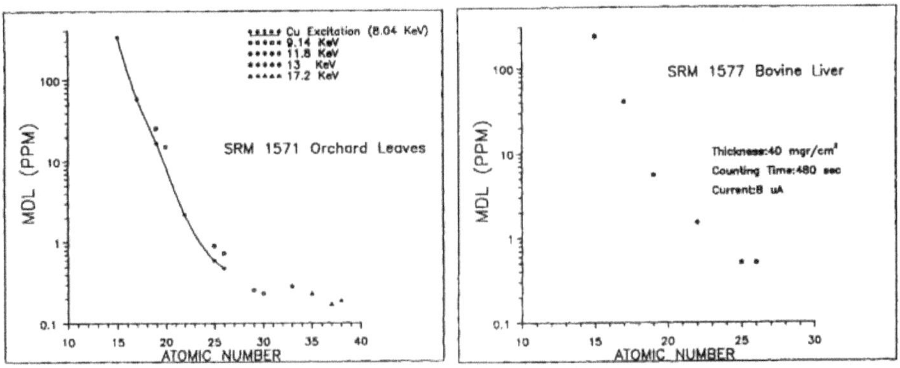

Fig. 4 Experimentaly determined MDL for the two standards at 8.04 KeV excitation. For the case of the 1571 standard, data obtained from monochromatized synchrotron radiation[7] at 9.1, 11.8, 13 and 17.2 KeV are also shown for comparison.

The use of the Ti production target in analyzing two geological samples are shown in figs 5, 6.

	This work	Chem. An.
Al(%)	32 ±3.2	30.49
Si(%)	1.08±0.1	1.4
S(%)	0.172±0.018	
K(%)	0.07±0.007	0.08

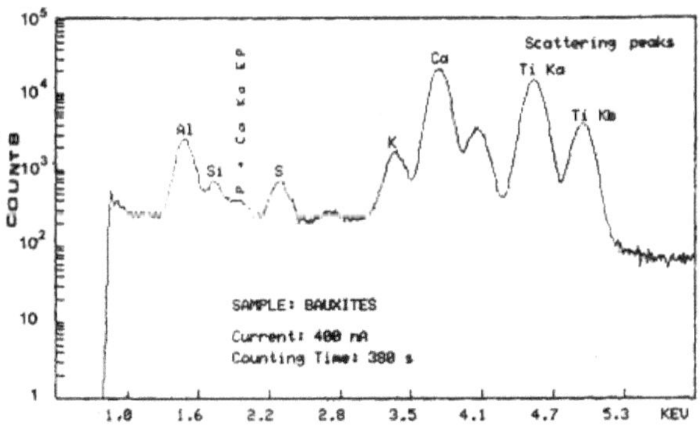

Fig.5 The results of the analysis of a bauxite sample using a Ti production target.

Mg(%)	30.7 ± 3
Al(%)	1.09 ± 0.1
Si(%)	13.3 ± 1.3
S(ppm)	62 ± 6
Cl(ppm)	242 ± 24
K(ppm)	115 ± 11
Ca(ppm)	358 ± 36

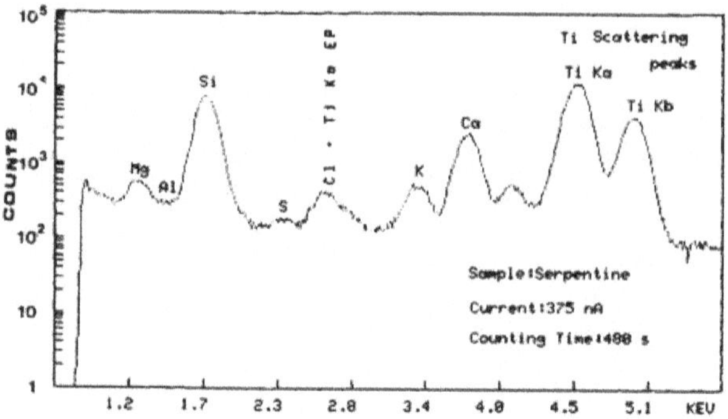

Fig.6 The results of the analysis of a serpentine sample using a Ti production target.

CONCLUSIONS

The monoenergetic photon beams produced by photon bombardment of medium weight targets, can be utilized for the analysis of biological and geological samples with a good sensitivity. The high intensity of the beam, its spectral quality, the possibility of avoiding interferences in the spectra by choosing an appropriate target, the simplicity in quantitative analysis due to the fundamental parameters approach, are the basic advantages of the method, which appear to be extremely valuable to the analyst.

REFERENCES

1. Gihwala, D., Olivier, C., Peisach, M., and Pineda, C.A., "The determination of chromium in steels by PIXE induced XRF" J. Trace Microprobe Tech., **1**, 199(1982).
2. Quynh, L.H., Demeter I., and Szökefalvi-Nagy Z., "PIXE induced XRF analysis of trace amounts of iron in pure copper matrix", Nucl. Inst. Meth. **B49**, 566(1990).
3. Sparks, C.J., Jr "X-ray Fluorescence Microprobe for Chemical Analysis" in Synchrotron Radiation Research p.459. Ed. by H. Winick and S. Doniach Plenum Press. New York 1980.
4. Grodzins, L., "Electron, Proton and Photon Induced X-ray Microprobes: Analytic Sensitivity Versus Spatial Resolution" Neuro Toxicology **4**, 23(1983).
5. Karydas, A.G., and Paradellis, T., "Coal XRF analysis using a proton induced copper x-ray beam" J. Coal Qual. **9(2)** 2669(1990).
6. Giauque, R.D., Goulding, F.S., Jaclevic, J.M. and Pehl R.H. " Trace elements determination with Semiconductor Detector x-ray spectrometers", Anal. Chem. **45** 671(1973).
7. Jones, K.W., Gordon, B. M., Hanson, A.L., Hastings, J.B., Howells, M.R., and Kraner, H.W. "Application of synchrotron Radiation to Elemental Analysis" Nucl. Inst. Meth. **B3** 225(1984).

PRECISION OF NEUTRON ACTIVATION ANALYSIS IN ARCHAEOMETRIC STUDIES OF CERAMICS

H. Mommsen, T. Beier and J. Weber
Institut für Strahlen- und Kernphysik, University Bonn, Nussallee 14–16, 5300 Bonn

ABSTRACT: The long term performance of the Neutron Activation Analysis (NAA) procedure and data evaluation in Bonn using different reactors and different detectors is checked. Concentration values of the well known Coal Fly Ash standard material (NBS 1633a) are presented and their standard deviations discussed. For 11 elements with statistical errors $< 2\%$ the average over all systematic error is found to be $\leq 1.9\%$ as obtained by a best relative fit of the data. Not this small error, but the natural inhomogeneity in clay limits provenance determinations of archaeological ceramics by NAA-'fingerprinting'.

INTRODUCTION

The principles of Neutron Activation Analysis (NAA) are well known. We use this analytical method routinely in Bonn to measure minor and trace elements in ceramic sherds found in archaeological excavations. The archaeometric aim is a more refined classification of pottery types than can be achieved by archaeological inspection only. Since the elemental concentration patterns in pottery depend on the raw materials, the clays and tempers used by the ancient potters, it is characteristic for the pottery workshop which normally exploited a claybed nearby. Therefore, if the chemical composition of ceramics produced at a certain pottery workshop is known, the assignment of pottery of unknown provenance showing the same elemental pattern to this place is possible. This 'fingerprinting' of pottery is a well accepted archaeometric method in provenance studies. The more elements are detected and the better the compositions are determined, the safer is the assignment.

In practice, several difficulties oppose this simple principle. Every analytical method has a measuring error. All clays as erosion products of rocks have a more or less similar elemental composition. The element pattern in clay from a certain claybed is not totally fixed but concentration values may vary slightly due to a natural occuring inhomogeneity. Therefore, to our opinion, to succeed in provenance studies of ceramics, a high analytical precision is necessary. Since NAA offers this property of high precision and has furthermore high sensitivity for many trace elements and is feasible for automation, we have chosen to use this method. In this paper, we will investigate the different sources of errors and show, that the precision attainable with NAA is not limiting provenance studies of pottery.

ANALYTICAL PROCEDURE AND DATA EVALUATION

Our experimental procedure is a modified version of that given by Perlman and Asaro[1)] and is described at length in ref. 2. We now are able to measure routinely over 30 elements using the Bonn ceramic powder as standard which was calibrated against the Berkeley pottery standard[1)]. A set of 24 samples and 4 standard samples in the form of pressed pills to guarantee fixed measuring geometry and containig each about 80 mg of material is irradiated for 90 min with a flux of $5.5 \cdot 10^{13}$ neutrons/(cm^2 s). After about 4 days used for transportation of the samples from the reactor to our laboratory, each sample is measured 5 times in the time range 4–24 days in different γ-energy regions, determining many elements several times independently. All the γ-spectra of the set are evaluated together with the help of a computer code, which was programmed to perform all necessary line intensity corrections automatically to achieve the high analytical precision wanted. These are

- general peak intensity corrections concerning the whole spectrum:
  - dead time (electronic losses)
  - peak broadening due to count rate effects
  - wrapping (Al-foil wrapping of sample pills to be subtracted)
  - matrix effects (absorption in sample for E_γ <50 keV)
- individual peak intensity corrections:
  - natural γ-radiation background (blanc spectrum to be subtracted)
  - intensity interferences due to underlying or partly overlapping peaks:
    1. invariant: corrected by undisturbed monitor line of the same decaying nuclide, detector efficiency dependent
    2. geometry and/or detector dependent: pile up-, scatter-, escape peaks, corrected like type 1 interferences

3. variable, dependent on time, flux or abundance: nuclear reactions, U-fission correction, corrected by time dependent monitor line from different nuclide

A complete list of all γ-lines, interferences and corrections is available on request.

PRECISION OF NAA

Missing or wrongly determined corrections result in systematic errors. Large systematic errors can be found by interlaboratory comparisons measuring standards of known composition. This work is in progress. During the last 5 years, in which we had to use 4 different reactors due to shut- and break downs and different detectors also, we included from time to time in the pottery sample sets as checks samples of the well known standard material Coal Fly Ash (CFA 1633a, National Bureau of Standards, Washington). Alltogether 70 samples of CFA, 19,32,15 and 4 respectively for the different reactors have been measured. Since the intensity corrections are detector and reactor (neutron flux distribution) dependent, the scatter of these data gives information about the precision achieved.

No significant differences between the reactor and detector subgroups are found. In table 1, we show in column 1 the total average concentration values M and their root mean square deviations (spreads) σ (in % of M) for 32 elements in CFA. The spread squared is: $\sigma^2_{total} = \sigma^2_{stat} + \sigma^2_{app} + \sigma^2_{shift} + \sigma^2_{remainder}$.

In column 2, the average errors $\sigma^2_{stat,elem}$ due to statistics only are given. A best relative fit of the data allows an estimation of some of the error contributions. Such a fit is obtained by multiplication of all concentration values of one sample with a constant factor. This corrects for constant batch- and sample differences like weighing errors or flux inhomogeneities. In column 3, average values obtained after such a fit calculation of the individual sample data to the values of column 1 using the 24 elements indicated are shown. Shifts of data of a few percent are found. The fit results in nearly unchanged average values, but decreases the spread values considerably. In the following, we will deduce the different error contributions averaged for the 11 values of elements Ce, Co, Cs, Eu, Fe, Hf, La, Sc, Sm, Th and Yb (all with $\sigma_{stat,elem} < 2\%$). The average statistical error σ_{stat} is 0.8%. Measuring one sample several times gives the average apparative error $\sigma_{app} = 0.2\%$. For the fitted data $\sigma_{shift} = 0$. With $\sigma_{total,fit} = 2.1\%$ $\sigma_{remainder} = 1.9\%$ is obtained. It contains the

Table 1: NAA concentrations of 32 elements of a group of 70 coal fly ash samples (NBS 1633a): averages M in ppm or percent and spreads σ in % of M

	raw values		av. stat. error	values after fit		given values	
	M	σ(%)	%	M	σ(%)	M	σ(%)
Ba *	1347.	(5.3)	2.9	1354.	(5.0)	1500.	
Ca%	1.40	(42.)	30.	1.34	(36.)	1.11	(0.9)
Ce *	176.	(6.8)	1.1	177.	(3.4)	180.	
Co *	43.9	(5.1)	0.46	44.1	(2.2)	46.0	
Cr *	197.	(7.8)	0.71	197.	(5.6)	196.	(3.1)
Cs *	10.6	(4.2)	1.4	10.6	(1.6)	11.0	
Eu *	3.60	(4.8)	1.2	3.62	(2.0)	4.00	
Fe%*	9.45	(4.9)	0.37	9.49	(2.0)	9.40	(1.1)
Hf *	7.31	(4.8)	1.3	7.35	(2.0)	7.60	
La *	82.1	(3.3)	0.47	81.3	(2.1)	—	
Lu *	1.13	(6.4)	2.9	1.14	(4.2)	—	
Na%*	0.17	(2.9)	1.9	0.17	(3.0)	0.17	(5.9)
Nd *	76.2	(7.3)	3.2	76.5	(4.2)	—	
Ni	159.	(20.)	15.	160.	(21.)	127.	(3.1)
Rb *	143.	(4.7)	2.6	144.	(2.6)	131.	(1.5)
Sb *	6.50	(5.7)	3.0	6.53	(3.7)	7.00	
Sc *	41.4	(4.5)	0.15	41.6	(1.5)	40.0	
Sm *	15.9	(4.7)	0.50	15.8	(2.8)	—	
Ta	1.77	(8.7)	1.8	1.81	(6.1)	—	
Tb *	2.32	(5.3)	3.2	2.33	(5.3)	—	
Th *	25.7	(4.4)	0.48	25.8	(1.6)	24.7	(1.2)
Ti%	0.75	(19.)	11.	0.75	(17.)	0.80	
U *	9.61	(6.1)	2.2	9.51	(4.6)	10.2	(1.0)
Yb *	7.76	(2.8)	1.4	7.68	(2.3)	—	
As *	153.	(5.8)	0.53	153.	(5.1)	145.	(10.)
Br	3.10	(33.)	7.8	3.08	(33.)	—	
Gd	14.9	(10.)	11.	15.0	(11.)	—	
K %*	1.99	(3.5)	2.4	1.97	(1.9)	1.88	(3.2)
Se	10.1	(13.)	6.4	10.2	(11.)	10.3	(5.8)
W *	6.19	(12.)	4.8	6.08	(10.)	—	
Zn *	218.	(11.)	1.7	219.	(11.)	220.	(4.5)
Zr	379.	(23.)	12.	380.	(22.)	—	

* : 24 elements used for best relative fit

elemental inhomogeneity in the CFA samples and the remaining systematic error of the determination of the concentrations of the 11 elements considered. If the CFA samples are supposed to be homogeneous, $\sigma_{remainder}$ gives the upper limit for the average long term systematic error. From $\sigma_{total,unfitted} = 4.6\%$ $\sigma_{shift} = 4.1\%$ is calculated. This error contains the weighing errors, the neutron flux inhomogeneity and other batch- and sample difference errors like varying water content (the CFA samples have not been dryed), which are taken out by the fit. In column 4 the certified (with error) and the informal values of CFA are quoted for comparison.

A comparison of the error contributions of the CFA samples with those of a group of 61 ceramic samples from a pottery in the Northern Argolid probably near Mycenae in the Bronze age[3)] shows, averaged over the same 11 elements, an increase in σ_{shift} from 4.1 to 4.3% and in $\sigma_{remainder}$ from 1.9 to 4.5%. Here σ_{shift} contains additionally the spread induced by pottery making practices like varying dilution of the clay by e. g. sand. The increase in $\sigma_{remainder}$ reflects the minimal spread due to the average natural inhomogeneity of the clay used for this pottery group: 4.1%. This shows, that the precision of our NAA measurements of ceramics is so high, that it adds not considerably to the spread in the data due to the always present natural inhomogeneity of the clay. Therefore, provenance determinations by NAA 'fingerprinting' are not limited by the precision obtainable with this method. Mycenaean pottery sherds found in *Amarna*[3)] in Egypt and one sherd in *Spain*[4)] have been identified to originate from the already mentioned Mycenaean workshop in the Argolid.

Acknowledgements: We thank the staff of the reactors in Jülich, in München and of the two reactors in Geesthacht for the irradiations.

REFERENCES

1. Mommsen, H., Kreuser, A., Weber, J. and Büsch, H., 'Neutron activation analysis of ceramics in the X-ray energy region', Nucl. Instr. Meth. in Phys. Res. A257, 451–461 (1987)

2. Perlman, I. and Asaro, F., 'Pottery analysis by neutron activation', Archaeometry 11, 21–52 (1969)

3. Mommsen, H., Beier, T., Diehl, U., Steffes, B. and Podzuweit, Ch., 'Provenance of Mycenaean sherds from tell el Amarna by NAA', to be published in Proc. Int. Symp. Archaeometry, Heidelberg (1990)

4. Mommsen, H., Lambrecht, D., Diehl, U., Pantenburg, F. J. and Weber, J., 'Eine mykenische Scherbe in Spanien: Bestätigung ihrer Herkunft mit der NAA', to be published in Prähistorische Zeitschrift (1990)

NUCLEAR FUEL ANALYSIS BY NEUTRON ACTIVATION

N. N. PAPADOPOULOS

National Center for Scientific Research "Demokritos"
15310 Aghia Paraskevi, Attikis, Greece

ABSTRACT

Nuclear fuel in various forms can be analyzed by fission-based neutron activation methods like the delayed fission neutron counting technique. High analytical accuracy is achieved by cyclic activation to improve the counting statistics. The flexibility of the nuclear analyzer has been enhanced by a grid-plate extension. By irradiation of bare and cadmium covered samples, different thermal to epithermal neutron flux ratios are achieved which even permit the determination of the U-235 isotope abundance besides the uranium element analysis. Comparisons with other methods gave satisfactory results. The principle of the technique can be extended also to multi-element neutron activation analysis to reduce matrix interference problems.

INTRODUCTION

Nuclear fuel has a wide concentration range in the nuclear fuel cycle, therefore the analytical system has to be flexible and adjustable to cover the total range. Special fission-based neutron activation methods and particularly the delayed-fission neutron-counting technique have been developed to meet other analytical requirements also, like high accuracy, high throughput, non-destructive sample preparation, low matrix interference, low analytical cost with automation a.o. [1-3]. In order to determine both the uranium element concentration and the U-235 isotope abundance in the sample, two measurements with different experimental conditions are needed, which can be achieved by differentiating the neutron flux spectrum. The same principle can also be applied to resolve overlapping peaks of gamma-ray spectra in multi-element neutron-activation analysis.

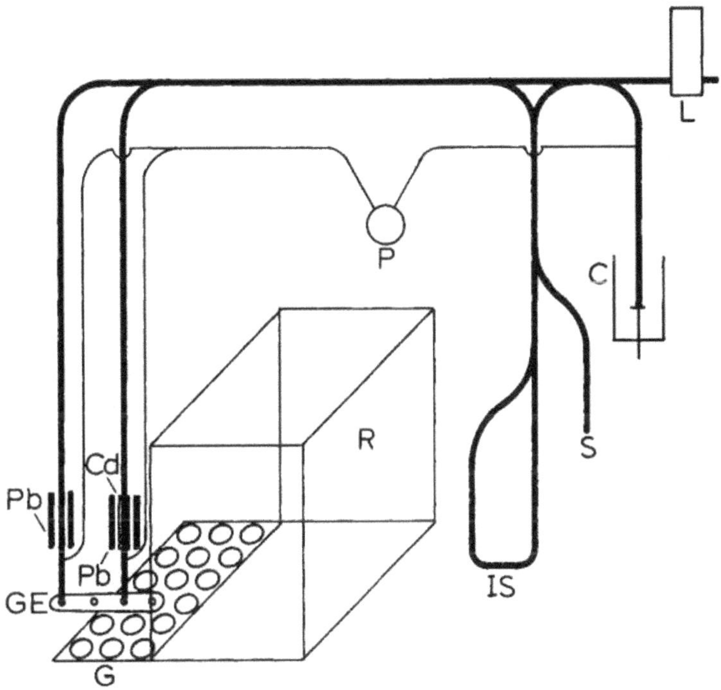

C	-	Counting Station	L	-	Loader
Cd	-	Cadmium Sleeve	P	-	Pump
G	-	Grid Plate	Pb	-	Lead Sleeve
GE	-	Grid Plate Extension	R	-	Reactor Core
IS	-	Intermediate Storage	S	-	Storage

Figure 1. <u>Nuclear Analyzer</u>

EXPERIMENTAL

The new flexible nuclear analyzer (Figure 1) includes a pneumatic transfer tube system with two movable terminals at the irradiation position close to the reactor core. The closer one is for enhanced epithermal neutron flux by cadmium coverage and the other one for prevailing thermal neutron flux by using only a lead sleeve for capsule protection from gamma irradiation. Cyclic activation with intermediate sample storage is foreseen to improve the counting statistics and thus the accuracy and sensitivity of the technique. Special care has been given to keep the aluminium tubes clean, by use of cotton in addition to the double sample containment to avoid capsule contamination and background increase. A grid plate extension has been used for nuclear safeguards analysis of samples with high uranium concentration to keep the irradiation end far enough away from the reactor core, for counting rate optimization. For geological samples of low uranium concentration, the tube terminals can easily be moved close to the reactor core to improve the counting statistics.

In multi-element neutron-activation analysis, two measurements with differentiated neutron spectrum, by use of bare and cadmium-covered samples, provide the necessary data for the resolution of overlapping gamma-ray peaks.

RESULTS AND DISCUSSION

The technique has been applied to natural uranium samples as well as to enriched and depleted uranium samples. Thorium samples can also be analysed by the differentiated neutron spectrum technique. Two standard reference material samples with different total uranium concentration and U-235 isotope abundance or thorium concentration are necessary to calibrate the analytical system. By applying the formula

$$aX + bY = C \qquad (1)$$

where X, Y = the U-235 and U-238 concentrations or the total U and Th concentrations

a,b = coefficients, and

C = counting rate per mass unit.

to the standards and the unknown samples in bare and cadmium-covered conditions, and by solving the resulting systems, the total uranium concentration and the U-235 isotope abundance or thorium concentration can be determined.

The International Atomic Energy Agency test samples gave good results for the U-235 isotope abundance, while some discrepancies in the total uranium concentration are under investigation.

An unknown powder sample for nuclear safeguards analysis has been compared with XRF and gamma-ray[4] spectroscopy with the following results:

DNC:	U-235/U-238 = 0.0071 ± 0.0002,	U = (80.5 ± 2.0)%
XRF:		U = (80 ± 4)%
GS :	U-235/U-238 = 0.0070 ± 0.0003	U = (81.7 ± 2.0)%

The results are in good agreement and show that the sample was natural uranium with element enrichment. In the case of multi-element neutron activation analysis Eq. 1 can again be applied where X and Y will be the concentrations of the two elements at the overlapping peaks and C will be the corresponding counts per mass unit.

CONCLUSION

From the preliminary results, it can be concluded that the new neutron activation technique offers new possibilities in accurate nuclear fuel analysis and that multiparameter analytical problems can be solved by multi-parameter nuclear analytical techniques.

ACKNOWLEDGEMENTS

The author would like to thank G. Lefkopoulos, C. Gatsos, N. Malliaros and the Reactor and Machine Shop staff for their contribution to the work.

REFERENCES

1. Papadopoulos, N. N., J. Radioanal. & Nucl. Chem., Articles, V.110, No. 1, 253, (1987).
2. Papadopoulos, N.N., J. Trace and Microprobe Techniques, V.6, No.1, 97, (1987).
3. Papadopoulos, N.N., Proc. ESARDA Symp. on Safeguards & Material Management, Luxembourg, 459, (1989).
4. Paradellis, T., Priv. Communication, (1990).

APPLICATIONS of MUONS IN CONDENSED MATTER PHYSICS

E. J. ANSALDO
Department of Physics
University of Saskatchewan
Saskatoon, Canada S7N0K3

ABSTRACT

An overview is given of muon spin rotation-relaxation results on selected condensed matter topics of great current interest. Of technical importance are measurements of the magnetic properties and the interplay of magnetism with superconductivity in the new high temperature superconductors, the location and electronic structure of hydrogenic centers in semiconductors, and the quantum diffusion properties of charged particles in metals and insulators.

INTRODUCTION

The availability of intense beams of muons at the so-called Meson Factories has allowed the full development of the μSR (muon spin rotation-relaxation-resonance) spectroscopy, of wide applicability in materials science and chemistry.[1] As a spin-$\frac{1}{2}$ light proton ($m_\mu \approx 1/9m_p$), the positive muon provides a unique microscopic probe of the electronic structure and dynamics of, for example, metallic systems, magnetic and superconducting materials, semiconductors and insulators (including organic and molecular systems). In insulators and semiconductors the μ^+ may capture an electron to form Muonium (Mu, μ^+e^-), a polarized, radioactive isotope of Hydrogen. In a typical experiment, the beam line transports (into the selected sample) muons produced by pion decay, with a kinetic energy of 4.1 MeV and 100% spin aligned in the direction of the linear momentum for pions initially at rest. A timing counter signals (t=0) the arrival of the muon at the sample, and a series of counters detect the anisotropic decay positrons as a function of angle and time. The depolarization of the muon spin, caused by its interaction with the effective fields at the implanted site in the material, is followed over typically 10 μs with nanosecond resolution. The technique is similar to the venerable perturbed angular distributions-correlations using excited nuclei, but it is more instructive to consider it as an analogue of proton NMR in which no external polarising field is required. In the μ^+ case the probe is a true elementary particle in interstitial locations of the lattice.

Since the decay positrons are of fairly high energy, the counters may be placed outside the magnets, cryostats, furnaces, etc. used to vary the sample environment , but a thin window is required for the incoming muons. In what follows, the power of the technique will be illustrated with a few examples of recent work on condensed matter topics of current interest, carried out mainly at the TRIUMF Cyclotron in Vancouver, Canada.

SUPERCONDUCTORS

In the particular case of the high-T oxide superconductors, the stopped muons are bound to oxygen ions, where they sense the internal field distributions in the mixed state. In the type-II superconductors, the magnetic field penetrates as flux quanta (fluxons), which form a triangular flux lattice. The resulting field distribution depolarizes the muon signal due to the well-known inhomogeneous broadening effect (T_2 relaxation). The depolarization rate is then $\sigma(\mu s^{-1}) \propto \lambda^{-2}$, where λ is the magnetic (London) penetration depth, a fundamental quantity whose temperature dependence has a bearing on the pairing mechanism. The μSR studies have shown that the new materials behave like conventional –s-wave paired– materials, an important result. In addition, a remarkable correlation was found between the (T=0K) value of σ and the critical temperature T_c, shown in Figure 1.

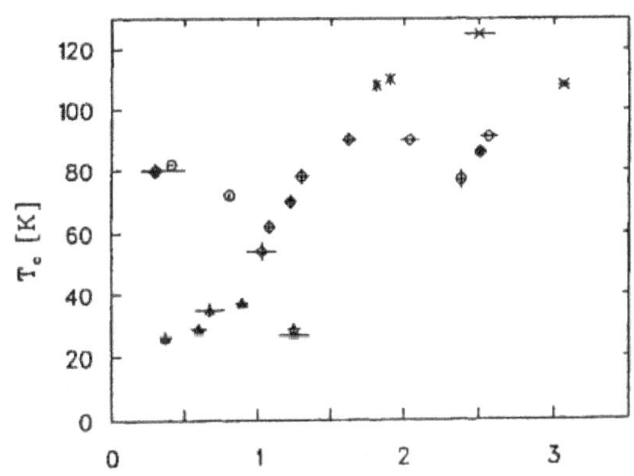

Figure 1. Plot of T_c *vs* relaxation rate $\sigma(0)$ for $La_{2-x}Sr_xCuO_{4-y}$ (triangles), $YBa_2Cu_3O_x$ (diamonds), Bi-Sr-Ca-Cu-O (circles), and (Bi-Pb-Tl)-2223 (asterisks). Except for x =0.15 and 0.20, samples of the $La_{2-x}Sr_xCuO_{4-y}$ system as well as $YBa_2Cu_3O_x$ samples with x below *ca.* 0.5 also display static magnetic ordering effects. The filled symbols are for crystalline samples. Note the substantially lower values of $\sigma(0)$ for the Bi-Sr-Ca-Cu-O family. This has to do with their peculiarly weak pinning forces, i.e. low critical currents.

Since $\sigma(0)$ is in turn a measure of the ratio of concentration to mass of the superconducting carriers, this correlation puts severe constraints on theories of the relation between electronic structure and T_c.

In magnetic materials, the depolarizing internal fields are due to electronic moments, and the measurements can take place in zero or in longitudinal applied field so that the type of magnetic order (e.g. antiferromagnetism *vs.* spin glass), and its dynamics may be determined. The cuprate oxide superconductors are known to develop from magnetic parents by hole doping, for example via trivalent Sr substituting for divalent La in La_2CuO_4. The magnetic phase diagram of $La_{2-x}Sr_xCuO_{4-y}$ was elucidated early by μSR in conjunction with neutron scattering experiments.[2] The data showed that random freezing of fixed moments supplants the antiferromagnetism as doping is increased, the system eventually developing metallic and superconducting properties while losing the magnetism. Similar magnetic effects also arise in $YBa_2Cu_3O_x$ for x varying from 6.0 to 6.5, with definite spin glass behaviour at $x = 6.45$. It is a matter of some controversy whether the apparent coexistence of magnetism and superconductivity is due to the intrinsic nature of the superconductive coupling mechanism or just due to microscopic inhomogeneities in the electronic structure.

HYDROGENIC ANALOGUE

The Mu atom usually forms as a defect centre in ionic and semiconducting crystals. The technical interest here is that Mu studies give information on the location and electronic structure of the analogous Hydrogen in those materials, of technical importance but difficult if not impossible to obtain by other techniques (ESR, ENDOR). Notable is the result that a dominant site for the hydrogenic defect in tetrahedral semiconductors is the bond center, i.e. the atom is located along a <111> direction in between two lattice atoms, involving a relaxation of the lattice parameters in the neighborhood.[3] The MuSR measurements, via the hyperfine parameters, also allowed the determination of the local spin density distribution. This resolved a long-standing controversy on the nature of such centres.

QUANTUM DIFFUSION

Another subject of considerable theoretical and practical interest is the mechanism(s) of diffusion of charged particles in metals and insulators. Here the muon also provides a unique probe. In this case the depolarization is due to the motion through the lattice. The depolarization rate σ in zero or applied fields may be related to the hop rate (diffusion constants) of the particle, whose spin experiences a fluctuating dipolar interaction with the nuclei as it tunnels from site to site in the host lattice. The main features of diffusion in metals are at present well understood, due in part to the theoretical effort to explain the μSR data of Figure 2.[4] In particular, the decrease in μ^+ hopping below 1K was at first surprising, but later explained by taking full account of the effect of conduction electrons. The theory was fully tested by experiments on ionic insulators (i.e. no conduction electrons), and in Al, where the onset of superconductivity actually "removes" conduction electron density and thus facilitates muon diffusion.

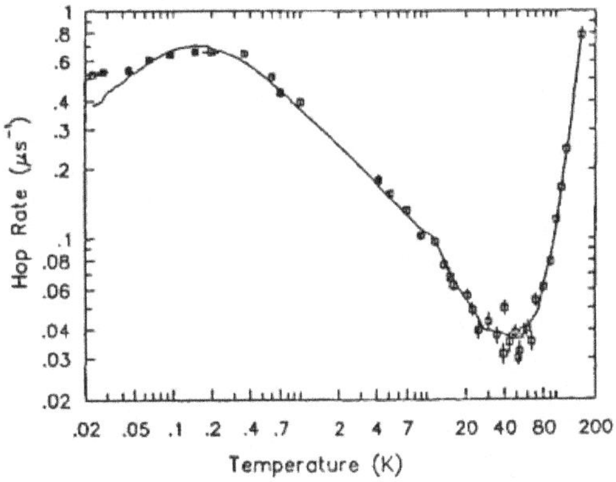

Figure 2. Muon hop rate in pure Copper.

CONCLUSION

Medium- and high- energy accelerators have provided a fertile ground for the new spectroscopy, illustrated only by a few selected "success stories" above. The future of the technique is bright, since μSR groups at several facilities worlwide (notably at TRIUMF in Canada, PSI in Switzerland, KEK in Japan, and RAL in England) are engaged in developing "user friendly" environments for outside scientists to obtain μSR results in their particular area of interest, using dedicated muon lines, as for example the M15 and M20 beam lines at TRIUMF.

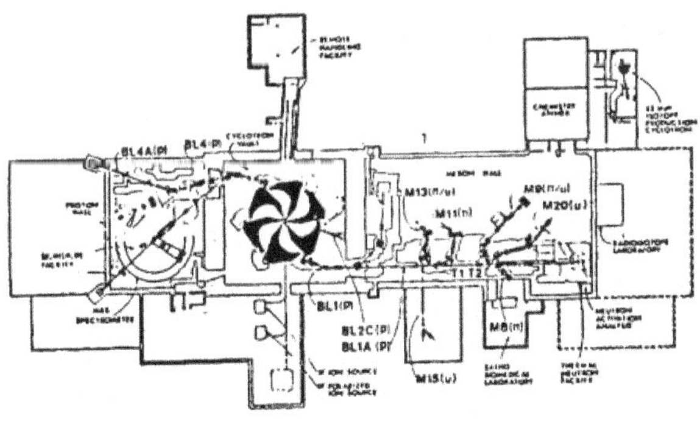

Figure 3. Plan of the experimental areas at TRIUMF. Beamlines M15 and M20 were developed for μSR work.

REFERENCES

1. For a description of the whole range of μSR applications, see the books:
   A. Schenck, *Muon Spin Rotation Spectroscopy-Principles and Applications in Solid State*, Adam Hilger Ltd. (1985), and
   D. C. Walker, *Muon and Muonium Chemistry*, Cambridge U. Press (1983),
   and the Proceedings of several specialized conferences Hyp. Int. **6** (1979), **8** (1981), **17-19** (1984), **31** (1986), and in press (1990).

2. G. Aeppli *et al.*, *Phys. Rev.* **B35**, 7129 (1987),
   D. R. Harshman *et al.*, *Phys. Rev.* **B39**, 851 (1989),
   Y. J. Uemura *et al.*, *Phys. Rev. Lett.* **62**, 2317 (1989).
   E. J. Ansaldo *et al.*, *Physica C* **162-164**, 259 (1989).

3. D. R. Harshman *et al.*, *Phys. Rev. Lett.* **63**, 1187 (1989).
   R. Kiefl *et al.*, *Phys. Rev. Lett.* **63**, 2136 (1989).
   J. H. Brewer *et al.*, *Phys. Rev. Lett* **60**, 1073 (1989).
   E. J. Ansaldo *et al. Phys. Rev.* **B40**, 2555 (1989).
   R. F. Kiefl *et al.*, *Phy. Rev. Lett.* **58**, 1780 (1987), and **60**, 224 (1988).

4. S. R. Kreitzman *at al.*, *Phys. Rev. Lett.* **56**, 181 (1986),
   R. Kadono *et al.*, *Phys Rev. B39*, (1989),
   R. Kiefl *et al.*, *Phys. Rev. Lett.* **62**, 792 (1989),
   G. M. Luke, Ph. D. Thesis, University of British Columbia (1989), and to be published Hyp. Int. (1990).

ION BEAM ANALYTICAL TECHNIQUES AT THE CISE 3.5 MV TANDEM VDG ACCELERATOR LABORATORY

C. Boni[1], A. Caridi[1], E. Cereda[1] and G. M. Braga Marcazzan[1,2]

1 Cise Tecnologie Innovative, P.O. Box 12081, 20134 Milan, Italy

2 Istituto Fisica Generale Applicata, University of Milan, Italy

ABSTRACT

The Cise 3.5 MV tandem VdG accelerator has been dedicated since the late seventies to the application of ion beam analytical techniques in the fields of environment and materials science.

In particular, PIXE (Particle Induced X-ray Emission), PIGE (Proton Induced Gamma-ray Emission), RBS (Rutherford Backscattering Spectrometry) and NRA (Nuclear Reaction Analysis) techniques have been set up and extensively applied.

In this paper, the overall features of the experimental set up are presented and an overview of typical applications is given.

INTRODUCTION

The home-made 3.5 MV tandem Van de Graaff accelerator has been operating since the mid seventies. First devoted to fundamental Nuclear Physics, it was then dedicated to the application of ion beam analysis.

PIXE technique was first set up and automated to be used also on an industrial scale[1)] and was then applied to environmental research[2-4)] and sometimes to biomedicine[5)]. PIXE difficulty in detecting light elements was recently overcome by setting up the PIGE technique to be used simultaneously with PIXE[6,7)] also for the analysis of thin samples. The PIXE-PIGE system is now extensively used in environmental applications[8] as well as in coal[9] and materials science[10].

RBS and NRA techniques were set up for material characterization[11,12] and for the analysis of carbon and oxygen in particulate matter[13].

EXPERIMENTAL

The technical characteristics of the tandem VdG accelerator are: 0.9-3.5 MV dynamic range of the terminal voltage; duoplasmatron ion source for protons or deuterons (in principle heavier ion beams could be injected but the solid stripper is rapidly deteriorated); the beam current ranges from a few nanoamps to some microamps; the energy resolution is within 5 keV with very good stability due to a liner system.

The ion beam, after passing through an analysis magnet, can be switched into three beam lines.

The PIXE-PIGE beam line, set up for in-vacuum analysis of both thick and thin samples, is extensively described elsewhere[6,7]. The peculiar feature is the simultaneous detection of X and gamma rays and for thin samples, the use of a proton beam with a suitable spread in energy by making the beam cross a rotating disc made up of six Al sectors of different thicknesses. The proton beam, whose primary energy is 3.52 MeV, gets a rectangular energy distribution about 270 keV wide resulting from the overlap of the six main components, partly superimposed by the straggling effect. In this way, the strong energy dependence of the prompt gamma reaction yields[14] is smoothed and the resulting gamma ray yield trends are reasonably constant over an energy interval larger than the energy loss in typical particulate matter samples up to 1 mg/cm^2 thick and made up of particles as large as about 20 um, making possible a straightforward quantitative analysis relative to standard reference samples.

As for the RBS and NRA set up, the beam is switched into the scattering chamber through a multiple tantalum slit beam collimator which makes the beam assume a rectangular shape. The samples are arranged in a ten position insulated target holder which, in order to improve the depth resolution, can be tilted with respect to the beam direction. The beam current is directly measured on the target holder, which is surrounded by an electron suppressor. The backscattered particles and reaction products are detected by a solid state detector connected to a standard electronic chain.

APPLICATIONS

Most of the beam time is dedicated to PIXE-PIGE in the context of atmospheric and coal-related applications.

As for environmental research, PIXE-PIGE is applied to the characterization of atmospheric aerosol samples collected in large urban areas and in sites near anthropogenic sources. Interesting results were obtained for Milan[2,7] where about 1300 aerosol samples collected in four sampling stations over one year allowed the determination of the main and trace elements, their distribution versus particle size, seasonal variations of composition, the effect of meteorological parameters on pollution levels, an estimate of the air quality and the identification of pollution sources by studying the correlations between elemental concentrations. Recently time-resolved measurements (streaker samplers with 1 hour resolution) in combination with meteorological data have yielded information on some anthropogenic source profiles and on the existence of a regional background of sulphates[4]. Similar studies were carried out in some Sicilian cities[3].

Research is carried out in order to assess the impact of coal and oil fired power plants. In a first step, an accurate characterization of the particulate matter released into the atmosphere was performed[15]. In a second step, studies on the diffusion of the emitted pollutants, on their fallout and on the air quality in the neighbouring areas are under way[2].

As for coal related technology, PIXE-PIGE is the principal screening technique combined with SEM, XPS, AES and XRD for the study of combustion processes and control devices, such as electrostatic precipitators[9], for fly ash abatement. The analysis of integral and size fractionated coal fly ash samples collected in different points of the power plants, leads to the estimate of the trace element pathways through the process and the control equipment and provides information about the ash particle formation mechanisms and enrichment phenomena for some elements.

RBS and NRA techniques have been set up for the analysis of carbon and oxygen in particulate matter[13] and for material characterization[11,12]. In particular, elemental concentration and depth profiles have been studied in implanted materials and interdiffusion phenomena, due to thermal treatments, have been characterized in multilayer structures.

CONCLUSIONS

The ion beam analysis carried out at Cise proved successful in many research fields. The setting up of an external beam line for PIXE-PIGE analysis of thin and thick samples as well as large objects and the development of DIXE-DIGE for the stoichiometry determination of thin oxide films are in progress.

The applicability of a proton microprobe to the characterization of coal fly ashes is being investigated in cooperation with the Zagreb Boskovic Institute group.

REFERENCES

1) Caruso, E., Cristofori, F., Marcazzan, G. M. and

Redaelli, P., Nucl. Instr. and Meth. 174,195(1980).

2) Braga Marcazzan, G. M., Caruso, E., Cereda, E., Redaelli, P., Bacci, P., Ventura, A. and Lombardo, G., Nucl. Instr. and Meth. B22,305(1987).

3) Boni, C., Caruso, E., Cereda, E., Lombardo, G., Braga Marcazzan, G.M. and Redaelli, P., J. Aerosol Sci. 19,7, 1271(1988).

4) Annegarn, H.J., Zucchiatti, A., Cereda, E. and Braga Marcazzan, G.M., Nucl. Instr. and Meth. B49,372(1990).

5) Caruso, E., Braga Marcazzan, G. M., Redaelli, P., Bonucci, E., Ballanti, P., Mazzaferro, S. and Coen, G., Biol. Trace Elem. Res. 10,123(1986).

6) Boni, C., Caridi, A., Cereda, E. and Braga Marcazzan, G.M., Nucl. Instr. and Meth. B47,133(1990).

7) Boni, C., Caridi, A., Cereda, E., Braga Marcazzan, G.M. and Redaelli, P., Nucl. Instr. and Meth. B49,106(1990).

8) Boni, C., Caruso, E., Cereda, E., Marcazzan, G. M. and Redaelli, P., J. Aerosol Sci. 19,7,1263(1988).

9) Boni, C., Caridi, A., Cereda, E. and Braga Marcazzan, G.M., Nucl. Instr. and Meth. B45,352(1990).

10) Boni, C., Caridi, A., Cereda, E., Braga Marcazzan, G. M., Parmigiani, F. and Scagliotti, M., Nucl. Instr. and Meth. B50,243(1990).

11) Boni, C., Braga Marcazzan, G.M., and Cereda, E., Proc. Int. Symp. on Three Day In-Depth Review on the Nuclear Accelerator Impact in the Interdisciplinary Field, eds. Mazzoldi P. and Moschini G., 71(1984).

12) Cereda, E., Braga Marcazzan, G. M., Parmigiani, F. and Scagliotti, M., Nucl. Instr. and Meth. B50,48(1990).

13) Boni, C., Braga Marcazzan, G.M., Caruso, E., Cereda, E. and Redaelli, P., Nucl. Instr. and Meth. B15,382(1986).

14) Boni, C., Cereda, E., Braga Marcazzan, G. M. and De Tomasi, V., Nucl. Instr. and Meth. B35,80(1988).

15) Bacci, P., Caruso, E., Braga Marcazzan, G. M. and Redaelli, P., Nucl. Instr. and Meth. B3,522(1984).

DEVELOPMENT AND USE OF NUCLEAR METHODS AT THE C.E.R.I. FOR RESEARCH AND APPLICATIONS

G. BLONDIAUX, J.L. DEBRUN

Centre d'Etudes et de Recherches par Irradiation (C.E.R.I.),
3A rue de la Férollerie, 45071 ORLEANS CEDEX 2 (France)

ABSTRACT

A variable energy cyclotron (protons up to 34 MeV) and a Van de Graaff (3 MV) are used at the C.E.R.I. for research and applications. Charged particles beams and fast secondary neutrons are used in many different fields: materials characterization, high energy ion implantation or modification of materials, thin layer activation, archaeometry, radiobiology, labelled molecules, radio-isotope production, neutron therapy and nuclear medicine. Users are either C.E.R.I. staff members, or outside users to whom the laboratory provides particle beams.

INTRODUCTION

The C.E.R.I. belongs to the chemistry division of the "Centre National de la Recherche Scientifique". There is a total of 30 C.N.R.S. persons working at the C.E.R.I., and 3500 C.N.R.S. persons in the chemistry division. The C.E.R.I. is a research laboratory, but it also provides charged particle beams to outside users, either free of charge for Universities and other C.N.R.S. laboratories, or on a commercial basis for other people. There are two particle accelerators in the laboratory: a variable energy cyclotron and a single ended Van de Graaff. Charged particles beams and fast secondary neutrons are used in many fields of research and application, with the notable exceptions of nuclear and atomic physics.

I. THE ACCELERATORS

a) The Cyclotron

This machine is a CGR-MeV 680; its characteristics are given in Table I.

Table I: Characteristics of the CGR-MeV 680 at the C.E.R.I.

PARTICLE	ENERGY RANGE (MeV)	MAXIMUM EXTRACTED INTENSITY (μA)
PROTON	5 - 36	50
DEUTERON	5 - 25	50
HELIUM 3	10 - 60	10
HELIUM 4	10 - 50	10

There are 4 beam lines:

- one beam line is for radioisotopes production; it is equiped with a 27°30 magnet, so that two irradiation positions are available. One is for the production of radioactive gases and is connected to the near-by nuclear medicine unit. The other position is for the production of radioisotopes by irradiation of solids; a pneumatic transfer system connects this position to hot cells in a high activity laboratory.

- two beam lines are used for activation analysis or for high energy ion implantation. Activation can be done in the vacuum or in the air, directly with ions, or with fast secondary neutrons.

- one beam line is for neutron therapy, and goes directly into the neutron therapy unit. 34 MeV protons are used; the horizontal beam is bent down by 90° and sent on a beryllium target.

b) The Van de Graaff

This accelerator is a vertical KN-3000 from HVEC, which has been almost totally rebuilt in 1986/1987. Its characteristics are given in Table II.

Table II. Characteristics of the KN-3000 at the C.E.R.I.

BEAMS	$H^+, {}^2H^+, {}^3He^+, {}^4He^+$	3.3 MV Maximum terminal voltage used ($N_2 + CO_2$ at 20 bar)	INTENSITY ON TARGET 10 μA maximum

BEAM LINES	1	2	3	4
	U.H.V. $5\ 10^{-10}$ torr + goniometer for channeling	"mini-PIXE" $30 \times 30 \mu m^2$	RBS or NRA with detection of ions	PIGE or activation

Tritons should soon be available on the machine. It is also planned to install a penning ion source for multicharged ions, and to routinely use higher terminal voltages by using 2% of SF_6 in the $N_2 + CO_2$ mixture.

II. VARIOUS USES OF THE MACHINES: GENERAL FIGURES

a) Cyclotron

Table III gives the present figures for the uses of the cyclotron.

Table III: Repartition of the beam time (1988-1989)

Research and Applications, non-medical service activities	Neutrontherapy (service activity)	Nuclear Medicine (service activity)
65%	30%	5%

The non-medical service activities represent a small fraction of the beam time; they consist in a few activation analyses and in irradiations for thin layer activation. Protons or HH^+ represent ~80% of the beams used, deuterons ~15% and helions 3 or 4 about 5%.

b) The Van de Graaff

The Van de Graaff has been used exclusively for the characterization of materials, until now.

Over 90% of the beam time was for research and applications, the rest was for service activities (mainly RBS analyses).

Protons were used 40% of the time, $^3He^+$ 25%, alphas 20% and deuterons 15% (1988-1989).

III. DESCRIPTION OF THE ACTIVITIES

A. Research and applications done by C.E.RI. staff members

a) Materials studies

Studies are done on semiconductors, in collaboration with other research laboratories; these studies have to do with crystal growth and with the influence of trace elements on the electronic properties of semiconductors. The elements of interest are mainly C and O, which are best analyzed using charged particle activation methods. Semiconductors that have been studied include: polycrystalline Si, Czochralski grown GaAs and InP and GaAlAs grown by vapour phase epitaxy on GaAs. A method combining charged particle activation and channeling has been developped: it provides unique information on the lattice location of C and O at sub-ppm levels. Studies on the stopping powers in channeled directions are being pursued, in connection with the above method. Charged particle activation methods are also developped and used in the frame of research contracts, for instance: trace analysis in fluoride glasses, in Li, Na, P, Al(Li) alloys, certification of trace elements in reference materials. Some work is also done in the field of thin layer activation, in collaboration with "Electricité de France".

To be complete, we have to mention that the laboratory is engaged, since 1987, in studies on resonance laser ionization mass spectrometry (RIMS) applied to ultra-trace and surface analysis.

[References: 1 to 13]

b) Modification of the characteristics of electronic devices by implantation

Silicon based devices can be improved by irradiation/implantation with charged particles; for instance, faster diodes are obtained, or smaller power losses can be achieved in the thyristors.This is due to local decrease of the lifetime of minority carriers: the resistivity reaches high values in the zone corresponding to the range of the particles (and remains high even after annealing). The laboratory is engaged in systematic studies on various devices, in collaboration with two industrial companies.

[Reference: 14]

c) Radioisotope production

Small-scale production of radioisotopes is studied, taking into account the specificities of the cyclotron. ^{7}Be, ^{56}Co, ^{57}Co and ^{127}Xe have been produced. ^{127}Xe, of interest in nuclear medicine, is obtained by irradiation of natural xenon via ^{127}Cs with a very high purity (> 99%) and with a yield of 3.7 MBq $(\mu A.h)^{-1}$.

[Reference: 15]

d) Labelled molecules

Part of the staff of the nuclear medicine unit of the hospital in Orléans, is doing research on/with labelled molecules, in association with the C.E.R.I. Most of the radioisotopes used are produced with the cyclotron: ^{15}O, ^{81m}Kr, ^{123}I, ^{201}Tl. [^{123}I and ^{201}Tl are not currently produced in Orléans].

Current research includes studies on the effect of drugs and on compounds for breast tumour imaging.

[References: 16-21]

e) Radiobiology

As for the nuclear medicine unit, part of the staff of the radiotherapy unit of the hospital is doing research in association with the C.E.R.I. (and with other laboratories). The goal of this research in radiobiology is to compare the effect of fast neutrons on living matter, with the effect of photons or electrons. In particular, studies have been performed on D.N.A. and on subgranular cells.

[References: 22-25]

B. OUTSIDE USERS

a) Neutron therapy

The neutron therapy unit belongs to the radiotherapy unit of the hospital. It was opened in 1981, and is the only neutron therapy unit in France, at the moment.

[For information contact: N. Breteau at the C.E.R.I.]

b) Nuclear Medicine

The nuclear medicine unit mainly uses commercial radionucleides. There is no PET camera. Radioisotopes made in Orleans with the cyclotron are: ^{15}O (CO^*_2, O^*_2 for brain and cardiovascular examinations) and ^{81m}Kr (for lung and cardio-vascular examinations).

[For information: contact F. Therain at the C.E.R.I.]

c) Materials characterization, archeometry, radioisotopes

Various studies or applications undertaken by outside users will be briefly mentionned; all persons in charge can be contacted at the C.E.R.I.

Geochemistry: activation analysis of platinoïds in geomaterials: (M. Volfinger).

Archaeometry: activation analysis of ancient glasses and coins: (J.N. Barrandon).

Biomaterials: activation analysis of coral for implantation in the body: (J.L. Irigaray).

Metallurgy: diffusion of radioisotopes in Tantalum: (J. Merini)

Production of ^{230}U for studies on decay by ^{14}C emission (M. Hussonois)

Production of $^{194,195,196,198}Au$ for studies on hyperfine structure: (J. Sauvage)

Thin layer activation on a commercial basis: (B. Janneau)

REFERENCES

1. Use of channeling in association with charged particles activation to study the position of light elements at trace level in crystals: the case of carbon in GaAs prepared by MO-VPE.
M.A. Misdaq, G. Blondiaux, J.P. André, M. Hage Ali, M. Valladon, C.J. Maggiore, J.L. Debrun
Nucl. Instr. and Meth. B15 (1986) 328.

2. Recent progress in the study of semiconductors using charged particle activation.
M.A. Misdaq, G. Blondiaux, N. Bordes, A. Giovagnoli, M. Valladon, L.C. Wei, M. Hage Ali, C.J. Maggiore, J.L. Debrun
J. of Radioanal. and Nucl. Chem. 110 (1987) 441.

3. Study of the incorporation and repartition of boron and carbon in Czochralski grown GaAs using charged particles activation.
L.C. Wei, G. Blondiaux, A. Giovagnoli, M. Valladon, J.L. Debrun
Nucl. Instr. and Meth. B24/25 (1987) 999.

4. Trace analysis using laser postionization of sputtered neutral atoms: preliminary results obtained with a Nd-YAG pumped dye laser and a quadrupole mass spectrometer.
O. Gobert, B. Dubreuil, P. Gelin, J.L. Debrun, R.L. Inglebert
Proceeding of the VI International Conference on SIMS, Versailles, septembre 1987. SIMS VI, p. 845, J. Wiley (1988).

5. Characterization of trace elements in semiconductors using charged particles activation analysis.
G. Blondiaux, A. Giovagnoli, M.A. Misdaq, C.J. Maggiore, M. Valladon, L.C. Wei, J.L. Debrun
Proceedings of the 12th International Symposium on the application of ion beams in materials science, Tokyo, september 1987, Hosei University Press, p. 85 (1988).

6. Studies on the analysis of trace metallic elements in semiconductrors using RIMS
P. Gelin, O. Gobert, B. Dubreuil, J.L. Debrun, R.L. Inglebert
Resonance Ionization Spectroscopy 1988, Institute of Physics Conference Series, n° 94, p. 201 (1988).

7. Study of the corrosion-erosion phenomenon in nuclear power plants using the thin layer activation method.
M. Bouchacourt, C. Marsigne, A. Dubail, G. Blondiaux and J.L. Debrun
Nuclear Instruments and Methods B40/411, p. 1199 (1989).

8. Proton activation analysis for the determination of trace elements in high purity ZrF4 used for the fabrication of low loss optical fibers.
M.F. Barthe, M.T. Bajard, A. Giovagnoli, G. Blondiaux, J.L. Debrun, M. Lesergent, Y. Tregoat and J.Y. Barraud
Nuclear Instruments ands Methods, B40/41, p. 1202 (1989).

9. Channeling and microactivation of materials.
C.J. Maggiore, J.D. Blasic, G. Blondiaux, J.L. Debrun, M. Hage Ali, E. Mathez, M.A. Misdaq, M. Valladon
Nuclear Instruments and Methods, B40/41, p. 1193 (1989).

10. Resonant ionization of sputtered neutral atoms for trace analysis in high purity materials.
P. Gelin, J.L. Debrun, O. Gobert, R.L. Inglebert, B. Dubreuil
Nuclear Instruments and Methods, B40/41, p. 290 (1989).

11. Analysis of carbon at trace level by ion sputtering and laser resonant postionization.
P. Gelin, M.F. Barthe, J.L. Debrun, O. Gobert, T. Gibert, R.L. Inglebert, B. Dubreuil
Nuclear Instruments and Methods, B45, 580 (1990).

12. Fast analysis of oxygen in fluoride glasses (ZBLAN) by charged particle activation $[^{16}O(d,n)^{17}F]$
M.F Barthe, A. Giovagnoli, G. Blondiaux, J.L. Debrun, Y. Tregoat, J.Y. Barraud
Nuclear Instruments and Methods, B45, 105 (1990)

13. Resonant multiphoton ionization of sputtered P atoms: application to the detection of phosphorus in silicon samples.
O. Gobert, B. Dubreuil, R.L. Inglebert, P. Gelin, J.L. Debrun
Physical review A, Vol. 41, n° 11, 6225 (1990)

14. J. Briaud et al.
Ann. Phys. (Fr), 14, 6C2, 597 (1989)

15. A. Mouadili et J. Vernois
Nucl. Inst. and Meth., to be published.

16. Labelling of oestrogen receptors in breast cancer
F. Thérain, J. Gros, G. Picaper
Nucl. Med. Biol., Vol. 13, n° 2, 141-146 (1986).

17. Synthesis of radio iodinated 16Iodo 11Methoxy 17Estradiol
J. Gros, F. Thérain
The Journal of Steroid Biochemistry, Vol. 25, Suppl., 45 (1986).

18. Dosimetrie de l'iode 123 chez le nouveau-né et l'enfant
M.T. Guilhem, F. Thérain
J. Biophysique et Biomécanique, Vol. 11, Suppl. au n° 2, 177-179 (1987).

19. Intérêt de la scintigraphie au 81M krypton dans l'étude de l'insuffisance coronarienne (résultat obenus chez 16 patients explorés également à l'aide de 201 thallium)
F. Thérain, P. Potin, D. Debrun, M. Valouiis
J. Médecine Nucléaire et Biophysique, Vol. 12, 350 (1988).

20. Assessment of myocardial perfusion using Krypton 81m in patients with coronary artery disease (CAD)
F. Thérain, P. Potin, D. Debrun, J. Gros, M. Valois.
Eur. J. of Nuclear Medicine, Vol. 14,, n° 5/6, p. 270 (1988)

21. 16α(^{131}I) Iodo 11β Methoxy-17bEstradiol: synthesis, in vivo distribution and preliminary clinical results.
F. thérain, J. Gros, D. Debrun, G. Picoper
Eur. J. of Nuclear Medicine, Vol. 15, n° 8, p. 542 (1989).

22. Study of several factors in RNA-protein cross-link formation induced by ionizing radiation within 70S ribosomes of E. coli MRE 600
B. Ekert, N. Giocanti and R. Sabattier
Int. J. Radiat. Biol., Vol. 50, N°3, 507-525 (1986)

23. Unusual alcaline elution pattern induced in mammalian cell DNA by fast neutrons P(34) + Be.
B. Ekert, R. Sabattier, M. Pironin
J. Radiat. Biol., Vol. 54, 39-53 (1988).

24. RBE of p(34) + Be neutron beam (Orleans) for intestinal crypts regeneration in Mice. Comparison with p(65) + Be neutron beam (Louvain-La-Neuve).
B. Destembert, J. Gueulette, M. Leclerc, R. Sabattier, N. Breteau
International Symposium on Particle Therapy, Meeting of the E.O.R.T.C. Heavy Particle Therapy Group, S.I.N. Villigen, 12-14 mars 1987. SIN Medical Newsletter, n°9, p. 44-47.

25. Fast neutrons irradiation on one rabbit cerebral hemisphere and comparative histological study with the other one. G. Gueneau, R. Sabattier, N. Breteau, V. Baille and L. Court
Proceedings of International Congress of Radiation Research, Edinburgh,n° 19-24 July 1987, Vol. 1, p. 318.

APPLICATIONS OF NUCLEAR TECHNIQUES IN THE ACCELERATOR LABORATORY OF THE UNIVERSITY OF HELSINKI

J. Keinonen

Accelerator Laboratory, University of Helsinki, Hämeentie 100, SF-00550 Helsinki, Finland

ABSTRACT

Applications of nuclear techniques in studies of slowing down of energetic ions in semiconductor materials and metals, ion implantation induced effects on them, interaction of defects, host atoms and implanted atoms, migration and solubility of doping atoms, elemental analysis and characterization of sample structures will be discussed. The appplications described with examples include the use of the following techniques: the Doppler shift attenuation method, nuclear resonance broadening, nuclear reaction analysis, and proton induced gamma-ray emission combined with the use of low-energy positron beam technique (not in our laboratory), Rutherford backscattering spectrometry and channeling, and partricle induced X-ray emission.

1. INTRODUCTION

The regular use of the laboratory's first accelerator, a 2.5 MV electrostatic accelerator VdG, began in 1956. The accelerator is still put to intense use. The operation of the laboratory's next main accelerator, a tandem accelerator EGP-10-II, started in 1982. The equipment of the laboratory is listed in Table 1. Ion beams in the energy region from 0 eV to tens of MeV can be produced by the accelerators of the laboratory. Ion beams from the EGP-10-II accelerator, isotope separator and 10 kV accelerator can be used to bombarding same target positioned in an ultra-high-vacuum chamber.

Table 1. Equipment of the Accelerator Laboratory of the University of Helsinki.

- 5 MV EGP-10-II tandem accelerator
- 2.5 MV electrostatic accelerator (VdG)
- 100 kV isotope separator
- 10 kV accelerator
- facilities for nonlinear laser spectroscopy
- facilities for production of diamond-like coatings
- facilities for research in the physics of molecules

The equipment installed at the laboratory, has enabled the implementation of a research programme in materials physics and applied physics based on the use of ion beams and applications of nuclear techniques. The scientific programme of the laboratory is listed in Table 2. In the following sections I shall briefly describe the research where nuclear techniques are employed and use examples of our recent studies to illustrate the combination of different techniques in studies of materials.

Table 2. Scientific programme.

Research based on utilization of ion beams

Materials physics

- interaction and slowing down of energetic ions in solids
- ion implantation induced effects on the structure of matter
- interaction of defects, host atoms, and implanted atoms
- migration and solubility of doping atoms
- production of new materials
- analysis of ion-implanted and irradiated materials

Applied physics

- elemental analysis
- radionuclide production

Nuclear physics

- gamma-ray spectroscopy of light nuclei
- astrophysics: cross sections of thermonuclear reactions

Laser physics

- nonlinear laser spectroscopy of ion beams

Research based on utilization of laser beams

- nonlinear laser spectroscopy of gases
- development of stabilized lasers
- physics of molecules

---

2. MATERIALS PHYSICS

In materials physics our principal subject of research is the physical processes taking place in solid matter during and after implantation of energetic ions.

For elemental analysis and structural characterization of implanted and ion-irradiated materials, several methods have been improved and employed. These are Nuclear Resonance Broadening (NRB), Nuclear Reaction Analysis (NRA), Particle Induced Gamma-Ray Emission (PIGE), Doppler Shift Attenuation Method (DSAM), Rutherford Backscattering Spectrometry (RBS), RBS/channeling, and Particle Induced X-Ray Emission (PIXE).

The damage produced in keV and MeV ion implantation of semiconductor materials Si and GaAs is the main subject of interest at the moment.[1,2] Nuclear techniques to study slowing down of ions in solids have been used and developed in the laboratory[3-5].

Ion beam techniques have enabled research on short migration distances, to obtain diffusivities from the changes in range profiles in previously uninvestigated temperature regions relevant to applications[6,7].

Very recently, the use of nuclear techniques has enabled us to make a basic experiment in ion beam mixing, i.e. to study relocation of Si atoms in keV and MeV Si-ion irradiation of crystalline Si (Ref. 8).

Ion beam analysis of high T_c superconductors with different nuclear techniques has been recently discussed by us[9].

3. APPLIED PHYSICS

External beam arrangement is usually used in elemental analysis with PIGE and PIXE. Serum and samples of tooth, bone and blood tissue[10], and the elemental composition of oil colour pigments used by various artists have been analysed[11].

A project launched at the laboratory has been to produce positron-active ^{18}F for positron emission tomography in medical diagnostics.

4. NUCLEAR SPECTROSCOPY

Nuclear spectroscopy has had an essential impact to the applications and development of nuclear techniques in our laboratory.

In nuclear spectroscopy research has been concentrated on the γ-ray spectroscopy of nuclei lighter than calcium. The main subject has been measurements of short nuclear lifetimes (1 fs - 1 ps) by the DSA method[3]. The development of sophisticated implanted targets, as well as the improved reliability and accuracy of the deduced lifetime values, stem from our work with implantation techniques, and ion implantation-induced effects in solids.

5. EXAMPLES ON APPLICATIONS OF NUCLEAR TECHNIQUES

5.1. Annealing Behavior of H Traps in Ne-Implanted Ta

In a recent work on the annealing behavior of H traps located at the interface of the Ta matrix and Ne precipitates, new information was obtained on the recovery stages of inert-gas-implantation induced defects in metals and their strong dependence on the implantation induced damage[12]. The concentration distributions of the implanted Ne atoms and trapped H atoms were measured by the NRB method through the reaction $^{22}Ne(p,\gamma)^{23}Na$ at $E_p = 1005$ keV and $^{1}H(^{15}N,\alpha\gamma)^{12}C$ at $E(^{15}N) = 6.385$ MeV, respectively. Atomic Ne densities within the solid Ne-precipitates were studied by means of the DSA method.

5.2. Stopping Power for Low-Velocity Ions in Semiconductor Materials

As a part of our systematic studies on the stopping power of semiconductor materials for low-velocity ions ($v \leq 4v_o$, where v_o is the Bohr velocity), the electronic stopping power for ^{26}Mg ions in Si, Ge, and GaAs has been recently studied in the energy region 0-1.0 MeV/nucleon by application of the DSA method[13]. An example of the results is shown in Fig. 1.

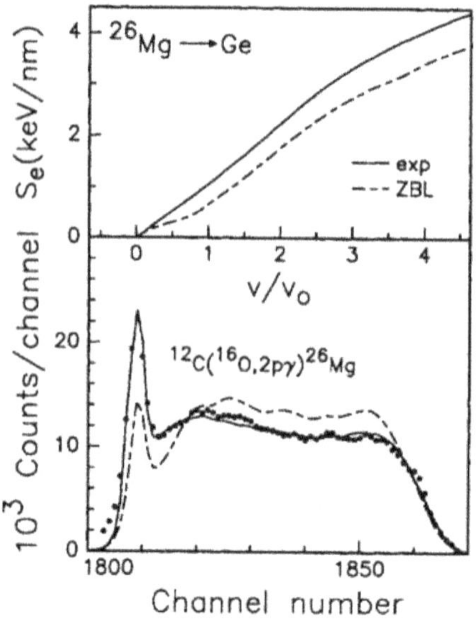

Fig. 1. Stopping power.

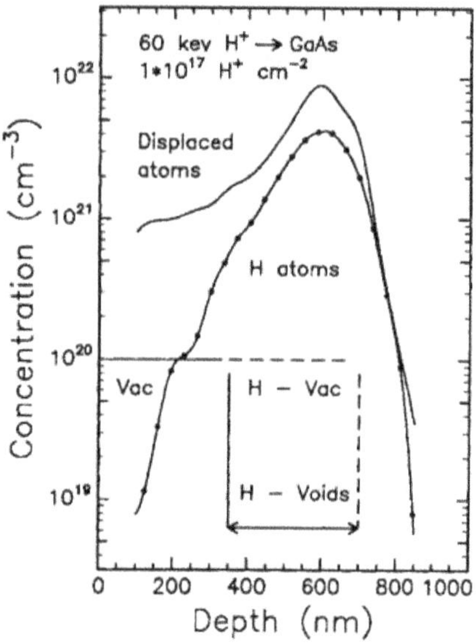

Fig. 2. Defects in GaAs.

Fig. 1. Portion of a γ-ray spectrum illustrating the measured 1.81 MeV γ-ray peak for ^{26}Mg ions slowing down in Ge. The energy of the ^{16}O beam was 20 MeV. The target was prepared by implanting the ^{12}C atoms into Ge. The solid line shows the line shape obtained in the computer simulation with the corrected electronic stopping power (exp in the upper part of the figure). The dashed line illustrates the line shape for the empirical electronic stopping power[14] (ZBL in the upper part of the figure). The velocity of the recoiling ions at the moment of γ-ray emission corresponds to the channel number as shown in the figure.

Fig. 2. Depth profiles of H atoms from NRB, vacancy defects from positron annihilation, and displaced Ga and As atoms from RBS/channeling in H-implanted semi-insulating GaAs.

5.3. Defects in H Implanted GaAs Studied by Nuclear Techniques

As a part of our systematic studies on the H implantation induced defects in semiconductor materials[1], NRB, RBS/channeling, RBS/PIXE and low-energy positron beam techniques have been recently used to study damage and implanted ion distributions in semi-insulating GaAs(100) after room temperature implantation of 10^{16}-10^{17} 60 keV H^+/cm^2 and 1×10^{16} 6.4 MeV $^{15}N^{2+}/cm^2$ (Ref. 2). The measurements with the low-energy positron beam were performed at the Technical University of Helsinki. The combination of the different techniques made it possible to draw solid conclusions from the annealing behavior of damage in GaAs. The redistribution of H after annealings was observed to depend on the migration of implantation-induced defects. Results are illustrated in Fig. 2. The combination of RBS/channeling and PIXE measurements have been employed to resolve the portions of the displaced Ga and As atoms in 6.4 MeV $^{15}N^{2+}$ irradiated GaAs.

REFERENCES

1) Keinonen, J., Hautala, M., Rauhala, E., Karttunen, V., Räisänen, J., Lahtinen, J., Vehanen, A., Punkka, E., and Hautojärvi, P., Phys. Rev. **B37**, 8264 (1988).

2) Saarinen, K., Hautojärvi, P., Mäkinen, J., Vehanen, A., Keinonen, J., Rauhala, E., Räisänen, J., Corbel, C., to be published.

3) Keinonen, J., in *Capture Gamma-Ray Spectroscopy and Related Topics 1984* (International Symposium, Knoxville, Tennessee), AIP Conf. Proc. No. 125, edited by S. Raman (AIP, New York, 1985) p.557.

4) Arstila, K., Keinonen, J., and Tikkanen, P., Phys. Rev. **B41**, 6117 (1990).

5) Kuronen, A., Räisänen, J., Keinonen, J., Tikkanen, P., and Rauhala, E., Nucl. Instrum. Methods Phys. Res. Sect. **B35**, 1 (1988).

6) Räisänen, J. and Keinonen, J., Appl. Phys. Lett. **49**, 773 (1986).

7) Räisänen, J., Keinonen, J., Karttunen, V., and Koponen, I., J. Appl. Phys. **64**, 2334 (1988).

8) Keinonen, J., Hautala, M., Koponen, I., and Erola, M., Phys. Rev. **B41**, May (1990).

9) Keinonen, J., Räisänen, J., and Rauhala, E., in *Studies of High Temperature Superconductors*, Vol. 4, edited by A. V. Narlikar (Nova Science Publishers, Inc., New York, 1989) p.239.

10) Räisänen, J., Nucl. Instrum. Methods Phys. Res. Sect. **B40/41**, 638 (1989).

11) Tuurnala, T., Hautojärvi, A., and Harva, K., Nucl. Instrum. Methods Phys. Res. Sect. **B14**, 70 (1986).

12) Keinonen, J., Karttunen, V., Räisänen, J., Bergmeister, F.-J., Luukkainen, A., and Tikkanen, P., Phys. Rev. **B34**, 8981 (1986).

13) Arstila, K., Keinonen, J., Tikkanen, P., Kuronen, A., to be published.

15) Ziegler, F., Biersack, J. P., and Littmark, J., *The Stopping Power and Ranges of Ions in Matter*, Vol. 1. (Pergamon, New York, 1985).

APPLIED NUCLEAR TECHNIQUES AT C.R.N. STRASBOURG

G. J. C. COSTA
Centre de Recherches Nucléaires C.N.R.S. - IN2P3 - U.L.P. F67037 Strasbourg Cedex (France)

ABSTRACT

A survey of some of the practical applications of nuclear techniques in Strasbourg will be presented. Three groups are concerned :

1. C.I.S. (Centre d'Irradiation de Strasbourg), dedicated to food conservation, using miscellaneous radiation processing equipment, both on a research and on industrial scale.
2. PHASE (PHysique et Applications des SEmiconducteurs), involved in semiconductor research using the RBS/NR/CPA/ERD and PIXE methods.
3. L.R.M.F. (Laboratoire de Recherche des Musées de France), associated with a group at the C.R.N., is devoted in the analysis of archaeological samples utilizing a 2 MV Pelletron Tandem Accelerator at the LOUVRE Museum (Paris).

Keywords : Food irradiation and dosimetry ; nuclear techniques on high TC superconductors ; archaeological characterization of art and objects.

1. C.I.S.

The methods of radiation treatment for food preservation by irradiation[1] have been conducted at the C.R.N. (Strasbourg) since 1988. This group has acquired an electron

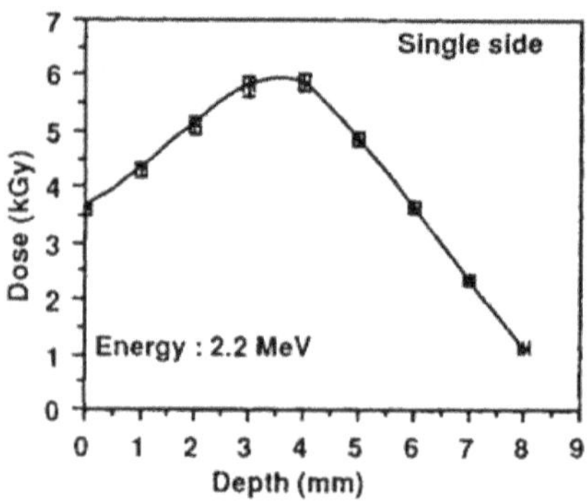

Fig. 1

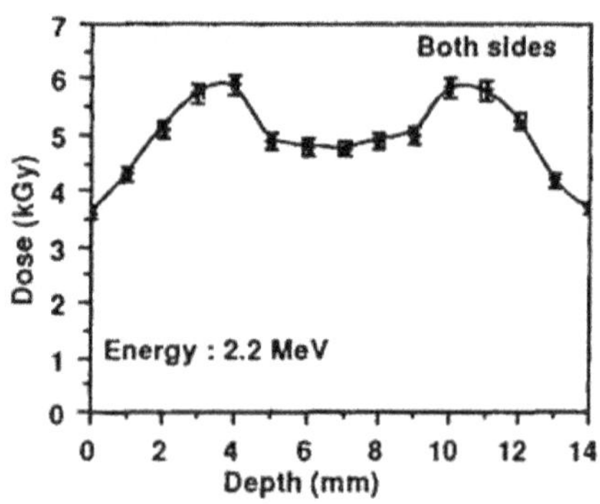

Fig. 2

accelerator of Van de Graaff type [2.5 MV - 150 μA]. Many studies and improvements of the conveyor system (optical coding and cycling) permit, through a X-ray converter target (W and Cu of 0.11 and 1.5 mm respectively) the suitable and optimal conditions for the homogeneity of the doses to be established, e.g. ***X-ray treatment conditions*** :

E_{e^-} : 2 MeV   I : 150 μA (maximum)   Type of treatment : on both sides

Length of a convoy cycle : 750 mm   Thickness of the matter (density = 1) : 80 mm

Fig. 1 shows that for 50 % of the dose (6 to 3 kGy), only a 6 mm thick product can be treated. On the other hand, a treatment on both sides ensures a sufficient depth (14 mm), while still keeping a 50 % variation of the dose [Fig. 2]. This is a physical process whereby energetic rays (γ, accelerated electron, X-rays) are used to improve the characteristics of a product. The ionisation technology is both flexible and powerful :

- ***in agro-food industry***
  - creation of variety
  - pasteurization, sterilization, sprout inhibition, etc...
- ***in other industrial sectors***
  - insect removal
  - medical and surgical equipment sterilization
  - packaging and basic products for cosmetics
  - physico-chemical modification of materials (grafting, linkage)

2. PHASE

For many years, the PHASE[2] group has applied nuclear methods to analysis and development using a wide range of ions at the 4 MV Van de Graaff and 18 MV Tandem accelerators. Starting from elemental analysis (C and O in Si, through C.P.A.) or profile concentration in amorphous semiconductors (Hydrogen through E.R.D.), this very active group uses some more analytical and sophisticated nuclear methods. The fact that σ is proportional to Z^2, permits them to use heavy ion beams like ^{35}Cl (30-50 MeV) or ^{127}I (240 MeV).

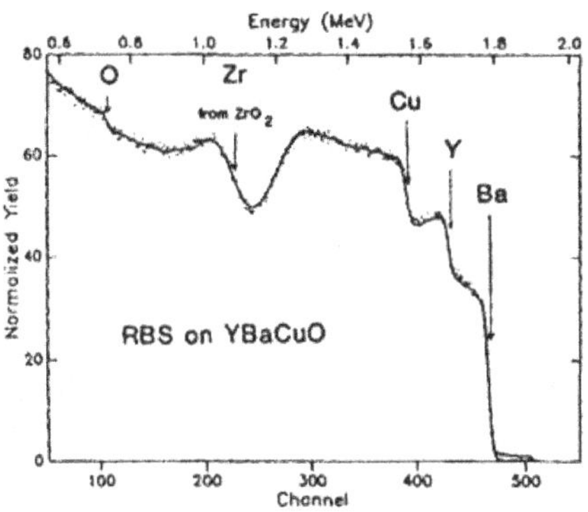

Fig. 3

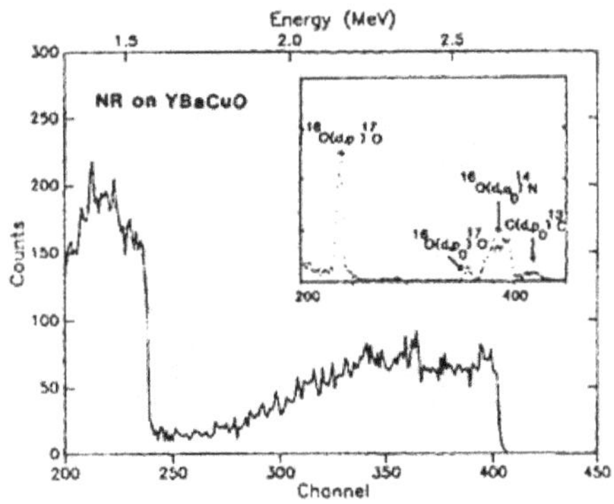

Fig. 4

Taking YBaCuO (a high Tc superconductor) deposits on YSZ, it is possible to obtain either, through R.B.S. (Fig. 3), N.R. (Fig. 4), E.R.D. (Fig. 5) and PIXE (Fig. 6), a high sensitivity analysis which is fundamental to semiconductor investigations.

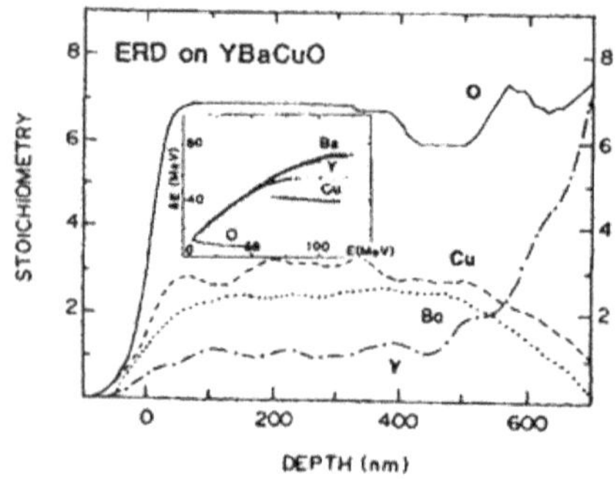

Fig. 5

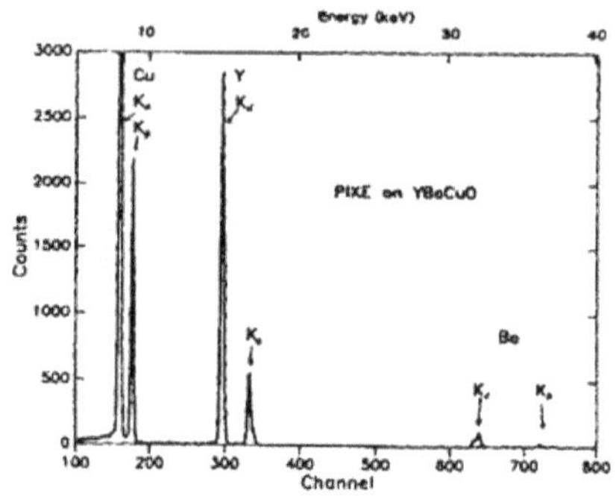

Fig. 6

3. L.R.M.F.

The L.R.M.F.[3], being mainly involved in the studies of art and archaeological objects, is utilizing a 2 MV Tandem Pelletron 6 SDH-2 from NEC at the Louvre (Paris). Through

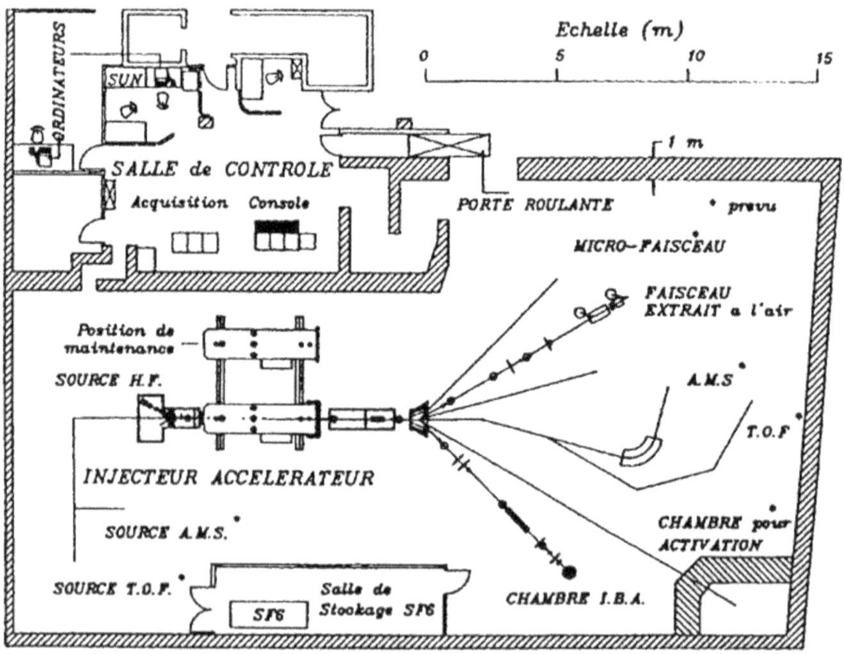

Fig. 7

different nuclear techniques (PIXE, PIGME and an external beam in air), this laboratory is able to obtain, with some confidence, data about the origin and composition of the samples. On Fig. 7, a general overview of the Louvre implantation is presented.

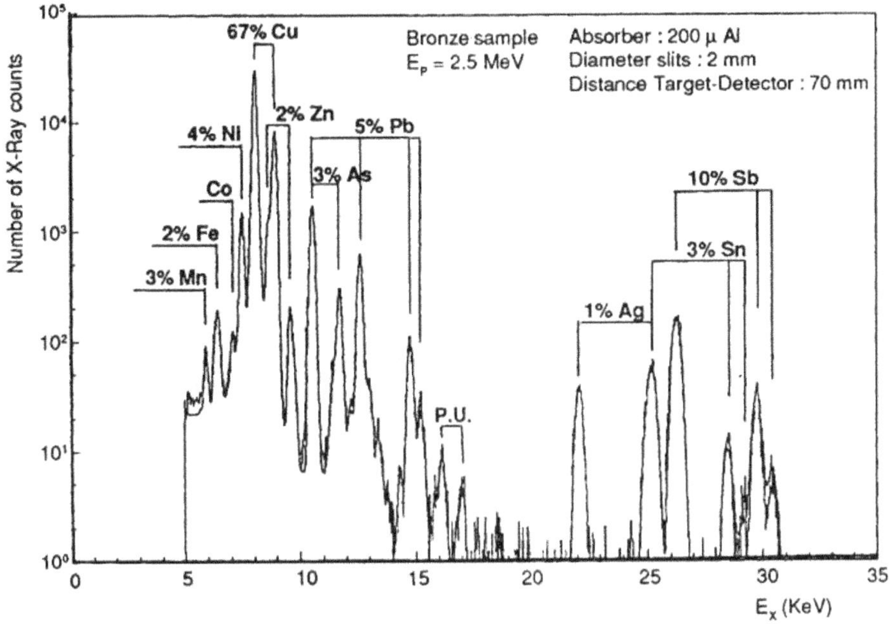

Fig. 8

Fig.8 represents the spectrum of a bronze sample. This spectrum has been analyzed with a specific X- ray computer programme and the element percentages are automatically calculated.

Acknowledgements

The author is grateful to all his colleagues for giving him, through their advices and papers, the opportunity to present this talk [International Conference on Applications of Nuclear Techniques CRETE - GREECE June 24-30, 1990].

References

1. *Rapport d'activité*, AERIAL Strasbourg 1989
2. *Rapport d'activité*, PHASE CRN Strasbourg July 1988
   Rapport d'activité, CRN Strasbourg 1988 & 1989
3. *Rapport d'activité*, LRMF Service de Physique 1988 - 1989

THE APPLIED NUCLEAR PHYSICS PROGRAM AT THE INSTITUTE OF NUCLEAR PHYSICS of the NCSR"DEMOKRITOS"

X. ASLANOGLOU

Institute of Nuclear Physics

NCSR "Demokritos", 153 10 Aghia Paraskevi, Greece

ABSTRACT

The applied Nuclear Physics program at the Institute of Nuclear Physics of the NCSR "Demokritos" is presented in this paper. The activities of the Laboratory are PIXE, PIGE, XRF, RBS and nuclear reaction analysis. Examples of ongoing projects are given for each of these activities.

INTRODUCTION

In the Institute of Nuclear Physics at "Demokritos" the applied Physics program involves five staff members and four graduate students. Most of the staff has a pure nuclear Physics background and has worked in Nuclear Spectroscopy and Heavy Ion Reactions, so the transition from basic research to applied was rather smooth and successfull.

On the facility side, the lab has a TANDEM TN/11 accelerator of terminal voltage 5.5 MV, a high current duoplasmatron source and an inverted sputter source, for heavy ion beams. Five beam lines depart from the accelerator and are dedicated to gamma ray experiments, proton intuced x-ray emission, H-detection and charge particle reactions, while the newest beam line is dedicated to neutron experiments, primarily for Astrophysics experiments.

A number of techniques have been developed and used for analytical perposes. These are PIXE, XRF, PIGE, RBS and Nuclear Reaction Analysis. In the following pages examples of projects involving these techniques will be given.

PROTON INDUCED X-RAY EMISSION

PIXE has been the most traditional activity in applied Physics in the Lab. It is usually done with 2.5 MeV protons using the external beam technique, through a Kapton foil. The problems usually studied are connected to Environmental Polution, Health Sciences, Archeometry etc.

An ongoing study in PIXE is the "Investigation of the shape of the Continuous Background in PIXE" by K. Potiriades (INP "Demokritos") and A. Katsanos (Univ. of Crete). The investigators are trying to calculate the continuous background analytically, based on calculation of Quasi Free Electron Bremsstrahlung (QFEB), Secondary Electron Bremsstrahlung (SEB) and Proton Bremsstrahlung (PB). Their data set consists of angular distributions of x-rays.

Fig. 1 shows the result of their calculation at 2.0 and 2.5 MeV, for a Mylar target ($(C_{10}H_8O_4)_n$). This calculation is based only on QFEB and SEB and the investigators believe that the agreement will improve if they include additional phenomena in the calculation. The agreement between the experiment and the calculation is better at the energy of 2.5 MeV, because the influence of SEB is stronger.

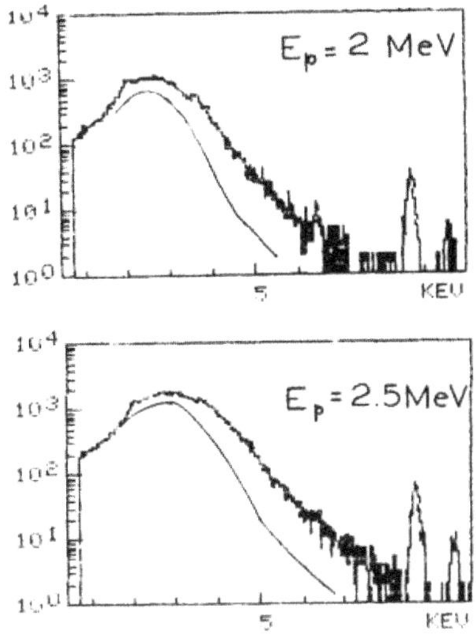

Fig. 1 Calculation of the background in PIXE, together with data from Mylar target

Another ongoing project in collaboration with the Center of Occupational Health and Safety of the Ministry of Labor is on the "Analysis of fumes in the environment of welding", by P. Kakanis (Dept. of Labor), M. Pilakouta and K. Potiriades (INP "Demokritos").

In this project, 60 workers were equiped with small pumps, pumping air in an area of 20 cm radius (the breathing sphere) and cellulose filters were used to trap the dust. The filters were then analyzed with PIXE. A typical spectrum is shown in Fig. 2, where peaks ranging from Ca to Pb are pronounced. The aim of this project is to correlate these data with blood analysis and epidemiological data in working environment.

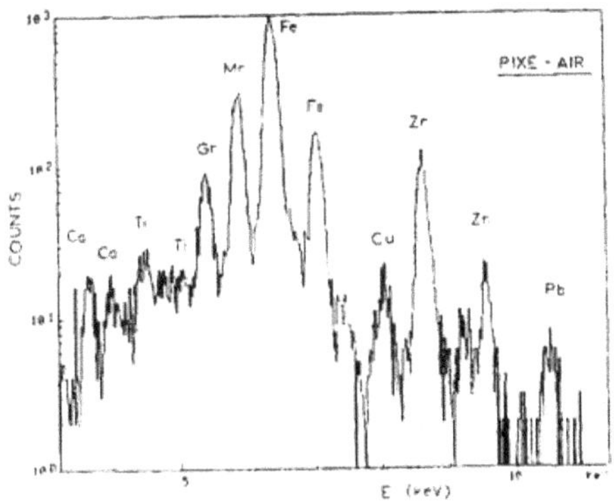

Fig. 2 PIXE spectrum of air in the environment of welding

PROTON INDUCED GAMMA RAY EMISSION (PIGE)

For quantitative analysis of light elements, (p, γ), (p, p'γ) and (p,αγ) reactions are used. This technique, when combined with PIXE or XRF measurements for heavier elements, gives a full analysis of the sample. The work on "Quantitative analysis of Al and trace elements in Bauxites" by A. Savidou et. al. is presented elsewhere in these proccedings.

Measurements of Carbon contamination in steel is a project which started after Steel Industry request. By measuring the 2365 KeV gamma ray following the $^{12}C(p,\gamma)^{13}N$ reaction, accurate quantitative analysis of C can be achieved. Fig. 3 shows a spectrum from this reaction, while Table I presents comparison between chemical and nuclear analysis of steel. The detection limit for these measurements was 0.1%.

TABLE I

Analysis of C in Steel

Chemical Analysis(%)	Nuclear Analysis(%)
0.8	0.88 ± 0.06
0.16	0.21 ± 0.02
0.20	0.30 ± 0.03
0.08	0.10 ± 0.01

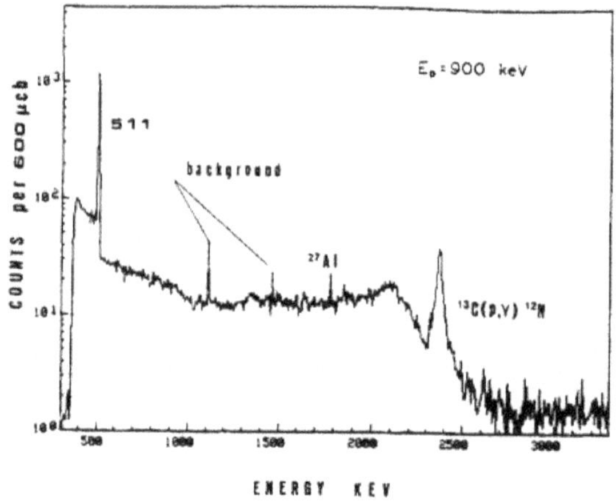

Fig. 3 Gamma ray spectrum of steel

Similar methods are routinelly used for detection of Li, B, O, S,P, elements which are usually found in ores.

RUTHERFORD BACK SCATTERING

In the RBS program beams of ^{7}Li and ^{12}C are utilized for determination of thickness of thin foils. In an ongoing collaboration between the Institute of Material Science and the Institute of Nuclear Physics at "Demokritos", the properties of thin Ti - Si films are studied by X. Aslanoglou, M. Pilakouta, V. Katselis (INP "Demokritos") and P. Aloupogiannis (IMS "Demokritos"). Titanium Silicides can be used in Very Large Scale Integrated Circuits (VLSI) as alternative material to Al, and the IMS at "Demokritos" is interested in the electrical properties and the thermodynamics of such materials.

In the process of manufacturing, Ti of thickness 400 - 1000 A is evaporated on Si substrate. Then, the Ti - Si compound is formed by annealing at 750^0 C and RBS is used to characterize the thickness and the exact nature of the compound. A typical spectrum of this material using a ^{12}C beam is shown in fig. 4.

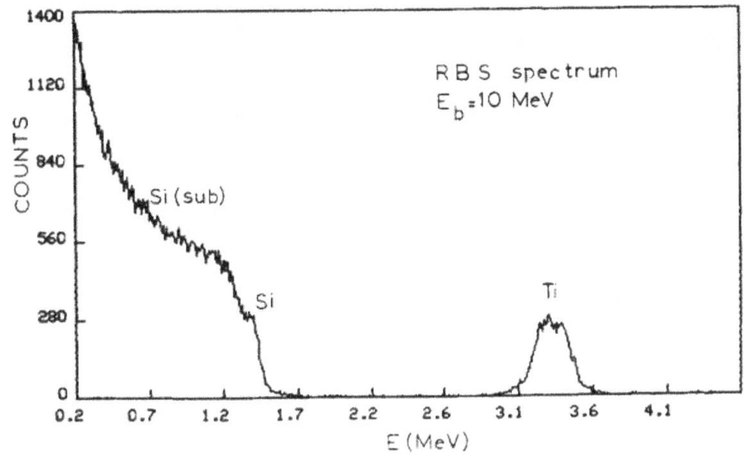

Fig. 4 RBS spectrum of $TiSi_2$ film on Si substrate using ^{12}C beam.

RBS is also routinelly used in thin foil characterization for a "thin foil Bank " that is under development at INP "D". These are produced by evaporation on Kapton substrate and used for XRF and PIXE calibrations.

X-RAY FLUORESCENCE

In XRF, sources of ^{241}Am and ^{109}Cd are used, coupled to a 100 mm^2 active area Si(Li) detector. With this set up, a elements limit as low as Sulfur can be measured. XRF has been another traditional activity of the laboratory for the last 15 years. The problems of interest have been analysis of biological and geological samples, polution problems, analysis of super-conducting materials, Archeometry and Medical Physics.

I will give two examples of recent projects, one from archeometry and one from Medical Physics.

Analysis of samples of ancient dyes used as either painting or cosmetic materials has been a long standing activity of the lab. By identifying the materials used in these dyes, conclusions can be extracted about comercial relations and cultural interactions between different places or the status of welth and civilization of a city. Most of them have been found to be inorganic materials.

In the case of "The analysis of the purple material found in Akrotiri, Thera" (E. Aloupi, Y. Maniatis, T. Paradellis) a purple coloured very light and fine powdered material was found during excuvations in the island of Thera. The XRF spectrum and the results of the analysis are given in Fig. 4. An unusually high content of Br (0.53%) is observed, which can not be attributed to sea water salt or any other obvious reason and points to an organic material, probably marine life. A subsequent XRD analysis showed strong Aragonite and Calcite phases of $CaCO_3$ indicating that the actual organic dye should not be more than 1.5% of the bulk material. Finally, the $CaCO_3$ phases were found to originate from sea shell powder, and the actual dye was identified to be purpura, a very expensive dye produced of the glands of molluscs.

Ca :	34.00 %
Fe :	1.50 %
Br :	0.53 %
Mn :	2300 ppm
Cu :	600 ppm
Zn :	600 ppm
Sr :	2900 ppm

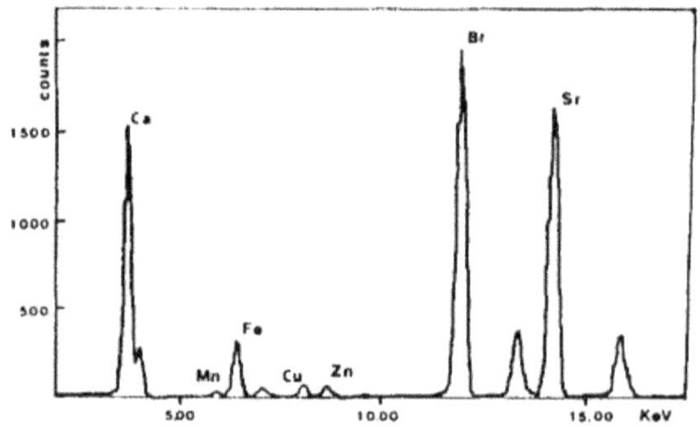

Fig. 5 XRF spectrum and concentrations of the purple material found in Akrotiri, Thera

In the medical Physics side, an investigation was done on the metabolism of Ca and Sr for patients with Chronic Renal Failure (CRF) under treatment with Continuous Ambulatory Peritonial Dialyses (CAPD), in collaboration with the School of Medicine, the University of Athens.

The results of XRF analysis of blood serum for normal people, patients with CRF and patients under CAPD are shown in fig. 5. In the case of Ca, patients with CRF show a deficit, which is fixed by CAPD. In the case of Sr though, the CRF group show an exess, which is not cured by the treatment. The reason is that, there is always a Sr contamination in Ca, because Ca and Sr are similar chemicaly and can not be adequatly separated. So, the Ca enriched CAPD solution, always contains an amount of Sr, which adds to the already existing Sr in the CRF patients. In addition, the observed increase of Sr in CRF parients is completely due to the fact that renal function is selective in discriminating Ca from Sr. Calcium is precious and reused while Strontium is extracted via urine. Patients on CRF do show increased Sr levels due to this insuficient renal discrimination.

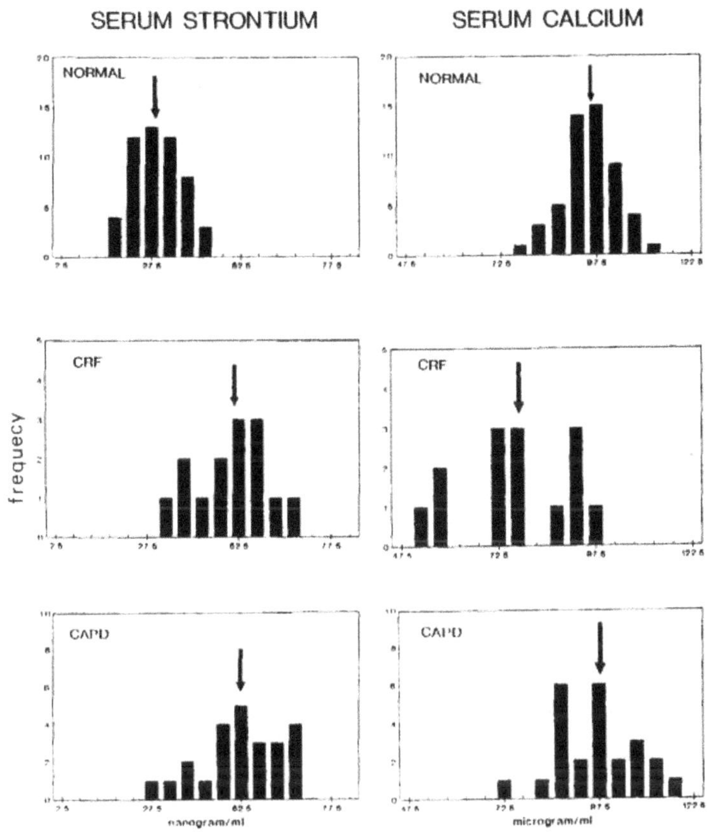

Fig. 6 Ca and Sr concentrations in serum for normal people, patients with Chromical Renal Failure and patients under CAPD

APPLICATION OF NUCLEAR TECHNIQUES IN THE CHEMISTRY DEPARTMENT OF THE UNIVERSITY OF THESSALONIKI, GREECE. - A Laboratory Report -

P. Misaelides

Department of Chemistry, Aristotle University,
GR-54006 Thessaloniki, Greece.

1. ABSTRACT

The Department of Chemistry of the University of Thessaloniki has established a Radiochemical Laboratory in order to cover the educational and training needs of the department and promote the research in the field of applications of nuclear techniques. The Laboratory has now, four years after its establishment, active training programmes and participates in a variety of research projects financed by national and international organizations.

Among the research activities of the Laboratory one could mention:

- the oxidation, corrosion and diffusion studies on material surfaces (metals, alloys, ceramics, marble) using Nuclear Reaction Analysis and Rutherford Backscattering techniques,
- the determination of trace amounts of light elements in semiconductor matrices using Charged Particle Activation Analysis,
- the environmental, archaeometrical and medical studies using Neutron Activation Analysis and radioanalytical techniques.

2. INTRODUCTION

The main research activities in the field of applications of nuclear techniques in the Department of Chemistry of the University of Thessaloniki are concentrated in the Radiochemical Laboratory established in 1986. The Laboratory, possessing no own accelerator or research reactor, has established strong collaborations with institutions having such experimental facilities (NCSR Demokritos, Nuclear Physics Institute of the University of Frankfurt, etc). These collaborations are funded by national and international organizations and agencies (General Secretariate of Research and Technology, VW-Foundation, IAEA, NATO etc).

The aim of this contribution is to provide information about the current research activities of this laboratory.

3. CURRENT RESEARCH ACTIVITIES OF THE RADIOCHEMICAL LABORATORY

3.1 Oxidation, corrosion and diffusion studies using ion beam analysis techniques.

A considerable part of the research activities of the laboratory is devoted to the study of oxidation, corrosion and diffusion processes on materials surfaces using ion beam techniques (Nuclear Reaction Analysis, Rutherford Backscattering Spectroscopy etc).

The high-temperature oxidation of stainless steels (AISI 321, 316 etc) in atmospheres containing sulfur oxides (simulation of the power stations process gases) is one of the current research projects of the Laboratory. The oxygen depth profiles are determined using the $^{16}O(d,p)^{17}O$ nuclear reaction at the 5 MV Tandem accelerator of the NCSR Demokritos. For the determination of the sulphur depth distribution in the surface layers of materials the $^{32}S(p,p'\gamma)^{32}S$ reaction has been selected. This nuclear reaction has in the energy region between 3.0 and 4.0 MeV two narrow resonances (E_R = 3379 and 3716 keV) allowing the determination of the depth distribution of sulphur in near-surface layers of materials. The cross sections of this reaction in the projectile energy region 3.0 to 4.0 MeV were determined using thin CdS-targets. In a series of recently started experiments the corrosion of ion implanted stainless steels in aggressive atmospheres at temperatures up to 750 °C is also under investigation. A complimentary study of the electrochemical behaviour of the ion implanted stainless steels in solutions of different acidities is also in progress. The ion implantations of the steels were performed using the 50 kV implanter of the Nuclear Physics Institute of the University of Frankfurt. As implants were chosen oxides forming elements increasing the passivation of the steels (Al, Y etc). Other ions (e.g. Si, B) as well as combinations of ions will be used for the further experimental work.

The study of the oxidation of zirconium and its alloys is a second research project of the laboratory. Although zirconium and its alloys are very important materials for the nuclear technology, their oxidation behaviour cannot be considered as completely understood. The data available in the literature show sometimes deviations of many orders of magnitude. The aim of this study is the understanding of the kinetics and mechanism of the oxidation of zirconium in the temperature region between 400 and 700 °C, that has not been systematically studied. For this purpose zirconium foils were oxidized in oxygen atmosphere and then annealed in vacuum at temperatures between 400 and 700 °C. The depth distribution of oxygen has been determined using the $^{16}O(d,p)^{17}O$ nuclear reaction. Of especial interest are also the determination of the diffusion coefficients of oxygen in oxidic layers using the $^{18}O(p,\alpha)^{15}N$ nuclear reaction and the influence of the presence of nitrogen on the oxidation rates of zirconium.

Finally a number of other corrosion and diffusion studies, e.g. the distribution of chemical matrix modifiers on graphite platforms for the electrothermal atomization atomic absorption spectrometry (ETA-AAS) using RBS, the diffusion of lithium in boron carbide and its interaction with stainless steels at high temperatures etc, are currently in progress.

3.2 DETERMINATION OF TRACES OF LIGHT ELEMENTS IN SEMICONDUCTING MATERIALS USING CHARGED PARTICLE ACTIVATION ANALYSIS.

The charged particle activation analysis has proven to be a valuable tool for the determination of trace amounts of light elements in semiconductor matrices offering extreme sensitivity and high selectivity. As part of the collaboration of the Radiochemical Laboratory of the University of Thessaloniki with the Nuclear Physics Institute of the University of Frankfurt the determination possibilities of trace amounts of boron, carbon, nitrogen and oxygen in silicon, GaAs and GaP by deuteron and helium-3 activation have been investigated. The relevant cross section data were determined using the 2.5 and the 7 MV van der Graaff accelerators of the University of Frankfurt.

The current activities of the Laboratory on this field include the determination of sulphur in GaAs matrices by means of the $^{32}S(^{3}He,p)^{34m}Cl$ and $^{34}S(p,n)^{34m}Cl$ nuclear reactions.

A dedicated activation chamber recently constructed will allow to perform proton and deuteron activations at the Tandem Accelerator Laboratory of the NCSR Demokritos.

3.3 ENVIROMNENTAL, ARCHAEOMETRICAL AND MEDICAL STUDIES USING I.N.A.A. AND RADIOANALYTICAL TECHNIQUES.

3.3.1 Environmental studies

After the monitoring of the fission products released during the Chernobyl nuclear accident in the environment of Thessaloniki and the studies on the adsorption and fixation of caesium-137 on the clay minerals vermiculite and montmorillonite, that followed, the laboratory started a three years programme on the determination of toxic concentrations of different elements (As, Sb, Cr, Co, Zn, Se, Cd, Hg, Pb, Mn, Br, etc) in roadside dust, air-particulates and plant samples from the urban, industrial and rural region of the city of Thessaloniki. The determination of the most elements was performed using Instrumental Neutron Activation Analysis. In the cases of elements, that cannot be determined by INAA (e.g. Pb, Cd etc), Atomic Absorption Spectrometry was used for the determinations.

The neutron irradiations take place at 5 MW Research Reactor of the NCSR Demokritos and the countings at Thessaloniki using two gamma-spectrometry setups with Ge(Li) detectors. Because of the transport of the samples only nuclides with half-life greater than 10 hours can be only counted in the Radiochemical Laboratory of the University of Thessaloniki.

Another project in the environmental sector is the determination of the natural radioactivity and the trace elements in mineral and spa waters of Northern Greece. A similar project dealing with therapeutic pelloids is also in progress.

3.3.2 Archaeometrical studies

The I.N.A.A. combined with high-resolution γ-ray spectroscopy has been applied in the study of a variety of archaeological samples including amber samples from the Aghia Paraskevi cemetery near Thessaloniki (6th c. BC), glass samples from Phillipi, bronze from Vergina, Bucchero ceramic from Lesbos and Izmir, byzantine ceramic from Cyprus etc. The most of these studies had as aim the localization of the provenance of the objects obtaining also information about the trade relations of the region, their chemical constitution and their way of fabrication. On this field there is a strong collaboration with the Archaeological Museum of Thessaloniki, the Department of Archaeology and History of Art of the University of Thessaloniki and the Nuclear Physics Institute of the University of Frankfurt. The neutron activations have been performed using the research reactors of the NCSR Democritos, the Research Centre Juelich and the University of Mainz in Germany.

4. SOME RECENT PUBLICATIONS OF THE LABORATORY

1. P. Misaelides, J. Krauskopf, G. Wolf and K. Bethge : Determination of carbon and oxygen in GaAs by means of ^{3}He-activation. Nucl. Instr.Meths. B18, 281 (1987)
2. V. Liebler, K. Bethge, J. Krauskopf, J. D. Meyer, P. Misaelides and G. Wolf: Determination of the excitation functions for carbon detection by charged particle activation analysis. Nucl.Instr. Meths. B36, 7 (1989)
3. K. Burkhard, A. Clouvas, H. Rothard, K. Kronenberger, R. Kirsch, P. Misaelides, A. Katsanos, C. Biedermann, J. Kemmler, K.O. Groenevelt: Secondary electron emission from thin foils under fast ion bombardment. Phys. Rev. B39, 6316 (1989)
4. M. Kessissoglou, E. Mirtsou, P. Misaelides, G. Wolf: Analysis of archaeological amber from the Aghia Paraskevi cemetery near Thessaloniki, by I.N.A.A. and I.R. spectroscopy. J. of Trace and Microprobe Techniques 7, 195 (1989).
5. P.Misaelides, C.Samara, M. Georgopoulos, Th. Kouimtzis, S. Synetos: Toxic elements in the environment of Thessaloniki, Greece. Part 1: Roadside dust analysis by I.N.A.A. and A.A.S. Toxicological and Environmental Chemistry 24, 191 (1989)
6. C. Tsartsarakos, P. Misaelides, A. Katsanos: Cross-sections of the $^{32}S(p,p'\gamma)^{32}S$ nuclear reaction used for the determination of sulphur on materials surfaces. Nucl.Instr.Meths. B45, 33 (1990)
7. F. Koehl, J. Krauskopf, P. Misaelides, G. Wolf and K. Bethge: Determination of nitrogen in semiconductors using the $^{14}N(p,\alpha)^{11}C$ and the $^{14}N(d,n)^{15}O$ nuclear reactions. Nucl.Instr.Meths. B50, 19 (1990)

ION BEAM FACILITIES AT CATANIA UNIVERSITY

Gaetano Foti
Dipartimento di Fisica, Corso Italia, 57,95129 CATANIA (Italy)

Introduction

Mev ion accelerators have been widely used in the sixty by nuclear physicist comunity to investigate nuclear matter.
In the last three decades, we have observed a progressive shift of the nuclear techniques to solid state or atomic physics. In our lab, in Catania a large community of nuclear and solid state physicist, about two hundred people, work with several accelerators:

Tandem - V.d.G.	13 MV	Experiments
V.d.G.	2.5 MV	Analysis
Ion Implanter	0.4.MV	Processing

Today high energy ion processing of materials may be considered one of the more scientific significant and technological developments; these include 1) microelectronics, 2) surface modification, 3) thin film technology, 4) ion implantation metallurgy, 5) nuclear track technology, 6) ion processing of biomaterials.

The understanding concerning the underlying mechanisms during ion-surface interaction requires the development of specific experimental methods: nuclear techniques are the most appropriate to analyse surface phenomenon occuring in the first monolayers up to 1 um depth.

Nuclear reaction analysis (NRA) and Rutherford Backscattering (RBS), combining with ion channeling technique are widely adopted to investigate on the depth profiles of impurities in solid, lattice location, phase formation, defect distribution, oxidation phenomena, ion mixing.

The big impact of Mev-ion beam analysis as diagnostic tool of solid matter was clear in mind of our chiefs, prof. E.Rimini and prof. J.W.Mayer of Caltech. In the early seventy, they organized a workshop of specialist to collect the most significant experimental data of the ion cross section energy loss, X-ray production yield, nuclear reaction data etc. in the KeV-MeV energy range. After the meeting, a handbook "Catania working data" was published and sent to many lab.s in the world. (1, 2).

About the activities of our laboratory up to day,

their can be divided in three groups:

I) Fondamental solid state physics - 40%

II) microelectronics - 40%

III biomaterials and interdisciplinary activities (astrophysics, medical, art) 20%

In the following section, I like to give same experimental results for each group.

I-Fundamental Solid State Physics

Ion beam induced epitaxial crystallization of surface amorphous layer or ion-induced annealing and/or amorphization of isolated damaged clusters in silicon has been extensive investigated in the last years (3). Crystallization occurs by a motion of the crystal-amorphous interface at temperatures as low as 500 K and depends on the energy deposited by the primary ions into elastic collisions.

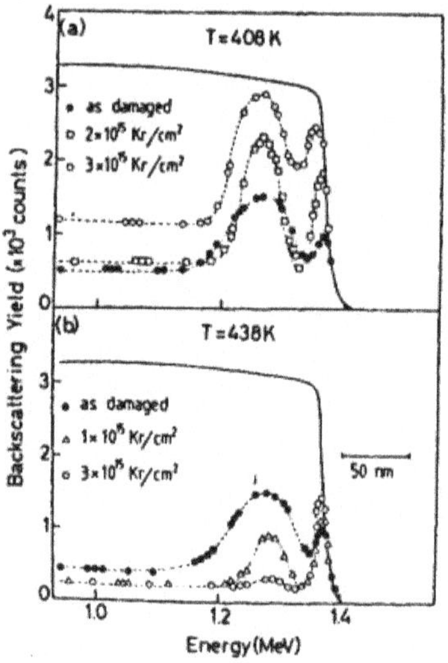

FIG.1 RBS spectra in random (continuos line) and along the (100) direction in as-damaged silicon crystals and after 600 KeV Kr ion irradiation, for two different substrate temperature

Fig.1 shows the RBS-channeling spectra with 2.0 MeV He beam recordered in silicon with amorphous clusters of 2.5 mm size after an irradiation with 600 KeV Kr2+ ions. By changing the substrate temperature from 408 K to 438 K, the Kr-ion irradiation produces an enhancement (low temperature irradiation) or a reduction (high temperature irradiation) of amorphous cluster sizeas a function of the ion fluence.

The measurement of the amount of disorder, as displaced atoms by RBS-channeling technique, allows to test a model for the phase transition induced by ion beam (4)

II -Microelectronics

Atom impurity introduced by ion implantation can be carefully controlled down 1 ppm of atomic concentration. Implantation and diffusion of gold in silicon single crystal is reported in Fig. 2, where the gold concentration has been measured as a function of the depth after very high temperature annealing.

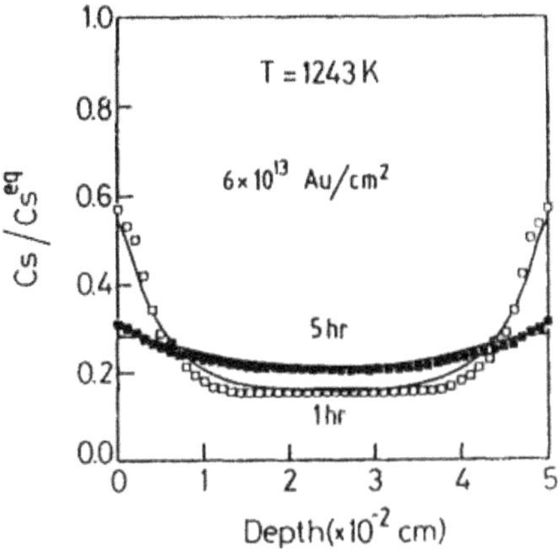

FIG. 2 Gold concentration profiles in ion implanted silicon measured with spreading resistance, after high temperature annealing (C_S^{eq}= $5x10^{15}/cm^3$)

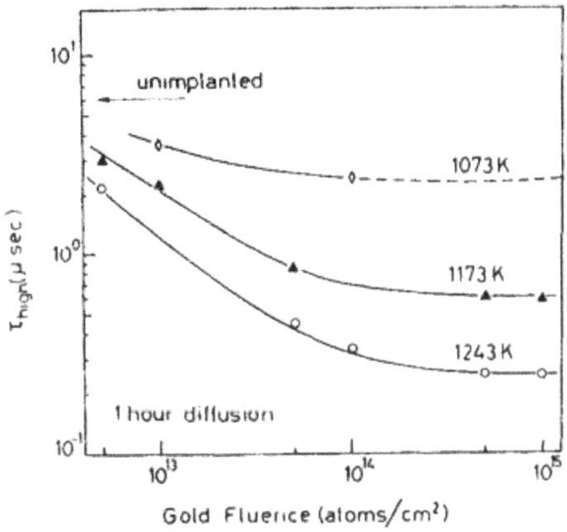

FIG. 3 Recombination lifetime v.s. gold fluence, measured by OCVD method, after high temperature annealing

Gold in silicon is an excellent way to control the high level recombination lifetime (high) as shown in Fig. 3 as a function of the ion fluence.

Ion implantation of gold produces unique concentration profiles at 1 ppm level, through the fine control of the amount of gold atoms in the silicon and depth implant control, the surface peak can be monitored with a high precision (5, 6).

Lattice location of boron in silicon after implantation and/or laser annealing can be otained with

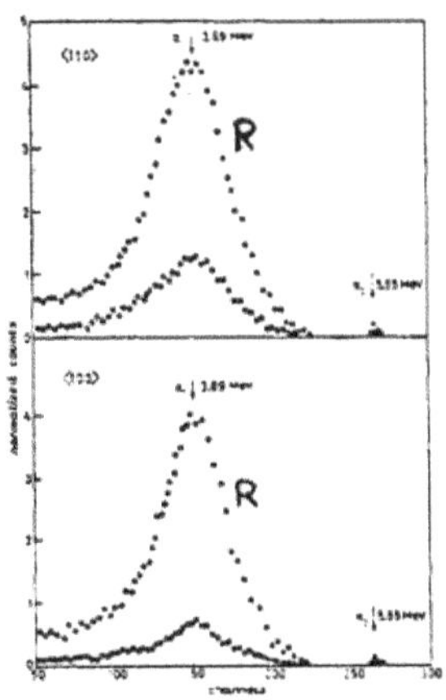

FIG. 4 NRA spectra in boron implanted silicon after laser annealing, recordered in random (R) and along the (110) and the (100) directions.

a specific nuclear reaction (,) as reported in Fig. 4, where the alpha yield are reported for random and aligned <100> incidence of MeV-proton beam. The fraction of boron located in the silicon lattice is just (1 - x), where x is the ratio of nuclear yield recorded along the <100> and random incidence. No models are required for the data analysis (7, 8).

III -Biomaterials

Ion etching of fluoropolymers is a very efficient way to get material removal, producing very small holes of micron size with smooth wall in a bulk material. Ion irradiations of polytetrafluoroethylene (PTFE) through metallic grids with MeV beams, are able to fabricate microfilters to drain biological fluid from the brain in hydrocephalus patients. PTFE ion etching rates are reported in Fig. 5 (a) for MeV proton and helium beams. High values of etching rate (10 mm/hr) make possible to remove large fraction of PTFE (8) as shown in Fig.5(b).

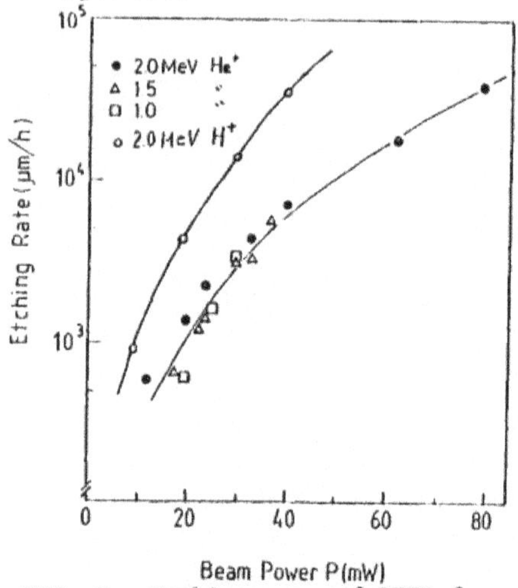

FIG. 5a Etching rate of PTFE for proton and helium irradiation for different ion beam power.

FIG. 5b SEM picture of PTFE irradiated with 2.0 MeV He beam through a metallic grid.

Ion beam analysis of natural hydroxyapatite has been used to measure the fluorine distribution in human teeth.

In Fig 6 is reported the depth distribution of fluorine in normal and fluorotic teeth by measuring, from the energy spectrum, the alpha particles emitted in the nuclear reaction (,). The surface enrichement is very high compared with the bulk value and becames more evident in fluorotic teeth. (9) By focusing the ion beam up 0.5x0.5 mm, it is possible to get a lateral distribution of fluorine atoms. In pathological teeth we observe a strong depletion of fluorine content in the decayed area, as shown in Fig. 7. (10).

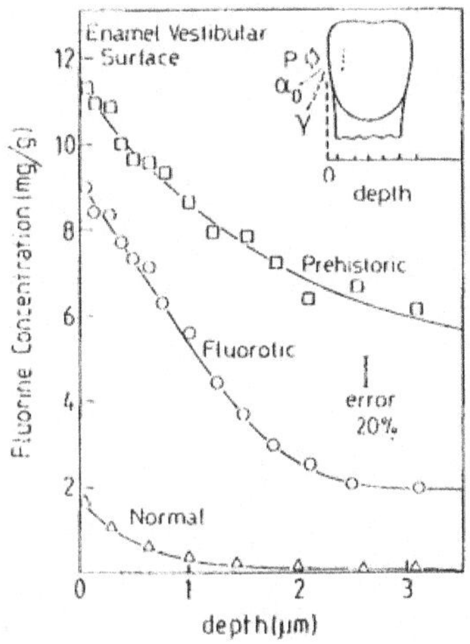

FIG. 6 Fluorine concentration v.s. depth measured with NRA in normal, fluorotic and very old enamel vestibular molar tooth.

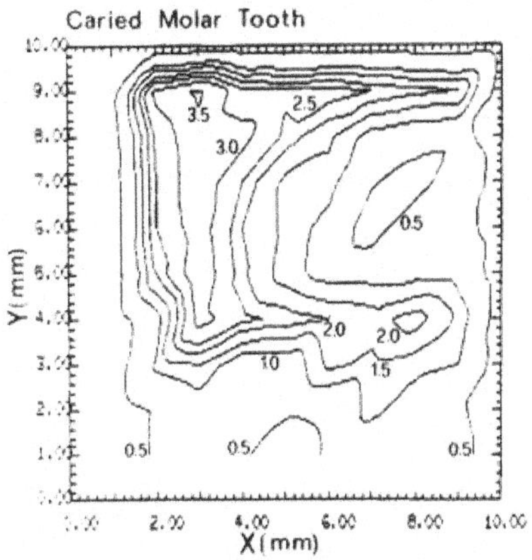

FIG. 7 Isoconcentration (mg/gr) fluorine map in decayed molar tooth.

Conclusions

Ion beam processing and ion beam analysis are powerfull techniques to modify and to analyse the first 1um-layer of materials. Reproducibility within ppm concentration values and submicron scale depth are the main factors which makes so actractive such ion beam techniques in industry. Moreover isotopic unrichment of solid target can be easely produce by ion implantation in the weight range 1-10 ugr/cmm and can be easely used to calibrate nuclear reactions.

References

(1) "Catania Working Data" USA - Italy Seminar on Ion Beam Analysis of Near Surface Regions (Catania June 17-20, 1974)

(2) Ion Beam Handbook For Material Analysis Ed. by J.W.Mayer and E.Rimini Academic Press Inc., New York, 1977

(3) A.Battaglia, F.Priolo, E.Rimini and G.Ferla Appl.Phys.Lett.56, 305(1990)

(4) F.Priolo, A.La Ferla and E.Rimini J.Mater.Res. 3, 1212(1988)

(5) S.Coffa, L.Calcagno, S.U.Campisano, G.calleri and G.Ferla J.Appl.Phys. 64, 6291(1988)

(6) S.Coffa, L.Calcagno and S.U.Campisano "Control of gold concentration in silicon" J.Appl.Phys. (1990) in press

(7) G.Della Mea and G.Foti Lett.N.Cimento 15, 127(1977)

(8) L.Torrisi and G.Foti "ion Beam Etching of PTFE" J.Mater.Res.(1990) in press

(9) L.Torrisi, G.Foti and S.U.Campisano Clinical Materials 5, 150(1990)

(10) L.Torrisi Phys.Med. 3, 213(1988)

Activities of the Radionuclides Metrology Group of the Central Bureau for Nuclear Measurements

T. Altzitzoglou, G. Bortels, B. Denecke, G. Grosse, D. Mouchel, D.F.G. Reher and W. Bambynek

CEC-JRC Central Bureau for Nuclear Measurements
B-2440 Geel, Belgium

ABSTRACT

A brief report on the past and present activities of the Radionuclides Metrology Group of the Central Bureau for Nuclear Measurements (CBNM) of the Joint Research Centre (JRC) is presented. Three of the recent projects are discussed in more detail.

INTRODUCTION

The Radionuclides Metrology Group is part of the Physics Division of the Central Bureau for Nuclear Measurements (CBNM) of the Joint Research Centre (JRC) of the Commission of the European Communities (CEC). CBNM at Geel has the role of promoting European standards and contributing to the harmonization of reference methodologies and materials and executes the program on nuclear measurements and reference materials.

The activities of the Radionuclides Metrology Group in particular lie in the areas of Nuclear Data for Standards and Radionuclide Metrology, with specific interests in the development and improvement of measurement techniques, the collection of nuclear decay and associated atomic data, the preparation of radioactive samples, special standards and reference sources, and the related fundamental research.

Over the past years, the group determined the decay scheme, the half-life and the life-time of excited levels of a variety of radionuclides, measured the probability of photon and electron emission, electron capture, alpha decay and internal conversion of a series of radionuclides, and compiled and evaluated data on the half-life of actinides, the experimental internal conversion, the K-shell fluorescence yields, the orbital electron capture, data for the International Directory of certified radioactive materials, etc.

In parallel, the group improved source preparation and counting techniques and provided services to third parties (reactor fluence measurements, fission foil mass determination by α particle counting, measurement of particulate matter in air).

The group participated in international comparisons of activity concentration measurements of selected radionuclides. Finally, studies have been performed on subjects like the internal ionization after nuclear decay, nuclear structure effects of l-forbidden transitions, radiative electron capture, orbital electron capture by the nucleus, X-ray fluorescence yields, Auger and Coster-Kronig transition probabilities, satellite peaks in α spectrometry, and peak fitting and non-linear energy response in α spectrometry.

From the more recent projects of the group, three were selected to be presented in more detail.

1. INTERNATIONAL COMPARISON OF ACTIVITY MEASUREMENTS OF A SOLUTION OF ^{125}I

The Bureau International des Poids et Mesures (BIPM) organized first a trial comparison in 1987 and then, in 1988, a full scale international comparison of activity measurements of a solution of ^{125}I. This nuclide has been selected for its importance in nuclear medicine, its difficulty to be measured precisely with the SIR system (Système International de Réference) and the discrepances in the half-life values reported in the literature.

Seven laboratories participated in the trial comparison and nineteen in the full scale comparison of activity measurements of a solution of ^{125}I, which demonstrates the interest in such an exercise. CBNM participated using four methods; (1) the total emission counting [1], (2) the sum peak coincidence counting [2], (3) the photon-photon coincidence and efficiency extrapolation [3] and (4) the photon-photon coincidence counting methods [4].

The weighted mean value based on 38 communicated results [5] is evaluated as (1425.3 ± 1.4) $Bq \cdot mg^{-1}$ on June 15, 1988 00h UT. Accordingly the unweighted mean value is (1429.9 ± 2.6) $Bq \cdot mg^{-1}$. The total range of the results is 7.2%. The results communicated by CBNM are given in Table 1.

Table 1. CBNM results on the activity determination of the ^{125}I solution

METHOD	Activity Concentration ($Bq \cdot mg^{-1}$) on June 15, 1988 00h UT	Combined Uncertainty ($Bq \cdot mg^{-1}$)
Sum peak coincidence counting [2]	1422.3	7.0
Photon-photon coincidence counting [4]	1425.8	7.7
Photon-photon coincidence and efficiency extrapolation [3]	1443.2	6.3
Total emission counting [1]	1427.9	4.9

A literature review reveals that new determinations of the half-life of ^{125}I are needed, as indicated at the 1989 meeting of the IAEA-CRP on "X- and gamma-ray standards for detector efficiency calibration" [6].

Point sources of ^{125}I were prepared and measured using a NaI(Tl) integral-line detector system. The data, corrected for background, dead time and pile-up, were then fitted using a weighted least-squares fitting routine. The adopted value is (59.37 ± 0.06) d .

2. EFFICIENCY CALIBRATION OF Si(Li) DETECTORS WITH X-RAY REFERENCE SOURCES AT ENERGIES BETWEEN 1 AND 5 keV

Modern analytical techniques, like proton-induced X-ray emission (PIXE) and X-ray fluorescence analysis (XRF), make use of high-resolution Si(Li) detectors to measure X rays in the energy region of 1 to 20 keV. Most of the methods for efficiency calibration in the low end of this energy region are based either on relative measurements or on standardized sources.

Indirect, shielded fluorescence sources have been prepared, by electrodepositing or spraying a ^{55}Fe suspension onto an annular titanium backing foil; the Mn-KX rays emitted by this radionuclide produce the majority of the fluorescence KX rays in elements with $Z<25$. The ^{55}Fe deposit is sealed with an 8-μm thick Mylar foil and the whole is placed in a cylindrical nickel container. The fluorescer foil can be exchanged to produce X rays of various energies. The advantages of such a design over the transmission-type fluorescence sources are (1) that the detector is shielded

against the Mn-KX rays and (2) the possibility to vary the energies of the fluorescence KX rays.
The defined low-solid-angle method (0.0395 sr in this case) was used to standardize the sources; apart from the large geometry factor, only minor corrections have to be applied to the measured count rates. The standardization of the sources was performed with an end-window gas-flow proportional counter.
Fluorescence sources of Al, Cl, Ca and Ti have been prepared and their emission rate certified to an accuracy of better than 2%, namely 2% for Al and 1.5% for Ti-KX rays [7].

3. A QUALITY-CONTROL EXERCISE OF RADIONUCLIDE CALIBRATORS AMONG BELGIAN HOSPITALS

In modern medicine, the number of patients exposed to radiation for diagnostic imaging purposes or for treatment has been increasing. In recent years international bodies, such as the World Health Organization (WHO) and the European Council of the CEC have proposed measures for the protection of persons undergoing medical examination or treatment.
By monitoring the technical performance of equipment, quality assurance programs aim at obtaining optimum diagnostic information with a minimum radiation dose to individual patients and staff. Since most of the radionuclide calibrators consist of an ionisation chamber, it is necessary to have calibrated gamma-ray emitting radiation sources of appropriate energy, activity and shape, whose calibration should be traceable to appropriate standards.
On the initiative of the Belgian Association of Hospital Physicists (BAHP) and the collaboration of the CBNM, eleven Belgian Hospitals participated in a quality-control exercise of radionuclide calibrators [8].
For this comparison, the nuclide ^{57}Co has been chosen because the energy of its principal gamma radiation is close to that of ^{99}Tc^{m}, its relatively long half-life allows repeated measurements, and most of the modern radionuclide calibrators are calibrated for ^{57}Co.
The ^{57}Co solution has been distributed to the participants into standard flame-sealed ampoules. Prior to distribution, the solution has been

standardized by the $4\pi\beta$-γ coincidence method [9] and the total emission counting method [1].
The evaluation of the 20 results (some of the participants used more than one apparatus) shows that about 2/3 of the results were within ±5% of the activity concentration determined at CBNM, but 15% of the results were outside the ±20% range. These results compare well with those obtained in the first ^{57}Co comparison performed in 1980 in the U.K. [10] and they confirm the suggestion for frequent periodic checks of the radionuclide calibrators used in hospitals.

ACKNOWLEDGMENTS

The authors would like to thank Dr. C. Ballaux of SCK/CEN, Mol, Belgium, Dr. U. Wätjen of CBNM and Mr. P. Merlo of Hôpital de Jolimont, Haine-St-Paul, Belgium for their valuable collaboration is some of the projects presented here.

REFERENCES

1. Denecke, B., Appl. Radiat. Isot. A38, 823 (1987).
2. Eldridge, J.S. and Crowther, P., Nucleonics, 22, No.6, 56 (1964).
3. Schrader, H. and Walz, K.F., Appl. Radiat. Isot. A38, 763 (1987).
4. Taylor, J.G.V., "Standardization of Radionuclides", (IAEA, Vienna, 1967), p. 341.
5. Ratel, G., "International Comparison of Activity Measurements of a Solution of ^{125}I (May 1988)", BIPM Report, BIPM-90/3 (1990).
6. Christmas, P., Nichols, A.L. and Lemmel, H.D., "X- and Gamma-Ray Standards for Detector Efficiency Calibration", INDC (NDS)-221 (1989).
7. Denecke, B., Grosse, G., Wätjen, U., Bambynek, W. and Ballaux, C., Nucl. Instr. and Meth. A286, 474 (1990).
8. Reher, D.F.G. and Merlo, P., "A quality control exercise of radionuclide calibrators among Belgian hospitals", Eur. J. Nucl. Med., in press.
9. Baerg, A.P., Nucl. Instr. and Meth. 112, 143 (1973).
10. Woods, M.J., "Intercomparison of ^{57}Co and ^{125}I in U.K. Hospitals 1980/81", NPL Report RS(EXT)56 (1981).

PROCESS MONITORING USING GAMMA-RAYS IN THE POWER INDUSTRY

by

Ira Lon Morgan, Ph.D. - IDM Corporation
Tommy Thompson - TU Electric Co.

ABSTRACT

The concepts of digital photon counting, discrete detector arrays, limited angle data collection, and application specific real-time data analysis have been combined in a pioneering effort. The result of this effort has produced the IRIS(1)(2)(3) system (Integrated Real-Time Inspection System). In a joint effort, IDM and TUE Co. are developing methods to apply the IRIS technology to the Power Utility Industry. IRIS provides the unique capability of performing dynamic digital measurement and inspection of tubular products, in situ, such as on-line piping in the power utility industry. The range of possible applications for the technology is broad.

Damage of high temperature components in electric power plants include many mechanisms such as creep, fatigue, plastic deformation, corrosion, erosion, cavitation, corrosion pitting, and stress corrosion fractures(4). Life assessment and damage analysis procedures for these components require a multi-disciplinary approach involving the knowledge of design, material composition, material behavior, and non-destructive component monitoring techniques.

The IRIS system has current capabilities of detecting creep at 0.2% of OD and rate of creep of 0.2% per year. This may include density changes of less than 0.5% as related to creep. Plastic deformation relating to non-linear hysteresis can be measurable at 0.2% OD. Wall thinning due to erosion or corrosion wear may be measured to an accuracy of less than 1%. Corrosion pitting with a 4-1 aspect ratio, with a depth of 1% of wall, are detected. Stress corrosion cracking with a 4-1 aspect ratio are detected with a current crack width in the region of $0.^5$ +/- 0.02 mm under most conditions. Under ideal conditions, crack widths as small as 0.2 mm have been detected. Porosity and inclusions of weld material are detectible at the 2-2T, 2-4T levels.

INTRODUCTION

The electric power industry has used standard nondestructive testing techniques which include x-ray radiography, ultrasonics, mechanical dimensional gaging, and eddy current techniques(5). These techniques, in general, require the shutdown of the power unit, a period of cool down, removal of insulation, testing, reinstallation of insulation, and start-up. The cost to industry of this mode of inspection is hundreds of millions of dollars per year. The need for real time process monitoring during plant operation is required in order to measure real time use degradation, increase power plant efficiency, reduce inspection cost, and prevent catastrophic failures[6].

IDM has developed a high efficiency multi-detector gamma ray system which measures photon transmission to an accuracy of 0.10%. Recent improvements in high speed detector electronics, high speed array processors, and computer systems has permitted the development of completely digital gamma ray transmission systems operating at photon count rates in excess of 10 megahertz. The digital photon counting system provides wide dynamic ranges, excellent linearity, and high signal to noise ratio. Precision dimension and density measurements, in real time during power plant operation are required in order to measure which components may be defective.

Defects are due to faulty processing or degradation during harsh environmental and operating conditions. In Table 1[7] are listed the causes for failure that generate the requirement for real time process monitoring. The initial components placed in a power plant such as high pressure steam pipes, pumps, valves, defuzers, boiler tubing, and other components may be defective due to initial vendor processing. These defects may be detected initially such that improper components are not employed. Use degradation is a longer term process and occurs during the operation of components at high temperature, high pressure, and corrosive environments. Table 1 also provides a listing of those properties which occur through use degradation. The IRIS systems, through precision and dynamic dimensional measurements, are capable of directly measuring many of the processes which cause reduced electric power plant efficiency and determine potential failure sites.

EXPERIMENTAL METHODS AND APPARATUS

The precision gamma ray transmission system as shown in Schematic 1, contains high energy gamma ray sources of Cobalt-60 and Iridium-192 mounted in a shield container which produces a fan-beam of photons. Photons are transmitted through the power plant component into a detector array composed of 128 high speed scintillation counters. The mechanism for translating the gamma ray transmission system, along the power plant component, plus rotation of the system, permits various angular transmission views. Gamma ray transmission measurements are made each 0.5 cm along the longitudinal axis of the component and angular measurements are taken through 360°. The accuracy of the transmission measurements are 0.1% and the stability and repeatability of the system was approximately 0.6%.

The mobile system may be transported from power plant to power plant by means of a trailer, an environmental computer enclosure, and operated under all environmental conditions. The electronics are environmentally housed, completely air conditioned, and operate with normal AC power.

The system operates in four distinct modes as follows:

The rapid scan gaging mode provides translational speeds as fast as 30 cm/minute in which the dimensions of the components are determined in relation to the outside diameter, inside diameter, wall thickness, eccentricity, ovality, and any major area of erosion wear, corrosion wear, cavitation, pitting or welding.

A second mode of operation requires multi-angular detailed gaging in order to determine small defects such as corrosion pitting and stress corrosion cracking with speeds of approximately 5 cm/minute.

The third mode of operation, upon detections of defects exceeding a predetermined standard deviation, as related to ASME codes for various components, requires the detected defect to be scanned in the analytical mode. A 0.5 cm slice planes are observed throughout 360°, at various angular spacings, in order to provide a three-dimensional map of the defect[8].

The fourth mode of operation used was to determine the nature, location, and size of the defect through a very high resolution, multi-angular analysis and triangulation method.

EXPERIMENTAL RESULTS

The real-time dimensional gaging of high pressure steam pipe as shown in Figure 1 presents a graphic display of the OD, ID, and wall of the high pressure steam pipe. The variation in the "Y" axis is of the order of 0.5 mm with the longitudinal measurements performed each 0.5 cm. One observes the location of weld structures and the step variation in the wall thickness is related to corrosion wear. In Figure 2, seven angular mass projection profiles are shown of a pipe with defects. The seven mass projection profiles are taken at seven equally spaced angular regions about the pipe. The data set represents in excess of one thousand digital measurements of transmission through the pipe and its insulation.

Defects are observed in the wall of the pipe and the information is archived for later analysis. Detailed defect analysis of the data as shown in Figure 2, are computed by differential analysis of the difference between defect regions and average structure values. The defects are highlighted in red, which exceed the defect threshold. In Figure 3, the ray paths which intersect a defect in the interior or the external portion of the pipe may be located through a triangulation method. The intersection of the ray paths in the external portion of the high pressure steam pipe are due to a corrosion pit with a 4-1 aspect ratio and a depth of less than 1% of wall thickness.

CONCLUSIONS

The IRIS systems using high speed digital photon counting with an accuracy of better than 0.1%, are capable of measuring in real-time many of the mechanisms which provide damage or catastrophic failure in high temperature components of electric power plants. Table 2 shows a listing of the primary mechanisms and defects and the accuracy of the system for monitoring the size, location, and rate of development of such defects. This enables the electric utility to perform three-dimensional finite element analysis, calculate meantime to failure, provide timely maintenance and repair during planned outages, and produces increased electric power plant efficiency, while preventing catastrophic failure.

References

(1) Morgan et al, US Patent No. 4,284,895, European Patent No. 0,201,849, entitled, "Method and Apparatus for Tomographic Examination of an Object by Penetrating Radiation"; Patent No. 4,437,006, entitled, "Method and Apparatus for Measuring Radiation in Computer-Assisted Tomography and Radiographic Applications"; Patent No. 4,725,963, entitled, "Method and Apparatus for Dimensional Analysis and Flaw Detection of Continuously Produced Tubular Products".

(2) Morgan et al, Patent Pending, entitled, "An Automated System for Controlling the Quality of Geometrically Regular-Shaped Products During Their Manufacture".

(3) Morgan et al, Patent Pending, entitled, "An Apparatus and Method for the Examination of Fluid Transmission Systems While in Operation".

(4) R. Viswanathan, Damage Mechanism and Life Assessment of High Temperature Components, ASM 1989.

(5) ANSI/ASME B31.1 - 1983 Edition.

(6) TU Electric Private Communication.

(7) R. S. Sharp, Research Techniques in Non-Destructive Testing Vol. IV, pg 162, 1980.

(8) I. L. Morgan et al, High Resolution Gamma Ray Tomography, Application of Accelerators in Research and Industry, North Holland Press 1986.

TABLE 1

Failure Causes that Generate the Need for Process Monitoring

Faulty Processing	Use Degradation
Wrong composition	Altered microstructure
Inclusions, debris	Corrosion/chemical attack
Embrittling impurities	Excess deformation
Excessive grain growth	Local overheating effects
Wrong dimension	Fatigue/creep damage
Faulty heat treatment	Internal oxidation
Faulty case hardening	Decarburization
Faulty surface treatment	Stress corrosion
	Impact/shock damage

TABLE 2

Real-Time Measurable Degradation

Mechanism	Critical Effect	IRIS Accuracy
Creep	2.0% ΔX in/10^5hrs	0.2% ΔX
Rate of Creep	0.5% ΔX/Yr	0.2% ΔX
Plastic Deformation	2.0% ΔX	0.2% ΔX
Corrosion Wear	12.5% ΔW	1.0% ΔW
Cavitation Wear	12.5% ΔW	1.0% ΔW
Solid Particle Erosion	10% ΔW	1.0% ΔW
Corrosion Pitting	4/1 Aspect 15% ΔW	4/1 Aspect 2% ΔW
Stress Cracking	4/1 Aspect 30% ΔW	4/1 Aspect 10% ΔW
Inclusion/Porosity	2% ΔP	0.5% ΔP
Weld Structure	1% ΔP	0.5% ΔP

ΔX = Change in OD
ΔW = Change in Wall
ΔP = Change in Density

IRIS Mobile Unit, Major Components

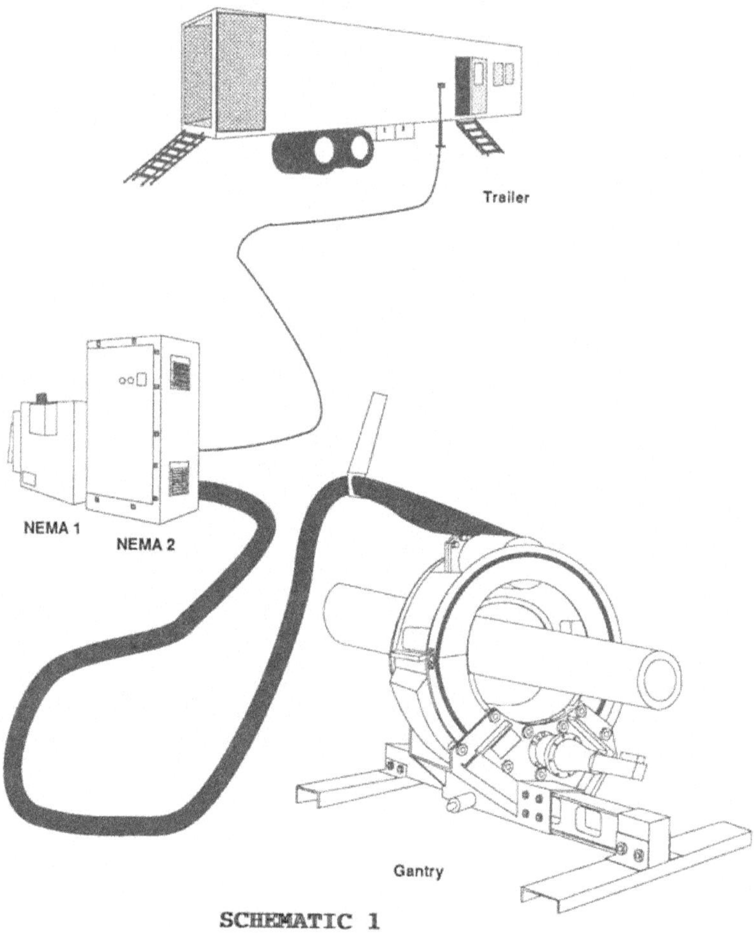

SCHEMATIC 1

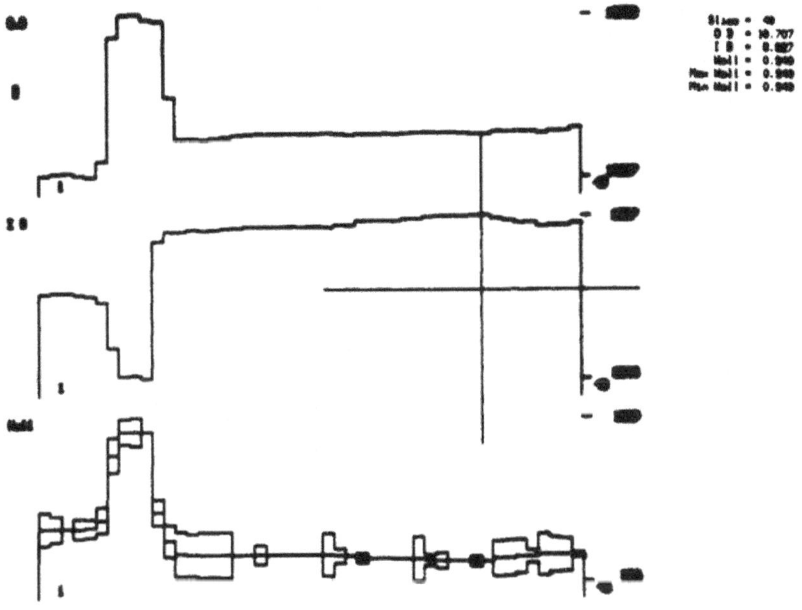

FIGURE 1. Operator dispiay of OD,ID,and Wall over 12" of pipe. The measurement was taken on sample "Stan" and extends for 12" over section C,starting in region B,2" behind the backing ring weld (weld #1).

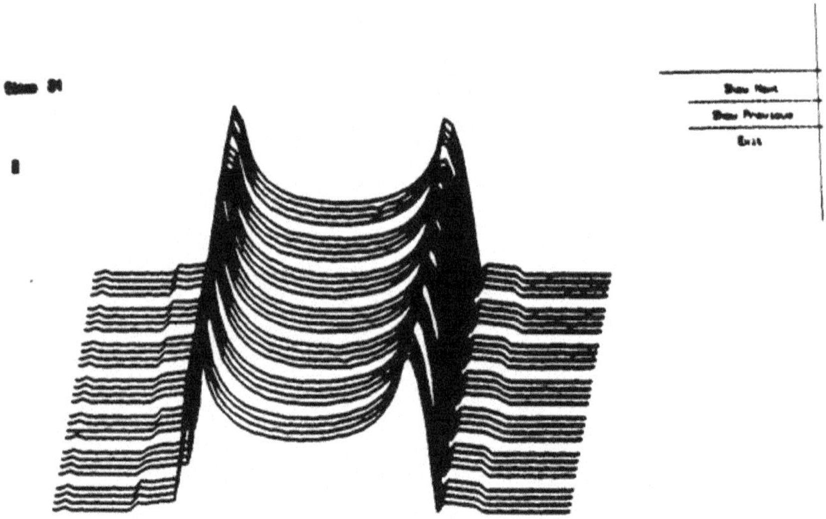

FIGURE 2. Measured pipe path length projections for a section of pipe sample "Stan" in section B at the 3/8" by .080" pit. The pit is not visible without further processing shown in Fig.3. The data was acquired as schematically shown in Figure 1. Note the presence of insulation extending to the left and right of the primary pipe profile.

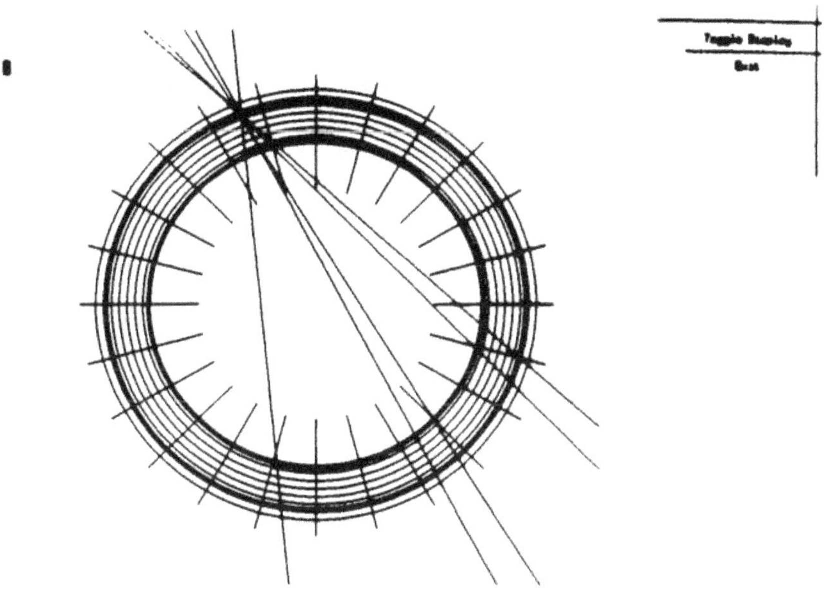

FIGURE 3. Triangulation display showing the location of the pit resulting from the intersection of the marked rays.

POTENTIAL OF NUCLEAR TECHNIQUES FOR ON-LINE BULK ANALYSIS IN THE MINERAL INDUSTRY

J.P. BARON* and L. DEBRAY**

* Laboratoire des Ponts et Chaussées, Centre de Nantes, BP 19 - 44340 BOUGUENAIS - FRANCE

** Ciments Lafarge, 3 et 5 Bd Louis Loucheur, BP 302 - 92219 ST CLOUD CEDEX - FRANCE

ABSTRACT

In the mineral industry, for commercial purposes and quality control operations, it is often necessary to have suitable quantitative data available concerning the chemical concentrations in products or materials. Until the recent years, this information was obtained by means of sampling techniques associated with conventional chemical analysis. These methods are ill-suited to plant automation and fabrication process monitoring systems. The nuclear techniques, owing to the penetration power of the radiations, can provide new control prospects that will eliminate the difficulties linked to the sampling techniques, hereby allowing direct on-line analysis of the material, without any preparation. The nuclear techniques liable to be used in the mineral industry are classified in four categories : natural radioactivity methods, gamma-gamma methods, gamma-neutron methods, pulsed neutron methods. The gamma-gamma method and the pulsed neutron method are based on analysis techniques by activation and neutron capture which have been used for more than 20 years in specialized laboratories to determine generally chemical elements in the form of traces in small samples. Based on these principles, the Laboratoire Central des Ponts et Chaussées (LCPC : Public Works Research Laboratory) has developed an application of these methods for concentration determination on site and in boreholes for the major chemical elements found in civil engineering materials. In collaboration with Ciments Lafarge, the LCPC has developed a methodology using nuclear techniques in order to control on-line the quality of raw materials, by irradiating the elements Silicon, Calcium, Aluminum and Iron. A device has been built and tested in a cement plant under real conditions. It includes a sampler placed on the conveyor belt of material feeding a cement plant at 1,000 tonnes per hour. This sampler is associated with an analysis hopper which includes the neutron irradiation system made up of a neutron generator tube and a NaI scintillator associated with a photomultiplier. The system is fed continuously by a controlled amount of raw material. The neutron generator permits the irradiation of this raw material for very short periods. Between these periods, the gamma detector analyzes the results of the nuclear reactions produced in the material just as it is present at the entry to the plant, that is to say without any preparation. The gamma-ray spectra given by this detector are analyzed with two distinctive types of measurement. One allows a utilization of the nuclear reaction due to the phenomenon of neutron capture. The other allows the utilization of the nuclear reaction due to the phenomenon of neutron activation. This device and this methodology have been tested on a cement plant in real conditions of utilization. Five pre-homogenization piles of 12,000 tonnes each have been analyzed, using this procedure. The information obtained under these conditions, in terms of real time, permits a measurement every two minutes of the four elements Silicon, Calcium, Aluminum and Iron, by a judicious exploitation of capture gamma spectra and activation. A comparison of these measurements was made with analysis of the same piles in parallel using a sampling tower giving every hour a sample analyzed by X-ray fluorescence.The comparison of the results obtained by these two methods of analysis shows the following advantages of neutron irradiation :

- greater frequency of analysis
- information obtained in terms of real time
- a greater quantity analyzed (around 50,000 times larger)
- a richer quality of information
- problems of representativeness of measurements practically eliminated.

The LCPC and Ciments Lafarge have patented the procedure and the installation of analysis using this technique and are planning now the construction of an industrial analyzer based on this innovative technique. In this way, in the future, it will be possible to obtain at the cement plant entry more homogeneous raw material permitting an easier optimization of crushing and burning operations to obtain a clinker having tricalcium silicate composition variations with a standard deviation in the order of 3 %.

1. INTRODUCTION - THE GENERAL REQUIREMENTS IN THE MINERAL INDUSTRY

In the mineral industry, it is often necessary to have suitable quantitative data on hand, concerning the chemical composition of the materials or final products. Such information is useful either during production to control the fabrication process, or at marketing stage to assess the quality of the materials or products. The conventional chemical analysis methods generally utilized in laboratories are suitable to obtain the information data needed. However, such methods can only be implemented on small samples. Now, in the mineral industry, the amounts of materials treated, either during the production or marketing operations, are huge, compared to the amount sampled and processed with the conventional chemical analysis methods. Until the 80's, these conventional chemical analysis methods have therefore been combined with more or less sophisticated material sampling techniques in order to obtain a sample liable to be processed in accordance with the conventional analysis methods. This procedure thus enables the desired information data to be obtained, but also generates two major problems :

* The sampling technique is often an operation which is very difficult to implement, requiring preparation of the material at certain stages of the sampling process (milling, drying..). The times needed for the sampling procedure, added to that for the chemical analysis, extends apreciably the time for obtaining the expected results.
* Despite the more or less high sophistication level of the sampling technique, the problem of sample representativity versus the tremendous amount of material from which it is issued, has still to be overcome.

Both problems make these control and analysis methods not well suited to the automation of the fabrication process currently expected in the mineral industry.

The nuclear techniques, owing to the deep penetration power of radiation, introduce new prospects in this major industrial field. Indeed, with the latter methods, it is possible to envisage new monitoring concepts that will considerably minimize the representativity problems of the sampling techniques by allowing their implementation directly on the total amount of material, without any specific preparation process. However, the mineral industry is an exacting field, and these new control concepts should not become a burden, further aggravating the economical factors affecting the sales of materials or products. Therefore, this aspect of the problem does not facilitate the task of the nuclear physicists who develop such methods. They are often compelled to use not all but only a part of the resources offered by the nuclear methods in order to develop simple tools, liable to be operated by unskilled personnel, in order to satisfy this major economical citerion.

2. THE NUCLEAR TECHNIQUES USABLE IN THE MINERAL INDUSTRY

The nuclear techniques most widely used in the mineral industry can be classed according to four essential categories, namely :

* The natural radioactivity methods : these methods make use of the gamma rays naturally emitted by the materials
* The gamma-gamma methods : they make use of the interactions between the gamma rays emitted by an isotopic source, and the atoms constituting the materials. These interactions are : the photo-electric effect, backscattering by Compton effect, and pair production.
* The gamma-neutron methods : these methods make use of the spectrometry of the gamma rays emitted during inelastic impacts or during the reactions generated by neutron capture by the nuclei of the atoms constituting the materials, and spectrometry of the gamma rays emitted by the isotopes issued from neutron activation reactions.
* The pulsed neutron methods : these are based on the same principle as the gamma-neutron methods, but the neutrons are emitted over only very short periods of time. Detection of the gamma rays resulting from the activation or capture phenomena only takes place between two successive neutron pulses.

This paper first briefly describes the principles of the first three methods, and presents typical application examples of these methods in the mineral industry. As the pulsed neutron methods are more recent, their principle will then be devèloped in further detail, and a typical application will be described for on-line analysis of the raw materials on the input line to a cement plant.

2.1 The natural radioactivity methods

2.1.1 Schematic description

This method is the most simple method currently available. It consists in recording the radioactivity of the rocks, resulting from the presence in infinitesimal quantities of three natural old-aged radio-elements : Potassium, Uranium and Thorium. Potassium emits a single gamma photon featuring an energy of 1.46 MeV, and is therefore easily identifiable. Uranium and Thorium feature complex gamma radiation spectra, but it is possible to detect only some characteristic rays in each of these radio-elements. A number of studies have demonstrated that the natural radioactivity of rocks or soils depends on their mineral composition and geological history. Recording the natural radioactivity of the materials in the mineral industry therefore constitutes a very simple and economical means of assessing certain characteristics.

2.1.2 Application examples

On-line analysis of the ash content of coal

Knowledge of the ash content of coal is an essential parameter during the extraction or utilization process. The chemical composition of ash includes silico-aluminates, present in variable quantities depending on the geology of the ground from which coal has been extracted. These silico-aluminates can be characterized by their natural radioactivity. PJ Mathew [1] has studied a system for continuous analysis of the ash content of coal, involving the measurement of the radioactivity in energy ranges relating to Potassium, Uranium and Thorium. Figure 1 shows the relationships obtained between the natural radioactivity and the ash content. It can be observed that these relationships are dependent upon the field from which the coal has been extracted. Calibration is therefore necessary, as a function of the various coal fields.

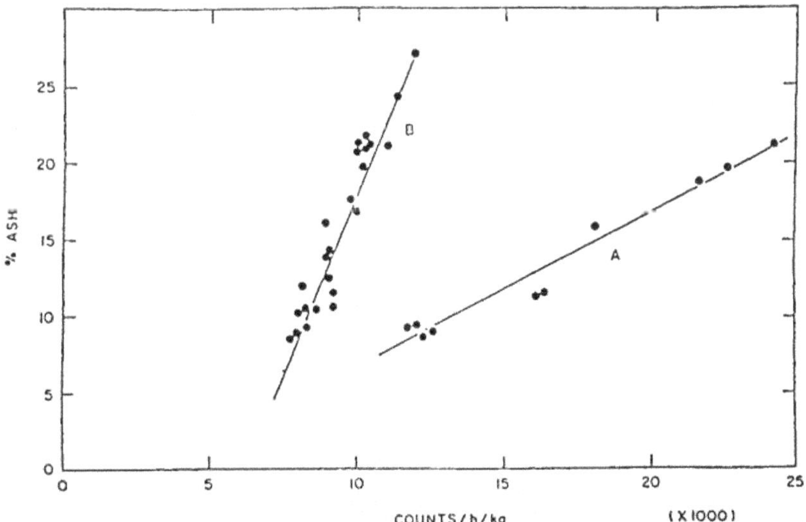

Fig. 1 - Relationship between the ash content and natural radioactivity for two types of coal (A) and (B) (from P.J. Mathew)

On-line analysis of iron ore schist content

Iron ores issued from sedimentary formations may be more or less contaminated by schists during the extraction process. Obviously, the commercial value of such an iron ore is not the same depending on whether it is slightly contaminated (higher grade), moderately contaminated (lower grade) or heavily contaminated (sterile). J.A MAYER and R.J. HOLMES have demonstrated [2] that contamination of iron ores by schists could be easily related to the natural radioactivity of the ore. They have developed a system designed for continuous measurement of the said natural radioactivity on a conveyor belt. This equipment enables automatic sorting out of the ore, as a function of the various grades.

Measurement of fly-ash enrichment during clinker milling operations in a cement plant.

During cement fabrication process, several additives such as gypsum or fly-ash are integrated into the clinker. It is essential to monitor the enrichment of the cement so produced in terms of fly-ash. L. BARANYAI [3] has demonstrated that it was possible to obtain a linear relationship between the natural radioactivity of the cement and the fly-ash content, and thus develop a simple system enabling such enrichment to be assessed either at the end of the production phase, or during clinker milling operations.

2.2 The gamma-gamma methods

2.2.1 Schematic description of the methods

"Gamma-Gamma" is a term generally used in nuclear techniques to indicate that the method is based on the utilization of a gamma ray isotopic source and a gamma ray detector. The characteristics of a medium are thus determined on the basis of the gamma ray scattering or absorption properties. Therefore, the interactions between the gamma ray and material are : pair production, backscattering by Compton effect, and absorption by photoelectric effect. Such interactions depend on the energy of the gamma radiation and chemical composition of the medium in which the said radiation propagates.

Pair production only takes place at gamma radiation energies higher than 1.022 MeV, and this effect varies as a function of the ratio Z^2/A (Z = number of electrons per atom and A = atomic mass), hereby meaning that it will be sensitive to the presence of heavy elements. Backscattering by Compton effect is essentially influenced by the energy of the gamma radiation and by the Z/A ratio of the medium, that is to say in fact the electronic density which is itself linked to the density of the medium. Absorption by photoelectric effect only takes place, practically, at energies less than 300 keV, and widely depends on the chemical composition of the medium (this effect approximately varies in the same manner as Z^5/A).

Depending on the energy of the gamma photons which will be used in the gamma methods, it can be seen that information either related to the density of the medium, or to the chemical composition of the medium, can be obtained.

2.2.2 Application examples

On-line analysis of iron ore

The method involving pair production is very easy to operate on materials conveyed through a conveyor belt. The technology is simple and a cost effective control method, now widely used in the mineral industry, can thus be obtained. Let us mention, as an example, the specific on-line iron ore analyzing system (figure 2) developed by R.J. HOLMES [4] of the CSIRO at Port Melbourne (Australia) for this application. The density variation effects of the medium have been taken into account by a judicious utilization of the energies of the spectrum of the detected gamma photons. Measurements are performed in two different energy windows : one corresponding to the Compton backscattering phenomena, the other to the pair production phenomena. However, care must be exercised to consistently maintain a minimum thickness of 15 cm iron ore on the conveyor belt.

The equipment is currently operational and performs control operations with an accuracy of ± 6% in terms of the iron content in the iron ore, when the measurements are integrated over time periods in the range of 10 minutes. Controlled material outputs can be as high as 6000 tonnes per hour.

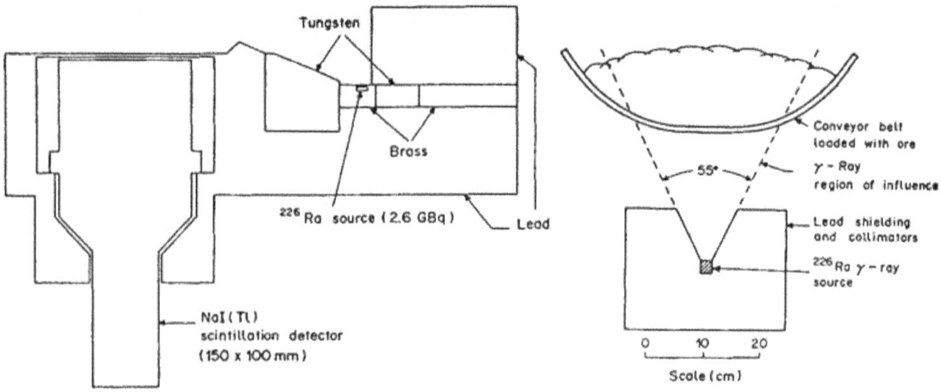

Figure 2 - Schematic sectional view of the equipment developed by CSIRO [4] for iron content assessment in iron ores, on a conveyor belt.

Density measurement in coal field mining applications

In the development of certain coal fields, it is useful to get prior knowledge of the position of certain layers consisting of clay schist, for example. Coal density varies between 1.2 and 1.5 ; that of schist may vary between 2.2 and 2.6 approximately. The difference in density is therefore high, and density measurement can allow both types of formations to be easily differentiated. C.G. CLAYTON and M.R. WORMALD [5] have demonstrated that it was possible to use a probe utilizing the gamma backscattering principle to determine the position of the schist layers in a coal field, by drilling process. The source used for gamma ray emission is Cesium 137. Two detectors are used to count the gamma photons backscattered by Compton effect. A detector, placed at long distance from the source (50 cm) enables information to be obtained on medium density. This information is but slightly perturbed by the drilling parameters (diameter variations, presence of water...). The second detector, arranged at short distance from the source (15 cm) gives, on the contrary, information which is highly influenced by the drilling parameters, but very little sensitive to medium density. A combination of the information from both detectors enables the density variations of the medium situated around the borehole to be measured with great reliability.

Technique of gamma ray transmission to two energies, applied to on-line analysis in the mineral industry

We have seen, in the schematic description of the gamma-gamma methods, that the interaction of gamma radiation is sensitive both to the energy of the said radiation and to the chemical composition of the medium in which interaction takes place. In complex media, it is sometimes difficult to correctly construe all interaction phenomena of the gamma radiation liable to exist. This is often the case in the mineral industry, where the processed materials are never simple. In such case, advantage can be taken of a technique based on the study of gamma ray transmission in such media, involving two different energies for the emitted gamma radiations. J.S. WATT and E.J. STEFFNER [6] have demonstrated the possibilities of such a technique for determining the concentration of elements featuring high atomic numbers present in a material the major elements of which feature low atomic numbers (this is often the case in the mineral industry). One selects a gamma radiation featuring an energy producing an absorption coefficient much higher for the element, the content of which is to be determined (high atomic number), than for the major elements of the tested material (low atomic number). The second energy of the gamma radiation is selected to obtain an absorption coefficient sensibly similar for all elements present in the material to be tested. The concentration of the investigated elements is then calculated by combination of the intensities of the gamma rays transmitted to both energies.

Thus, the application of this technique can be envisaged on conveyor belts for determining :

- the lead content in lead-zinc ore
- the ash content of coal
- the iron content and sulphur content of coal
- the iron content in high grade iron ore
- the tin and tungsten contents in ore, provided that the iron content is not too high.

2.3 Gamma-neutron methods

2.3.1 Schematic description of the methods

In the gamma-neutron methods, utilization is made of the phenomena linked to the nuclear reactions occurring between the neutrons and nuclei of the atoms constituting the medium. The analysis of these nuclear reactions and the results obtained have now been used for twenty years or so in specialized laboratories for the purpose of irradiating chemical elements to determine their concentration, generally under the form of traces in small samples. However, the specific character of the neutrons has rapidly caused the situation to turn toward other types of irradiation processes featuring extended variation.

The different types of nuclear reactions commonly used in the nuclear techniques applied to the mineral industry are described in the following.

Neutronic inelastic scattering : inelastic gamma method

When a neutron is highly energetical, it may, upon collision with the nucleus of an atom, rebound with a loss of energy. This is called inelastic scattering. The nucleus is then in an excited state and, its subsequent passage to the ground state takes place with the emission of a gamma photon. The energy of this gamma photon is characteristic of the nucleus excited in the alone reaction, and the reaction takes place over a very short time, immediately after the inelastic scattering.

Neutron capture reaction : capture gamma ray method, or prompt gamma ray method

When a neutron has a low energy (energy close to the thermal level), it may interact with the nuclei of the elements present and get captured by them. This phenomenon is referred to as neutron capture by the nucleus. This capture is spontaneously accompanied by the emission of a gamma radiation the energy of which is characteristic of the nucleus having initiated the reaction. This phenomenon is called the capture gamma radiation, or the prompt gamma radiation.

The activation reaction : analyzing method by neutron activation

In certain cases, the neutron may interact with the nucleus of an atom to produce a radioactive artificial isotope. During this reaction, a particle is emitted. Depending on the initial energy of the neutron, this particle could be another neutron, a proton, an alpha or a gamma radiation. Subsequently, the radioisotope so produced reverts to a steady state by emitting, in certain instances, a gamma radiation the intensity of which decreases according to the half-life of the radioisotope. The radioisotope produced and, possibly, its decreasing gamma radiation, are representative of the nature of the atom having generated this reaction.

Thus, it can be seen that the gamma spectrometry of a material subjected to neutron bombardment will yield highly valuable information concerning the chemical composition of the material. The essential problem is to extract this information from the material irradiated in bulk. This is a difficult operation, as this amount of highly valuable information is available without any differentiation. Optimized data can be obtained from this information, provided that differentiation could be made, that is to say if the separation of events of a same type could be kept under control. But this is not the only problem to be overcome, as the mineral material handled during the production, development or commercialisation operations always stands in highly diversified conditions in terms of density and humidity. Both parameters, independent of the chemical composition of the medium, do influence the characteristics of the neutron flux being used. Only full control of these parameters, that is to say density and humidity, can render efficient the analytical methods involving such principles of nuclear physics.

2.3.2 Overview of possible applications

For ten years or so, the mineral industry has followed the same approach as that undertaken by the oil industry concerning the application of the gamma-neutron methods. It has been rapidly understood that such methods could bring prospective solutions to problems hitherto ill-solved. As many solutions have been developed, it is rather difficult to give an exhaustive overview of all possible applications. Therefore, only a few typical applications, representative of the problems liable to be solved with these methods, are developed in the following .

Bauxite analysis

The prototype of a bauxite analyzer has been developed by CSIRO at Port Melbourne (Australia) (7). This analyzer enables the aluminum and silicon concentrations to be simultaneously determined in a bauxite sample of mass 3 kg. The measurement principle used takes into account the activation reactions with fast neutrons ^{27}Al (n,p), ^{27}Mg for aluminum and ^{28}Si (n,p) ^{28}Al for silicon. The sample is irradiated by the neutrons emitted by an Americium-Berylium source. The duration of a measurement is in the range of 15 minutes. The calibration of the method takes into account the influence of sample density and possible variation of the flux of fast neutrons in the sample. Thus, the aluminum content can be assessed with an accuracy of $\pm$ 0.7 % in the range between 15 and 45 %, whereas the silicon content can be assessed with an accuracy of $\pm$ 0.3 % in the range between 0 and 15 %.

Bulk analysis of sulphur, lead, zinc and iron in the development process of lead ores

J.B. CUNNINGHAM et al [8] have developed an on-line analyzer for continuous measurement of sulphur, lead, zinc and iron concentrations during the utilization process of lead ores. The technique uses the gamma radiation emitted during the inelastic scattering of the neutrons emitted by a source of Americium-Berylium on the said elements. Figure 3 shows a schematic section of the tested system. The study has demonstrated that measurements could be made with the following accuracy levels :

- ± 0.3 % on sulphur for an average content of 6 %
- ± 0.85 % on lead for an average content of 40 %
- ± 0.15 % on zinc for an average content of 6 %
- ± 0.3 % on iron for an average content of 12 %

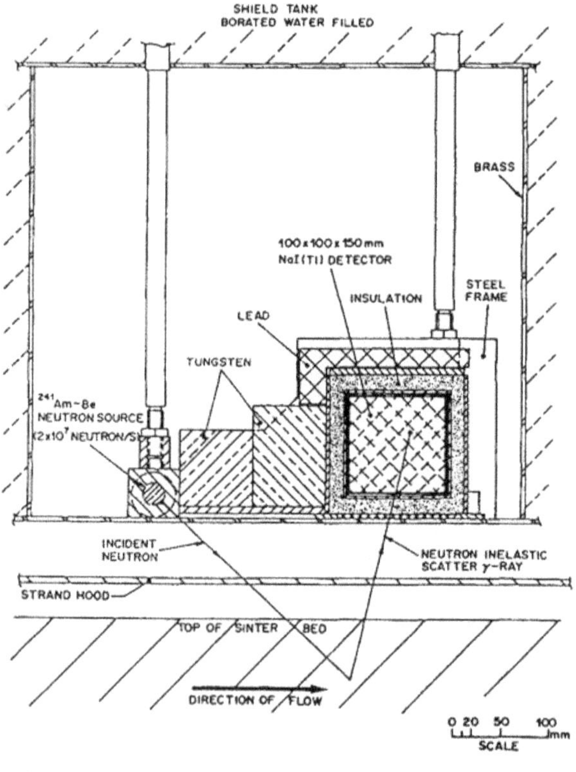

Figure 3 - Schematic section of the analyzer developed for determining the sulphur, lead, zinc and iron contents by neutron inelastic scattering (from J.B. CUNNINGHAM et al [8]).

However, the calibration of the method must take into account the influence of the water content to obtain the said accuracy levels. The information concerning the water content is obtained by measurement of the flux of thermal neutrons.

Geochemical applications

M.D. GLASCOCK, R.M. COVENEY Jr, C.W. TITLE, M.L. GARTNER and R.D. MURPHY [9] have studied the possibility of applying the captured gamma radiation analyzing process to the geochemical field. They consider that this technique allows 15 to 20 elements related to the geological materials to be determined.

In geochemical applications, this irradiation technique should render the work easier in the following fields :

- Study of matrix effects
- Bore and rare earth elements concentration determination in geochemical studies
- Exploration and development of mineral deposits.

On-line assessment of sulphur content in coal

G. VOURVOPOULOS [10] has demonstrated that on-line assessment of sulphur content in coal was possible, using the method of prompt gamma rays. Work has been conducted on medium size samples (13 kg) irradiated by neutrons produced by a source of Californium 252 and thermalized in paraffin. Detection of the prompt gamma rays is performed using a high purity germanium detector. The gamma spectra utilization technique requires an anti-Compton system and a calculation algorithm for the Gaussian curves of the photoelectric peaks. It has been demonstrated that the presence of trace elements absorbing the neutrons, such as chlorine, had no influence on the assessment of the sulphur content. Figure 4 shows that the assessment method is linear, and enables simultaneous measurement of sulphur and chlorine.

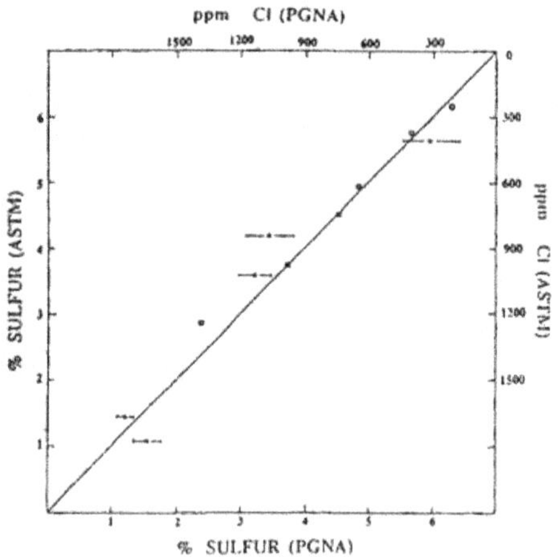

Figure 4 - Comparison of chlorine and sulphur contents determined by the ASTM method and prompt gamma method, for five different coals (from G. VOURVOPOULOS et al [10]).

2.4 *Pulsed neutron method*

2.4.1 Schematic description

The gamma neutron methods have high-lighted the great interest of their implementation in the mineral industry for the development of on-line analytical applications. However, their analytical performances are often impaired, as interpretation of the phenomena caused by the nuclear reactions is extremely complicated. The pulsed neutron method brings valuable help in the interpretation of such phenomena. Using a neutron generator (small miniature particle accelerator utilizing the deuterium-tritium reaction), it is possible to irradiate the material during very short time intervals. Advantage is then taken of the time interval between two successive neutronic pulses to operate the gamma ray detector. During this period, the detector is no longer perturbed by the interactions it may itself have with the neutrons. The signal-to-noise ratio is thus considerably improved. Besides, it is possible to select the moment at which the detector is operated after termination of the neutronic pulse. As further developed in the following, this feature allows utilization of the gamma ray spectra, leading to the separation of the phenomena caused by the neutron capture reactions and by neutron activation. The interpretation of these phenomena in terms of the chemical elements concentrations is then much more reliable, hereby enhancing the analytical performances of the gamma neutron methods.

2.4.2 Application to raw material on-line analysis in the cement industry

In a cement plant, keeping the quality of the cement under control is closely linked to the chemical composition of the raw material used for cement fabrication. Accurate determination of lime (CaO), Silica (SiO_2), alumina (Al_2O_3), and iron oxyde (Fe_2O_3) contained in the raw material, enables cement klinker to be produced.

Extracted from the quarries, the raw material has the aspect of stones with dimensions ranging between 0 and 40 mm. This raw material is conveyed from the quarry to the cement plant input line through a conveyor belt, at an average rate of up to 1,000 tonnes/hour. It is subsequently prepared in order to have the amounts of silicates and aluminates required in the final product. However, the disparity between the geological layers of the quarry causes fairly high variations in the composition of the four basic elements : Ca, Si, Al and Fe.

Currently, repeatability of component determination is obtained through pre-homogenization plants, where the material is arranged in multiple successive horizontal layers on a pile, and reclaimed in a plane normal to the pile in order to intersect all layers. Monitoring of pile average composition is performed through a sampling tower installed upstream of the pre-homogenization plant. The tower supplies a sample, theoretically representative of the raw material before its introduction into the milling plant, and subsequently into the kiln. After preparation in the laboratory, this sample is analyzed with X-ray fluorescence. Chemical corrections are then introduced at the milling stage, hereby ensuring correct irradiation of the raw material. Currently, information is available only at an hourly rate, in the best conditions. The processed sample is tiny, weighing a few grams and representativity problems may occur.

To gain full control over the raw material without any specific preparation process, the LCPC and Ciment Lafarge have conducted joint research aimed at studying the possibility of using neutrons as projectiles for direct on-line analysis of the chemical composition of raw materials. Indeed, owing to two essential characteristics associated with the neutron irradiation analyzing techniques :

- Utilization of high energy neutrons capable of deeply penetrating into the material (a neutron flux of 14 MeV is attenuated by 88% or so through 30 cm of raw material)

• The gamma rays emitted during neutron capture reactions feature a high energy : between 1 and 10 MeV (a gamma ray flux of 5 MeV is attenuated by 75% or so through 30 cm of raw material)

It is possible to design a neutron irradiation analyzing technique giving results on a quantity of materials 50,000 times as big as with the X-ray fluorescence technique associated with a sampling tower. In such conditions, the representativity problems are practically eliminated. Besides, this new technique can be implemented in order to give results on a quasi real time basis, thus enabling its direct insertion to be envisioned into the cement plant process, in the future.

2.4.3 Principles developed to implement the system

Preliminary laboratory tests conducted on raw material specimens and samples have enabled the definition and development of an instrumental neutron irradiation technique on the one hand, and a method on the other hand, in order to obtain signals capable of being used for the purpose of analysis and determination of the chemical elements involved in cement fabrication. On the above basis, an experimental system has been designed to demonstrate, in a cement plant, the feasibility of the continuous control process applied to raw materials in accordance with the neutron irradiation technique, during the constitution of the pre-homogenization piles. This system has been designed and installed in the Ciments Lafarge plant situated at Saint Pierre La Cour (France), where the experiments have been conducted.

Mechanical system

Having taken into account the structure of the sampling tower in the Saint Pierre La Cour cement plant, it was decided to install an analyzing hopper in order to monitor the raw material taken by a disk type sampler. The disk sampler conveys into the analyzing hopper 1/60 (16 tonnes/hour) of the raw material brought into the plant. Figure 5 shows a simplified schematic diagram of the system. The analyzing hopper consists of a small inverted steel cone, approximately 2 m high, with an average diameter of 0.60 m. The selection of such a diameter is based on the necessity of irradiating the complete flux of material within the range of the neutrons, having a radius of 0.3 m or so. The selected height eliminates the so-called "border" phenomena generated by the predictable variations of the level of material in the hopper. An 8 mm thick polyethylene covering applied on the internal wall of the hopper plays the role of "neutron reflector", that reduces the velocity of the neutrons, and reflects them toward the flux of material.

A 30 mm thick polyethylene tube is placed in the center of the hopper. This tubing is designed to introduce the analyzing sensor into the material core. This tubing acts as a neutron moderator, increasing the flux of thermal neutrons through the material and, consequently, enhancing the amount of capture nuclear reactions.

The analyzing hopper is fed continuously. A vibration discharge system ensures continuous, fairly homogeneous flow of material within the volume of investigation.

Analyzing sensor and associated instruments

The analyzing sensor is comprised of two essential elements :

- A 14 MeV neutron generator tube enabling irradiation of the material to be analyzed. This tube is now manufactured in France by SODERN.
- A gamma ray detector consisting of a sodium iodide crystal, 2 inches in diameter, 4 inches long. This crystal is associated with a photomultiplier. The crystal is placed at a distance of 30 cm or so from the neutron emission point. A shielding consisting of 30 % boron laden polyethylene is placed between the crystal and the neutron generator tube in order to protect the crystal from the neutron flux. So designed, this gamma ray detection system operates with acceptable performance.

To obtain the gamma ray spectra, a high performance multi-channel analyzer is associated with the analyzing sensor. A micro-computer of the PC-AT type ensures the piloting functions of the multi-channel analyzer, processes the obtained spectra, stores the collected information and displays the results on an almost real time basis.

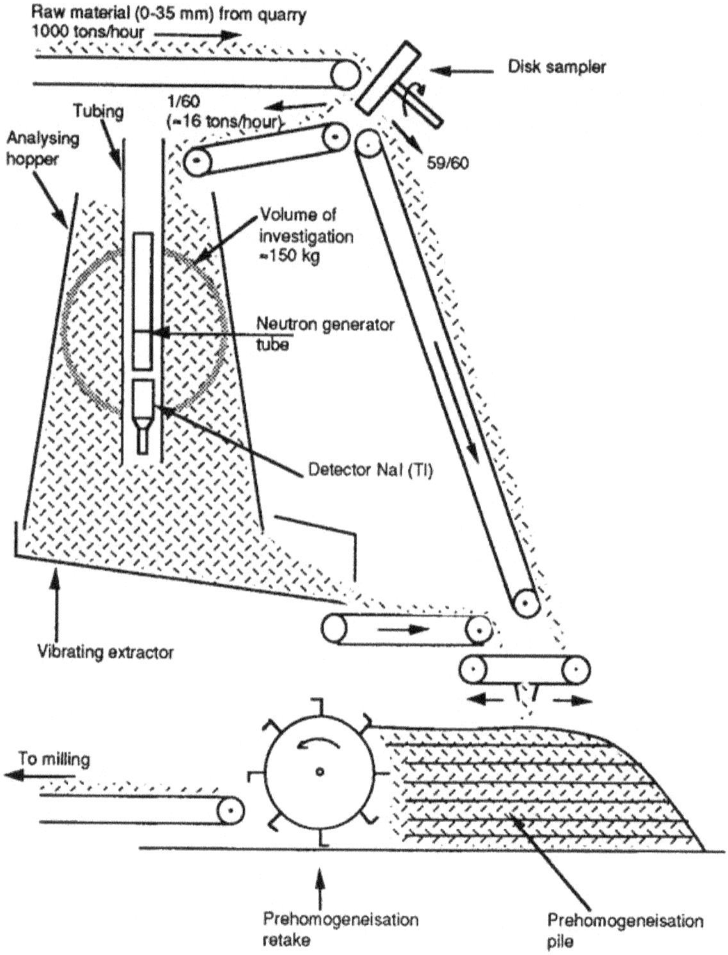

Fig. 5. Synoptic scheme of raw material analysing process by neutron irradiation

Operating principle of the system in the pulsed mode

Two possibilities are available to irradiate bulk material for analysis purposes : using an isotopic source, or a neutron generator tube. In the method jointly developed by the LCPC and Ciments Lafarge, the solution involving the neutron generator tube has been retained. This choice has been made on the basis of the criteria enabling the development of an analyzing method well suited to industrial applications, while featuring high analytical performance. Indeed, the utilization of an isotopic source has the drawback of requiring the enforcement of stringent rules concerning protection and safety, especially in industrial environments. In addition, the flux of neutrons supplied by an isotopic source decreases in time, and emission is continuous. A neutron generator tube involves less stringent safety rules when operated in an industrial environment, as neutron emission is controlled by electrical and electronic systems. The greatest advantage however of the neutron generator tube is the ability to operate in a pulse mode. In such a mode, the analyzing capabilities are more widely diversified than with an isotopic source, as differentiation is possible between the various types of nuclear reactions. Figure 6 shows how the irradiation and gamma ray detection phases are organized as a function of time.

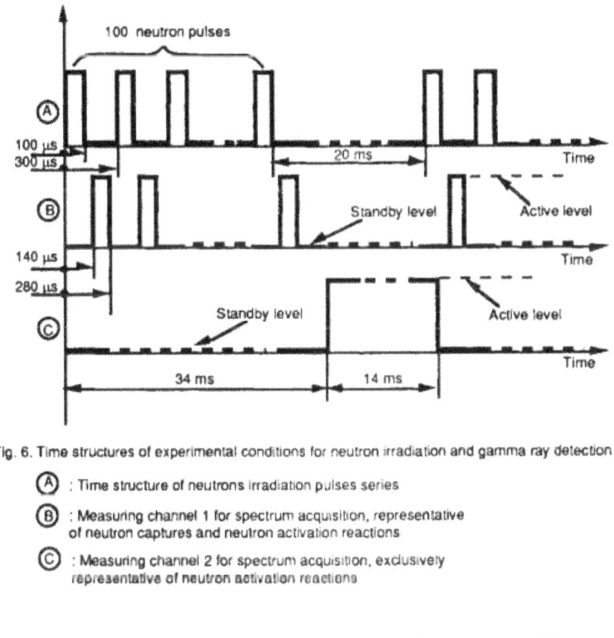

Fig. 6. Time structures of experimental conditions for neutron irradiation and gamma ray detection

(A) : Time structure of neutrons irradiation pulses series

(B) : Measuring channel 1 for spectrum acquisition, representative of neutron captures and neutron activation reactions

(C) : Measuring channel 2 for spectrum acquisition, exclusively representative of neutron activation reactions

The neutron generator tube is controlled by an electronic system enabling irradiation of the material to be obtained for periods of time of 100 microseconds at 300 microseconds intervals. Two analog to digital converters are connected to the gamma ray detector and are controlled in the anti-coincidence mode relative to the pulsed neutron emission. Then, the associated multichannel analyzer can be operated on two analyzing channels as shown on figure 6. Channel 1 is operative during each neutron pulse phase over a time period of 140 microseconds. A pulse counter enables the neutron emission to be stopped after 100 neutron pulses. During the interruption period, the duration of which has been fixed at 20 milliseconds, channel 2 becomes operative for 14 milliseconds.

This pulse irradiation method and the gamma detection technique between each such pulse enables two gamma ray spectra to be obtained simultaneously. The spectrum obtained on channel 1 is representative of the activation and capture phenomena which occured in the material so irradiated. The spectrum obtained on channel 2 is representative of the activation phenomena exclusively, as sufficient time (4 milliseconds) has been allowed to elapse after the trailing edge of the last pulse out of the 100 neutron pulses, to render this channel operative. The difference between the spectra obtained on channels 1 and 2 thus enables obtaining the spectrum representative of the capture phenomena, exclusively.

Thus, with the use of a neutron generator tube, it can be seen that a method, both original and innovative, can be developed, which does feature enhanced characteristics as compared to an isotopic source. Thus, access to the differentiation of the capture and activation phenomena is possible for interpreting the spectra.

Spectrum utilization method

Figure 7 shows the spectra obtained after differentiation of the capture and activation phenomena. Owing to the fact that a sodium iodide detector has been used, it appeared difficult to extract the photo-electric peaks from the Compton background in order to set-up a process enabling concentration determination of the chemical elements. Our interest has been preferentially aimed at extended spectral regions enabling :

- Characterization of the concentration of an element to be dosed (Ca, Si, Al, Fe)
- Taking into account the parameters having effects on the measurement

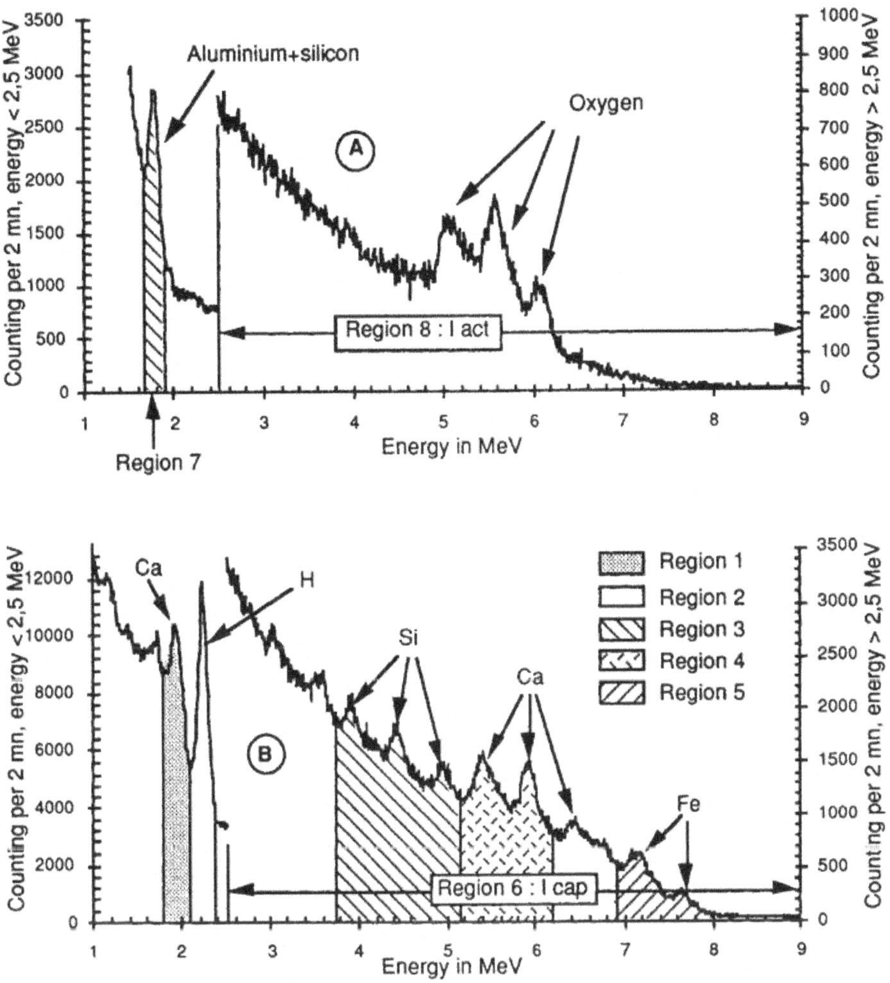

Fig. 7. Typical spectrums obtained in the analysing hopper with raw material

(A) : Spectrum exclusively representative of neutron activation reactions

(B) : Spectrum exclusively representative of neutron capture reactions

Studies have enabled regions of specific interest to be highlighted in the gamma ray radiation spectra, resulting from the capture and activation phenomena respectively. These regions of interest each encompass at least one characteristic photo-electric peak of one of the investigated elements, and are given in Tables I and II.

Table I - Regions of interest to utilize neutron capture phenomena.

Region	Element	Associated gamma-ray	Identification	Energy band
1	Ca	1,94 MeV	N_{Ca1}	1,81 - 2,09 MeV
2	H	2,22 Mev	-	-
3	Si	4,93 MeV	N_{Si}	3,75 - 5,13 MeV
4	Ca	6,42 MeV	N_{Ca2}	5,14 - 6,20 MeV
5	Fe	7,64 MeV	N_{Fe}	6,88 - 10,0 MeV
6	Region of interest associated to capture phenomena		I_{capt}	2,50 - 10,0 MeV

Table II - Regions of interest to utilize neutron activation phenomena.

Region	Element	Associated gamma-ray	Identification	Energy band
7	Al + Si	1,78 MeV	N_{Al}	1,65 - 1,90 MeV
8	Region of interest associated to activation phenomena		I_{act}	2,50 - 10,0 MeV

The studies of the said regions of the spectrum resulting from the capture phenomenon have demonstrated that the following system of equations could be established :

$$\begin{aligned} N_{Ca} &= f_1(C_{Ca}, C_{Si}, C_{Fe}, \phi_{th}, w, \rho) \\ N_{Si} &= f_2\,(C_{Si}, C_{Ca}, C_{Fe}, \phi_{th}, w, \rho) \\ N_{Fe} &= f_3\,(C_{Fe}, \phi_{th}, w, \rho) \end{aligned} \qquad (1)$$

and, for the spectrum resulting from the activation phenomena :

$$N_{Al} = f_4\,(C_{Al}, C_{Si}, \phi_{th}, \phi_{rap}, w, \rho) \qquad (2)$$

- N_{Ca}, N_{Si}, N_{Fe}, N_{Al} are the intregrals of the regions of interest retained in the spectra in association with the elements Ca, Si, Fe and Al
- C_{Ca}, C_{Si}, C_{Fe}, C_{Al} are the concentrations of the elements Ca, Si, Fe, Al present in the material subjected to the neutron irradiation process in the pulsed mode
- ϕ_{th} is the flux of thermal neutrons in the irradiated material
- ϕ_{rap} is the flux of fast neutrons in the irradiated material
- W is the water content of the irradiated material
- ρ is the mass per volume unit of the irradiated material

This system of equations shows that a region of interest of the spectrum resulting from the activation phenomena or capture phenomena is influenced by the contents of the various elements contained in the irradiated material and by the following parameters that affect the collected information data depending on the status of the irradiated material :

- Its water content W
- Its mass per volume unit ρ
- The flux of thermal neutrons ϕ_{th}, or fast neutrons ϕ_{rap}

For Calcium, we use two regions of interest : N_{Ca1} and N_{Ca2} (refer to Table I an Table II). At this stage, our research has demonstrated that the integrals of the regions of interest in the spectra (activation spectrum and capture spectrum), between 2.5 MeV and 10 MeV, did allow the effects of w, ρ, ϕ_{th} and ϕ_{rap} to be taken into account. Indeed, this integral is but little affected by the variations of the raw material chemical composition. On the contrary, sensitivity increases as soon as any of the above parameters increases, either independently from each other, or simultaneously.

The regions of interest being associated with the various variable factors in accordance with the indications given in Tables I and II, the contents of the investigated elements can be deducted by inversing the relations of this system of equations :

$$
\begin{aligned}
C_{Ca} &= g_1\,(N_{Ca1}, N_{Ca2}, N_{Si}, N_{Fe}, I_{cap}) \\
C_{Si} &= g_2\,(N_{Si}, N_{Ca1}, N_{Ca2}, N_{Fe}, I_{cap}) \\
C_{Fe} &= g_3\,(N_{Fe}, I_{cap}) \\
C_{Al} &= g_4\,(N_{Al}, C_{Si}, I_{cap}, I_{act})
\end{aligned}
\qquad (3)
$$

Where :

I_{cap} : integral ranging between 2.5 and 10 MeV, in the spectrum resulting from capture phenomena

I_{act} : integral ranging between 2.5 and 10 MeV, in the spectrum resulting from activation phenomena.

Our experiments have demonstrated that operation of this system was finally possible, with a linear relation per each chemical element in which the constant term represents the contribution of the other chemical elements contained in the material subjected to the analysis, and that of the related environment :

$$\begin{aligned} C_{Ca} &= (P_1 N_{Ca1} + P_2 N_{Ca2} + P_3 N_{Si} + P_4 N_{Fe})/I_{cap} + P_5 \\ C_{Si} &= (P'_1 N_{Si} + P'_2 N_{Ca1} + P'_3 N_{Ca2} + P'_4 N_{Fe})/I_{cap} + P'_5 \\ C_{Fe} &= (P''_1 N_{Fe})/I_{cap} + P''_2 \\ C_{Al} &= (P'''_1 N_{Al} + P'''_2 C_{Si}) \times I_{act}/I_{cap} + P'''_3 \end{aligned} \tag{4}$$

It should be noted that the determination of A1 by means of the spectrum resulting from the activation phenomena depends on the initial determination of Si, using the spectrum caused by the capture phenomena.

2.4.4 Method calibration - measurement accuracy

The calibration of the suggested method consists in determining the P_i coefficients of the above system of equations (4). This determination can be made by multiple linear regression on spectra obtained with various materials the chemical composition of which is known. Then, the proportional contents of any other unknown material can be assessed under the same conditions. To have such a method giving satisfactory results, it is therefore necessary, at the initial stage, to have materials available, whose composition is known exactly. This composition must be the same throughout the whole volume. On the basis of our experiments, it has been found that the latter condition is very difficult to obtain, taking into account the volume of investigations involved in the measurement method : a sphere of approximately 30 cm radius, that is a hundred liters or so, representing between 100 and 150 kg of material. We have attempted using a method consisting in taking three samples of correct size (60 kg) in such a measurement volume. The calibration parameters could be assessed by linear regression on 6 different materials featuring highly different chemical compositions. However, on the basis of the variations of the three samples taken for each of the 6 materials involved, we do believe that the calibration parameters so determined are impaired by certain errors. Then, inaccuracies will result on the determination of the elements made with such a calibration. Such inaccuracies have been studied by comparisons made between the reference analysis on samples, and those determined by the neutron irradiation technique. The typical standard deviations between the differences so observed between both methods can give an idea of the inaccuracies in question. For analyzing times in the range of two minutes, we have obtained for the standard deviation σ in % of weight absolute concentration :

- For CaO $\sigma = 0.96$ %
- For SiO_2 $\sigma = 1.75$ %
- For Fe_2O_3 $\sigma = 0.19$ %
- For Al_2O_3 $\sigma = 0.43$ %

These inaccuracies are such that it would appear difficult to rely on this new method to perform process control in a cement plant.

In order to actually test the qualities of the neutron irradiation technique applied to raw material control, we have taken advantage of the fact that, at the Saint Pierre La Cour cement plant of Ciments Lafarge, the sampling tower (in which we had installed our measurement system) was supplying a sample at hourly intervals, which was subsequently analyzed using the X-ray fluorescence method. We have therefore assumed that these hourly analysis using XRF could serve as reference in order to characterize the average variations that our study could record during the continuous measurements. On the ground of this new assumption, we have performed a new calibration based on the following :

- Each measurement performed according to the neutron irradiation process consists of the moving average of five successive measurements lasting 2 minutes each. As this operation is performed ininterruptedly, a fairly good image of the variations in the chemical composition of the raw materials is obtained while the pile is being built-up.
- The values obtained from XRF analysis during the corresponding period of time are associated with each measurement performed according to the neutron irradiation method.

Then, the calibration parameters are calculated by multiple linear regression from these pairs of values. Two hundred fifty six neutron irradiation measurements have been processed in this manner, which have enabled a much better calibration of our method to be obtained.

On the basis of the above calibration, we have conducted an experiment bearing on repeated measurements of the same raw material (material composition not varying in time). The purpose of these tests was to assess the precision of the measurements. Knowledge of this characteristic is of high interest, as it is an expression of the intrinsic stability of the analyzing method linked to the instrumentation and counting statistics.

Over an extended period of time (9 hours approximately) we have performed successive measurements of 2 min each, on a moving average value of 5 measurements at a time. These measurements have been used to determine the four elements Ca, Si, Fe, Al. The standard deviations σ obtained on these determinations are as follows in % of weight absolute concentration :

- For Cao $\sigma = 0.34\ \%$
- For SiO_2 $\sigma = 0.37\ \%$
- For Fe2O3 $\sigma = 0.06\ \%$
- For Al2O3 $\sigma = 0.11\ \%$

The same experiment was resumed two months later, and gave the same results. Under such conditions, it can be concluded that the performance of the raw material analyzing method by the neutron irradiation process is acceptable for process control purposes in a cement plant.

2.4.5 Control of pre-homogenization piles - results

Using the system and method described in the foregoing, we have examined the composition of three pre-homogenization piles of 12,000 tonnes each. The data collected on a quasi real time basis have enabled a determination of the four elements Silicon, Calcium, Iron and Aluminum to be obtained every two minutes. In parallel with the said tests, the sampling tower associated with the XRF system did supply a determination of the same elements at an hourly rate. A comparative representation of both types of measurements is given in Figure 8 for pile N° 2 (determination of Fe_2O_3 and Al_2O_3) and pile N° 3 (determination of CaO and SiO_2). These results have been selected as they are fairly representative of the data obtained on all piles controlled within the scope of this survey. It is noted that the results as a whole are in fairly good agreement with those obtained from the sampling tower and XRF system. Over all, they appear to be satisfactory for contemplating the development of an industrial analyzer based on the stated principles.

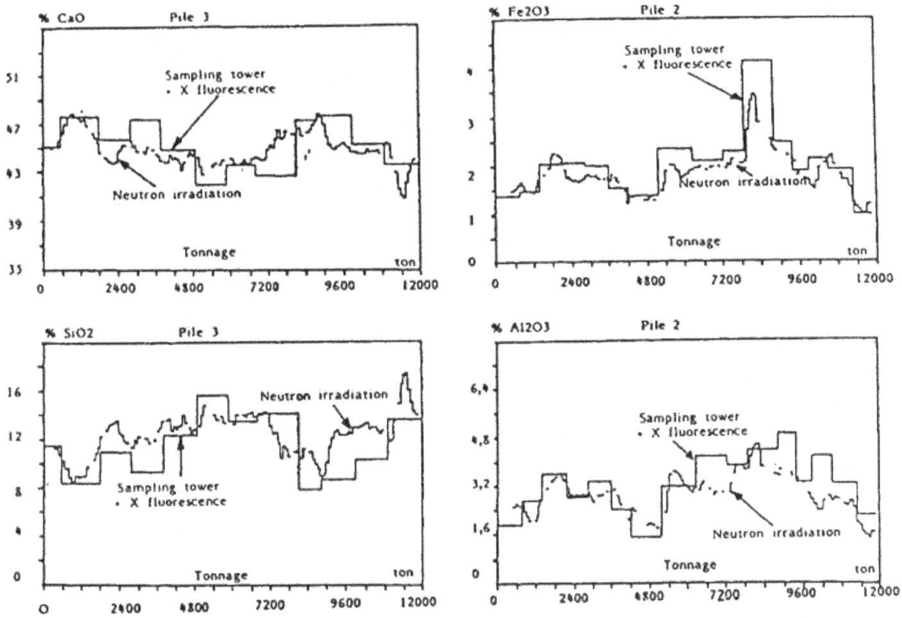

Fig. 8 - Typical comparative representation between the measurements by the sampling tower associated with X-ray fluorescence and the measurements by neutron irradiation.

2.4.6 Discussion

The tests conducted on the site at the Saint Pierre La Cour cement plant of Ciments Lafarge on the analysis of unprepared raw materials by the neutron irradiation method did constitute an important step for applying laboratory know-how to industrial process control applications. The analysis of pre-homogenization pile follow-up in real time gave positive results. However, we have noted that the results are widely dependent upon the quality of the calibration, that is the repeatability of the samples taken as reference. Within the scope of the development of an industrial analyzer based on these principles, a solution must be found concerning this problem.

Despite this drawback, the comparison between the results obtained from both analyzing methods (sampling tower - XRF system and neutron irradiation) highlights the noticeable advantages of the raw material analyzing method by the neutron irradiation process :

- Much higher volume of analyzed material (50,000 times larger)
- Measurement representativity problems practically eliminated
- Higher analyzing rates
- Data obtained in quasi real time
- Qualitatively richer information.

The LCPC and Ciments Lafarges have filed a patent both for the process and the analyzing system linked to this technique. They are now contemplating the construction of an industrial analyzer based on this innovative technique. Thus, in the future, it will be possible to obtain more homogeneous raw material at the plant input. This will enable easier optimization of the milling and burning operations for obtaining a clinker having Tricalcium Silicate composition variations with a standard deviation in the order of 3 %. The LCPC and Ciments Lafarge are thus in a position to propose a solution which is well adapted to product quality improvement for industrial concerns involved in cement production and mineral activities.

3. CONCLUSION

From the rather wide overview developed in the foregoing, it clearly appears that the nuclear methods do constitute a somewhat new and powerful potential for the control operations needed in the mineral industry. As this is the case for any new method, an adaptation period is obviously necessary, during which knowledge has to be gained on how to use the new tools generated by these methods. This is often a critical period. Nevertheless, the nuclear methods are highly diversified, hereby allowing, as described above the development of analyzing techniques resulting in extremely simple control methods (such as natural radioactivity) or highly sophisticated analyzing methods (such as the pulsed neutron method). Advantage has to be taken of this wide diversity to smoothly depart from the customary procedures and philosophies currently existing in the mineral industry, and make the profession understand the powerful prospects of this technique. It is only at a later stage that the complex methods will fully reveal all their possibilities and will become more easily integrated, to suit the often severe economical requirements of the trade.

For the future, the major advantage offered by the nuclear techniques lies in the capability of developing well suited control strategies, enabling on-line installation on the material itself, and not requiring any preparation of the material. Besides, advantage could be easily taken of the short response time needed by the nuclear methods in order to supply reliable data to the continuous monitoring systems of the fabrication process. These major features are essential as they will allow :

* Better utilization of the production tools
* Improvement of product quality
* Uniform product quality
* Energy saving in certain instances

REFERENCES

[1] MATHEW P.J., "On stream ash analysis of coal based on its natural gamma-ray activity, Proceedings Coal Research Conference - Wellington, New Zealand, 15-17 October 1985, Volume 2, paper n° 31.1

[2] AYLMER J.A., HOLMES R.J., "On stream shale detector for sedimentary iron ores", Proc. Austral. Inst. Min. Metall. n° 280 : 29-34, December 1981

[3] BARANYAI L., "Measuring of fly-ash in cement grinding on the basis of its natural radioactivity", Appl. Radiat. Isot., Vol. 37, n° 11, pp. 1111-1114, 1986

[4] AYLMER J.A., HOLMES R.J., RUTHERFORD L.CL, "Dynamic trial of an iron ore analyser based on pair production", Nucl. Geophys., Vol 1, n° 1, pp. 51-58, 1987

[5] CLAYTON C.G., WORMALD M.R., "Coal analysis by nuclear methods", Int. J. Appl. Radiat. Isot. Vol 34, n° 1, pp. 3-22, 1983

[6] WATT J.S., STEFFNER E.J., "Dual energy gamma-ray transmission technique applied to on-line analysis in the coal and mineral industries", Int. J. Appl. Radiat. Isot. Vol. 36, n° 11, pp. 867-877, 1985

[7] HOLMES R.J. et al, "A prototype bauxite analyzer based on fast neutron activation analysis", Nucl. Geophys. Vol 1, n° 1, pp. 41-50, 1987

[8] J.B. CUNNINGHAM et al "Bulk analysis of sulphur, lead, zinc and iron in lead sinter feed using neutron inelastic scatter gamma rays", Int. J. Appl. Radiat. Isot., Vol 35, n° 7, pp. 635-643, 1984.

[9] GLASCOCK M.D., COVENEY R.M. Jr., TITTLE C.W., GARTNER M.L. and MURPHY R.D., "Geochemical applications for prompt gamma neutron activation", Nucl. Instr. and Meth. in Phys. Research, B. 10/11 (1985) 1042-1436

[10] VOURVOPOULOS G., HUMPHREY D.L., SETTERS P.L. and LAMKIN K., "On-line elemental analysis of coal via high resolution PGNA", Nucl. Instr. and Meth. in Phys. Research, B 40/41 (1989) 853-856.

INDUSTRIAL APPLICATIONS OF POSITRON EMISSION TOMOGRAPHY

G.M. Field, P. Fowles and M.R. Hawkesworth

University of Birmingham, Edgbaston,
Birmingham, B15 2TT, United Kingdom.

ABSTRACT

An outline is given of the basic principles of positron emission tomography (PET) and its powerful application and potential for a wide range of current industrial and engineering problems.

1. INTRODUCTION

The application of nuclear imaging techniques to medicine is well established. Medical research into the functioning of the brain, in particular, has benefitted greatly from the use of these techniques, in which radioactive fluids and gases are introduced into the body and their distribution traced using position-sensitive detectors, or 'cameras'. For imaging involving gamma-rays, two types of camera geometry are used. Single-photon imaging (scintigraphy) has its major applications in the field of medicine where advantage is made of the relatively low energy gammas (technetium-99 emits 140 keV photons, for example) and an optimum camera-subject separation of about 100 mm.

The technique of positron emission tomography (PET), which also has its origins in medical imaging, has now been extended to encompass problems in engineering and industry. PET depends on the coincident detection of two collinear 511 keV gamma-rays arising from the annihilation of positrons emitted by suitable radioisotopes occupying the space to be mapped. The PET camera, therefore, requires at least two position-sensitive detectors, one on each side of the subject, or a single detector array encircling the subject. The penetrating power of 511 keV photons and the relatively weak dependence of resolving power on detector separation are attractive for industrial applications. For such non-medical work, radioisotope tracing can be used to greater advantage in industry, since the sources can be made more intense and the exposure time made longer to provide better, clearer images.

In the following section the positron camera is described in some detail. A number of the industrial applications of PET are then demonstrated, and finally, I shall discuss the new projects in which the Team at the University of Birmingham are involved.

2. THE POSITRON CAMERA

The two position-sensitive detectors of the University of Birmingham positron camera are based on Multiwire Proportional Counter (MWPC) technology, their design being developed from the Rutherford Appleton Laboratory (RAL) Mk.I MWPC camera [1]. The design is based on using lead plate cathodes to achieve adequate conversion of photons into electrons for detection, and delay lines are used to provide the X and Y coordinates of events.

A simplied diagram illustrating the operating principles of the RAL/University of Birmingham industrial positron camera is given in Figure 1. The most important single function of the readout electronics is the identification of the pairs of events associated with the coincidence gamma-rays. The signals from all (twenty) anode

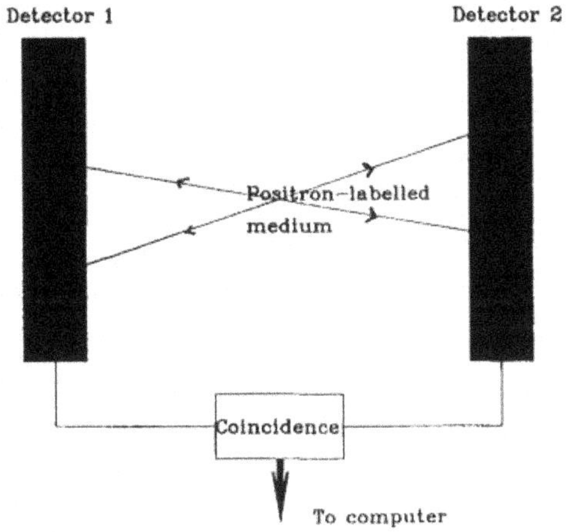

Figure 1. Schematic illustration of PET

planes in a detector are passed to a coincidence resolving unit. Only events occurring within 10 ns of an event in the opposite detector are accepted and this coincidence provides the trigger for the whole readout cycle. Two coincident sets of XYZ coordinates are then supplied and stored as raw data for subsequent data processing.

The simplest method of image production is to back-project each gamma-ray vector calculated from the raw data onto a series of Z planes lying parallel to the detector faces. Each plane contains equal contributions from all the data recorded but in varying degrees of focus depending on the distance of regions of activity from the plane.

These back-projection images, or tomograms, form the first stage of the 3D image reconstruction. A number of numerical techniques may then be applied to sharpen the image in each plane, so removing background effects. Alternatively the entire array of planes can be deconvolved with the experimental or theoretical 3D point source response function of the camera. Corrections for possible photon attenuation are also made, before the final tomograms are displayed. In engineering applications of PET, drawing overlays may be used to enhance the interpretation of the radioisotope tomograms.

3. DEMONSTRATIONS OF THE TECHNIQUE

Positron emission tomography was first applied to the study of oil lubrication in operating engines and test rigs, as part of a project carried out jointly with the University of Birmingham, the Science and Engineering Research Council, Rolls-Royce plc and Burmah-Castrol Ltd [2,3].

Figure 2 illustrates the application of PET to map the flow of oil in a bearing test rig, and shows the detailed and accurate information that can be obtained with

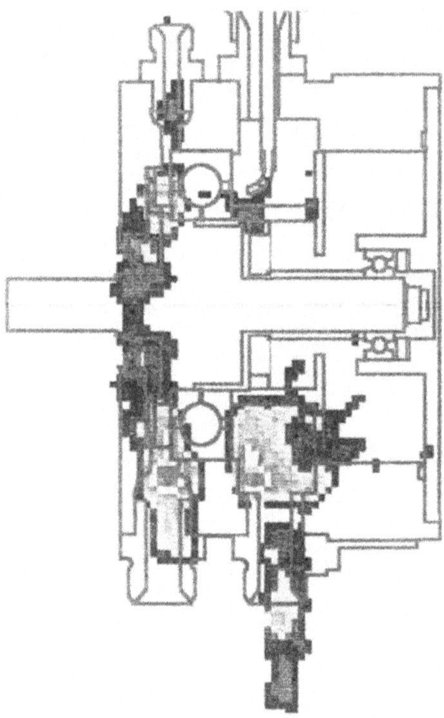

Figure 2. Flow mapping in a bearing test rig

this technique. Shown is the basic tomogram together with the same information in an engineering drawing overlay which locates the oil flow pattern precisely within the engine. The positron-emitting radiolabel used frequently in the engineering studies is ^{68}Ga with a half-life of 68 minutes. ^{18}F, with a half-life of 110 minutes, is typical of a range of useful positron labels that can be produced locally on the 10 MeV Nuffield Cyclotron at the University of Birmingham.

Since this initial exploratory stage, PET has been applied to numerous other problems of interest to Industry. A joint programme between the Departments of Physics and Chemical Engineering to study flow in fluidized beds, chemical hoppers and paste extruders was carried out, where PET was used to locate and track discrete positron sources, providing 3D maps of flow streams and their velocities [3)]. Some examples of the wealth of data produced by this technique are shown in the following figures.

Figure 3 shows the path followed by a particle, in this case a 10 mm diameter glass bead containing 50 μCi ^{22}Na, falling with sand through a rectangular-section (300 mm $\times$ 200 mm) egg-timer type hopper.

The movement of a heavy particle (12 mm glass bead, 50 μCi ^{22}Na) in a sand-filled fluidized bed with a sloping base distributor is illustrated in Figure 4. This is a simulation of non-combustible material, or slag (i.e. the bead), in a fluidized-bed furnace. It is seen that slag will be consistently transported to the top of the distributor where it can be conveniently removed without shutting-down the furnace.

Figure 5 is a superposition of five centre-line images of a paste (clay/bentonite) in an extruder 75 mm in internal diameter with 10 mm thick steel walls. This figure shows, it is believed for the first time using PET, the changing shape of a paste front as it enters the extrusion die. The line source label for the front was made by adding ^{68}Ga to a sample of the paste.

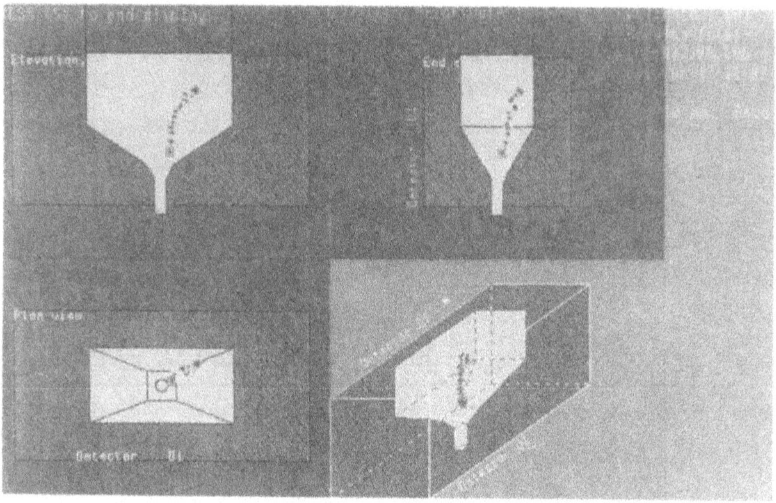

Figure 3. Flow streams in a sand hopper

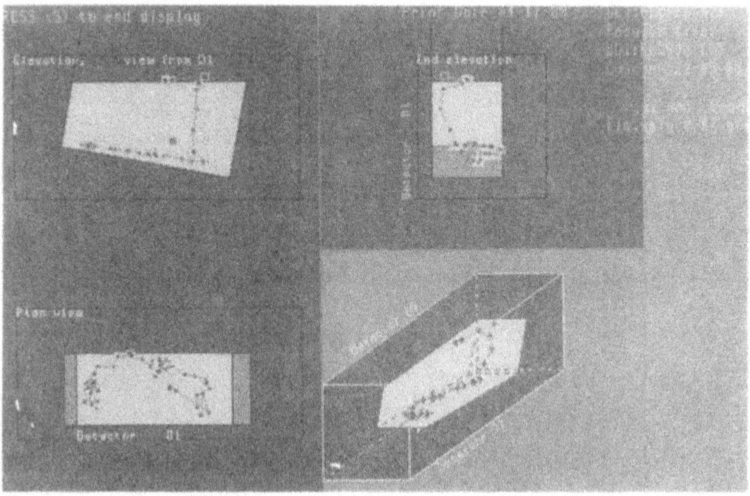

Figure 4. Particle tracking in a fluidized bed

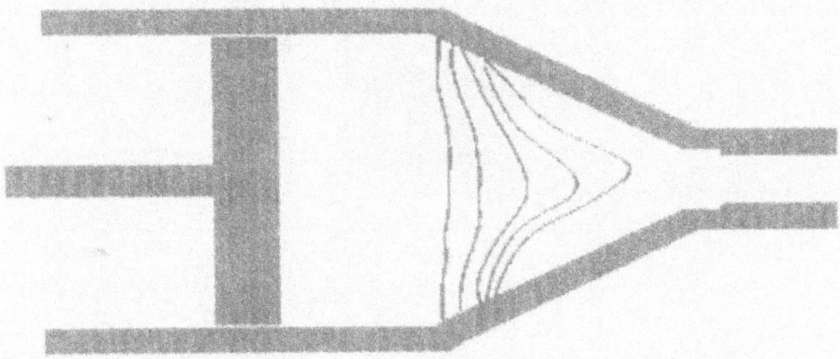

Figure 5. Paste Extrusion

4. NEW PROJECTS

Following these successes, a number of new projects involving PET have been initiated. One of these is a joint programme between the Schools of Physics and Chemical Engineering, this time to study the mechanics of powder mixing using positron emission tomography. The aim of the research is to understand, characterize and quantify the bulk flow of solids in mixing equipment, intially with a view to gaining an understanding of how mixing occurs. This study will then be extended to an understanding of the bulk flow of solids in mixing equipment be these free–flowing or cohesive and to develop means of following and describing the mixture. The non–invasive nature of PET should eventually allow quantitative data to be accrued to test the various current theories of mixing.

As a first step in the study, stream flow patterns using a single tracer particle have been investigated. For this purpose a domestic mixer has been used to demonstrate the mixing process. The mixtures considered in this preliminary investigation have been both free–flowing and cohesive. Starting from a powder

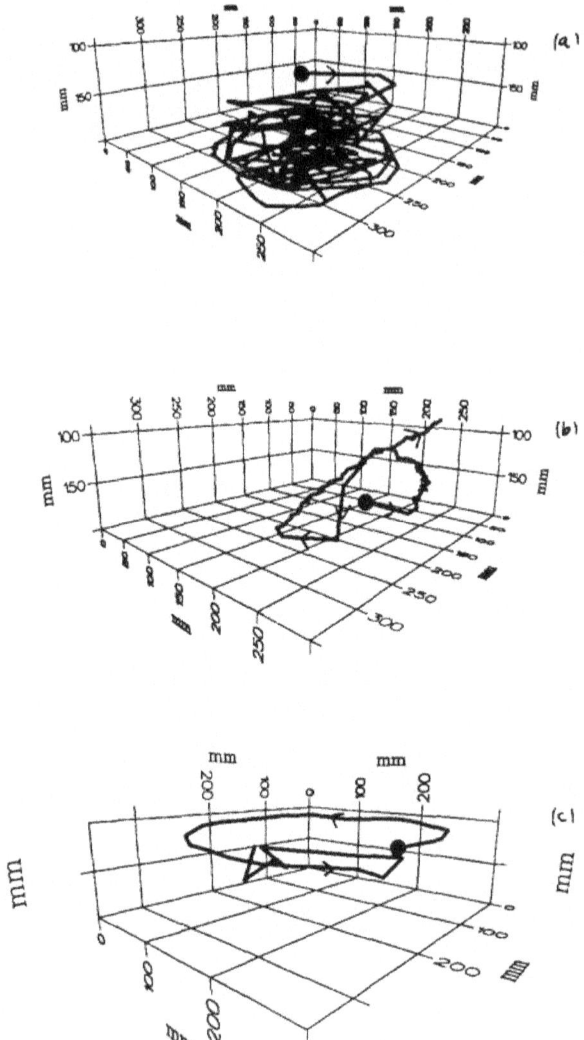

Figure 6. Mixing processes in (a) a powder, (b) a biscuit crumble, and (c) a liquid slurry

consisting of α–alumina and adding water, it has been possible to vary the consistency of the mixture from a powder, to a crumble (cohesive) to a liquid slurry (free–flowing). A single radioactive tracer particle of diameter 3 mm has been used to investigate the 3D flow patterns in each of the three cases. The results obtained in this preliminary investigation are shown in Figure 6. It is clear that the stream flow patterns are very different for the three cases considered, and results of this nature will provide a databank for the modelling and understanding of these mixing processes.

The results of this preliminary experiment has brought an enthusiastic response from a number of industrial mixing companies, with the promise of assistance in the form of the loan of industrial mixers and mixtures. Other industrial applications are currently being identified and solved in the food and petroleum industries with future developments anticipated in the study of the geological storage of radioactive waste.

5. CONCLUSIONS

PET has proved itself to be a successful example of technology transfer from the field of pure nuclear physics to applied physics and engineering.

6. ACKNOWLEDGEMENTS

The authors wish to thank the Science and Engineering Research Council for support for this work.

7. REFERENCES

1) Bateman J.E., Connolly J.F., Stephenson R., Tappern G.J. and Flesher A.C., Nucl. Instr. and Meth. **225** (1984) 209.

2) Hawkesworth M.R., O'Dwyer M.A., Walker J., Fowles P., Heritage J., Stewart P.A.E., Witcomb R.C., Bateman J.E., Connolly J.F. and Stephenson R., Nucl. Instr. and Meth. **A253** (1986) 145.

3) Hawkesworth M.R., Bemrose C.R., Fowles P. and O'Dwyer M.A., in '*Tomography and Scatter Imaging*', IOP Publishing, Bristol (1989) pp 67–79.

4) Bemrose C.R., Fowles P., Hawkesworth M.R. and O'Dwyer M.A., Nucl. Instr. and Meth. **A273** (1988) 874.

ION BEAM TECHNIQUES FOR MATERIALS MODIFICATION AND ANALYSIS

F.D. McDaniel, D.L. Weathers, J.L. Duggan, S. Matteson, and D.K. Marble

Department of Physics and Center for Materials Characterization
University of North Texas
Denton, Texas 76203
USA

ABSTRACT: The University of North Texas has developed a laboratory for the characterization and modification of materials employing atomic and nuclear physics techniques. Currently, materials may be characterized by nuclear reaction analysis, hydrogen forward scattering, particle induced x-ray emission, Rutherford backscattering spectrometry, ion channeling, and accelerator mass spectrometry of stable isotopes. Materials modification may be performed by ion implantation. A brief discussion is included of each of these techniques. Four accelerators are available for materials research and atomic collision physics, including a 200 kV Cockcroft-Walton, a 200 kV Nova implanter, a 2.5 MV, single-ended Van de Graaff, and a 3 MV National Electrostatics Corporation Tandem. The strong interactions developed with industry and future plans for the laboratory are discussed.

INTRODUCTION

For many materials, the technological limits of the material have been exceeded. Further progress in many areas of research and industrial applications requires the development of new materials or the improvement of existing ones. The National Academy of Science and Engineering has reported, "Perhaps what is most significant about materials....is that, in parallel with the development of social organizations and the art of language, it tended to be a limiting factor in determining the rate at which civilization could advance."[1] There is a present need to characterize trace element impurities in materials at levels of 100 parts-per-million to sub parts-per-billion. In many cases, the impurities have a large effect on the properties of the material, causing changes in the electrical conductivity of semiconductors and in the strength and corrosion resistance of metals, for example.

THE ION BEAM MODIFICATION AND ANALYSIS LABORATORY (IBMAL)

The Ion Beam Modification and Analysis Laboratory (IBMAL)[2] at the University of North Texas (UNT) utilizes a number of techniques for the characterization and modification of materials. Presently available techniques include nuclear reaction analysis (NRA),[3] hydrogen forward scattering (HFS),[4] particle induced x-ray emission (PIXE),[5] Rutherford backscattering spectrometry (RBS),[6] ion channeling (IC),[7] accelerator mass spectrometry

(AMS), [8,9] and high- and low-energy ion implantation.[10] The IBMAL is also used for atomic collision experiments to study ion charge-state dependence of x-ray and Auger electron production.[11]

The IBMAL houses four accelerators: a Texas Nuclear Model 9500, 200 kV, 10 mA Cockcroft-Walton; a 200 kV, 3 mA Nova Model NV-200 ion implanter; a 2.5 MV single-ended, High Voltage Engineering Corporation Model AK Van de Graaff; and a National Electrostatics Corporation Model 9-SDH2 3 MV tandem. The Cockcroft-Walton accelerator is located in a 78 m^2 laboratory while the other three accelerators are located in a 450 m^2 laboratory. See fig. 1. The 200 kV Cockcroft-Walton accelerator is used for low-energy, x-ray cross section measurements. The 3 MV tandem accelerator, which is the workhorse of the laboratory, was purchased with an rf charge exchange ion source (alphatross) and a Cs sputter-type ion source (SNICS). A third, ultra-clean, microbeam, raster-scanning, depth-profiling ion source is being constructed for the 3 MV accelerator for AMS. The 2.5 MV accelerator is being configured for routine PIXE and RBS. The 200 kV ion implanter is being configured so that the implanter ion beam may be injected into the same target chamber and/or end station as the ion beam from the 3 MV tandem accelerator in order to be able to implant into the same sample over a wide range of ion energies from 20 keV to 15 MeV.

At the present time, capital expenditures in the IBMAL total about $3,500,000. Yearly maintenance and operating expenses are approximately $500,000 and includes student and post doctoral stipends and faculty Summer salaries. Laboratory personnel, which include full time, adjunct, and visiting professors, post doctoral and industrial research associates, and student research assistants, total over 30 people.

Nuclear Reaction Analysis (NRA)

There are over 500 different nuclear reactions that may be used to study trace quantities of an element in a host matrix. Prompt nuclear reaction analysis[3,12] measures the average concentration over a relatively large range in the sample of 10^3 to 10^5 Angstroms and has a sensitivity of parts-per-million (ppm). Resonant nuclear reaction analysis occurs over a much narrower ion energy range and therefore allows determination of the impurity concentration as a function of depth in the sample with depth resolutions of 100 to 1000 Angstroms and sensitivities of 100 to 1000 ppm. The NRA apparatus has recently been used to measure H in thin ZnS and diamond films. The H is correlated with the device performance for the ZnS detectors and as a function of annealing conditions for diamond films. BGO and NaI detectors were used to measure gamma rays from the resonant reaction, H(^{19}F,alpha gamma)^{16}O, at 6.4 MeV.

Hydrogen Forward Scattering (HFS)

Hydrogen forward scattering (HFS) or elastic recoil detection analysis (ERDA) may be used to measure the depth distribution of light elements in heavy materials.[4,13] ERDA consists of detection, at forward angles, of the light elements recoiling after a collision with an incoming ion. The incoming ion will scatter the light ion from the solid into a Si surface barrier detector. Depth resolution of < 300 Angstroms may be obtained. ERDA, which is limited to relatively thin film targets, is available in the IBMAL.

Particle Induced X-Ray Emission (PIXE)

PIXE is a multi-element, nondestructive technique that can measure element concentration from B to U with a sensitivity of 1 ppm.[5,14] A Link Analytical Si(Li) detector with three window positions (Be, Formvar/Al, windowless) is used for the x-ray measurements. The Guelph PIXE software package[15] is used to fit the PIXE spectra and determine elemental concentrations. The PIXE apparatus has been used recently to measure trace element concentrations in stainless steel metal samples taken from turbines and tubing used in steam power plants.

Rutherford Backscattering Spectrometry (RBS) and Ion Channeling (IC)

RBS provides information on the mass and distribution of elements in the sample.[6] Heavy ions may be used to increase the depth resolution of the concentration profiles due to the increased dE/dx of the heavier ions. RBS has been used with ion channeling (IC) to also give information on the disorder in crystals. In IC, ions undergo correlated scattering from lattice atoms along the low index rows or planes of a crystal.[7] Recently, RBS has been used in the IBMAL to study thin, steam-aged, Pb-Sn solder-plated strips. H scattering was used to determine oxide thickness and B scattering was used to study the Pb-Sn stoichiometry.

Accelerator Mass Spectrometry (AMS)

Accelerator mass spectrometry was proposed in 1977 as a new technique for making precise measurements of radioisotope ratios for dating of archaeological or geological materials.[16,17] AMS is being used at UNT to determine impurity levels of stable isotopes.[8,9] The AMS technique for stable isotopes is similar to Secondary Ion Mass Spectrometry (SIMS)[18] except a particle accelerator accelerates the atomic and molecular ions to high velocities and passes them through a gas or foil stripper. The electron stripping causes the molecular ions to dissociate into atomic ions, removing the molecular interferences.[19] Momentum-per-charge and energy-per-charge discrimination, followed by total energy detection, provides unambiguous elemental identification.

The AMS facility shown in fig. 2 was custom designed and constructed through a collaboration between UNT, Texas Instruments Inc., the National Science Foundation, and the Office of Naval Research. The design objectives of the AMS facility are: (1) detect trace impurities in electronic materials and metals at sensitivities of ppt (1 part in 10^{12} or 10^{10} atoms / cm^3), (2) remove molecular interferences present in SIMS, (3) construct an ultra-low sample contamination ion source, (4) sputter depth profiling of impurity distribution in the sample, (5) develop microbeam imaging of lateral distributions of impurities in the sample, (6) measure secondary electron images of the sample, and (7) provide rapid sample turnaround utilizing computer control of equipment, data acquisition, and analysis.

In fig. 2, secondary negative ions of the sample material being studied are produced in an ultra-clean, Cs sputter-type ion source and are accelerated to an energy of 70 keV. The development of the new ultra-clean ion source, which is under construction at this time, is the subject of a paper by Matteson et al.[20] The secondary negative ions are momentum-per-charge analyzed with a 90° magnet (M/dM = 500 for a 2 mm slit width) and injected into the tandem accelerator. The ions are attracted to the positive potential at the center of the accelerator resulting in ion energies of several MeV where they pass through a gas or foil stripper. Electrons are removed through collisions with the stripper gas or foil. The resulting positive ions are then repelled by the positive potential resulting in even higher ion energies. After exiting the accelerator, secondary ions from the original sample are momentum-per-charge analyzed with a 40° magnet (M/dM = 350 for a 2 mm slit width) and energy-per-charge analyzed with a 45° electrostatic analyzer (E/dE = 700 for a 2 mm slit width), and detected by total energy measurements in a surface barrier detector. The electrostatic analyzer (ESA) has been discussed previously.[21]

The entire AMS facility is under computer control. A personal computer is interfaced to the accelerator, the 90° and 40° magnets, the 45° ESA, Faraday cup controllers, and particle counting electronics, so that masses may be scanned and ion current or particle counts may be accumulated automatically. User-friendly, menu-driven software programs have been written in TURBO PASCAL 4.0 to provide fast, reliable computer control of the AMS apparatus and electronics.[22]

For ion beam currents greater than 10^{-14} A at the detector, a Faraday cup is used to monitor the ion beam intensity. For lower intensity ion beams, particle detection is used. For some injected masses, molecular break-up produces some atomic ions that have the same charge-to-mass ratio and would not be discriminated against by the 40° magnet or the ESA. However, these isotopes having the same charge-to-mass ratio do not have the same energy and are discriminated against, by energy, in the Si surface barrier detector.

At the present time, the AMS facility is operating and preliminary mass scans are being obtained using the conventional ion source technology that came with the accelerator (i.e., the National Electrostatics Corporation - SNICS Cs sputter-type ion source). This source provides many elements that contaminate the samples of interest and limit the sensitivity of the AMS facility. Fig. 3 shows a computer controlled, molecular-interference free mass scan of stable isotopes from a float zone Si sample in the mass range from 1 - 80 u for an accelerator terminal voltage of V_t = 2 MV and an ion charge state of Q = +3. Negative ions were not extracted from the sample for elements with zero or negative electron affinities. Fig. 4 shows an AMS mass scan of the same float zone Si sample (V_t = 2 MV, Q = +3) in the mass range of 1 - 80 u for "silicide" dissociation products. Negative molecular ions of the form MSi^- are extracted from the sample and injected into the accelerator. The dissociation fragments of mass M^{+3} are detected.

Fig. 5 presents a bar graph of calculated sensitivities for the data displayed in figs. 3 and 4. SIMS relative sensitivity factors[18] for Cs sputtering of Si were used to calculate the sensitivities and do not correct for charge state variations. At present, the sensitivities for some elements such as Al, K, Ti, Cr, Mn, Fe, Ni, Cu and Zn are limited by sample contamination in the SNICS ion source. These sensitivities should dramatically improve with the completion of the ultra-clean microbeam ion source. Appropriate choice of injected atomic or molecular ion and ion charge state, optimization of ion transmission, and completion of the ultra-clean, microbeam ion source should allow ppt sensitivities to be obtained for essentially all elements in the periodic table.

Ion Implantation

Ion implantation may be used to modify the surface and near-surface regions of almost any material.[10] In semiconductors, ion implantation has rapidly replaced diffusion techniques because of the reliability and precision of the process and because it is the key to miniaturization, vertical device structures, shallow junction formation, and isolation of devices. In metals, this method can result in increased hardness, wear, and corrosion resistance.[10] An excellent review paper has recently been published on the many industrial applications of ion implantation.[23]

The IBMAL has the capability of either/both high- and low-energy ion implantation into the same chamber from the Nova, 200 kV ion implanter or the National Electrostatics Corporation, 3 MV tandem accelerator. In addition, the 200 kV Cockcroft-Walton accelerator may also be used for ion implantation.

Atomic Collision Physics

The atomic collision research has focussed on charge-exchange and direct-ionization processes in ion-atom collisions[11] and fundamental measurements of x-ray production cross sections.[24] These types of measurements are important for the accèlerator mass spectrometry and PIXE research.

Future Plans for the IBMAL

Future plans for the IBMAL include a PIXE ion microprobe, that is external to the vacuum system, and a creep furnace, that will be attached to a beam line, for testing metal samples, subjected to varying temperature and stress.

INTERACTION BETWEEN INDUSTRY AND ACADEMIC LABORATORIES

The industrial interactions with the IBMAL, which were begun in the late 1970's, have continued to increase dramatically over the past few years. One of the reasons for the close interactions with industry has been the International Conference on the Application of Accelerators in Research and Industry, which is held at UNT in November on even numbered years.[25] In addition, an NSF Industry/University Cooperative Research Center has recently been formed to foster interactions with industry and the transfer of technology to industry.

SUMMARY

The Ion Beam Modification and Analysis Laboratory at the University of North Texas has four accelerators involved in fundamental and applied research in atomic and nuclear physics. A number of research projects are associated with the modification or characterization of materials, including nuclear reaction analysis, hydrogen forward scattering, particle induced x-ray emission, Rutherford backscattering spectrometry, ion channeling, accelerator mass spectrometry, and ion implantation. Characterization techniques such as accelerator mass spectrometry extend capabilities beyond any instrument in existance today. The personnel of the IBMAL have developed strong collaborations with industry that have resulted in joint research projects and increased research funding. Future plans include the development of an external microbeam PIXE system and a creep furnace to measure dynamic changes in metals as a function of temperature and stress.

ACKNOWLEDGEMENTS

Work supported in part by the National Science Foundation Grants No. DMR-8812331 and ECD-9003099, the Office of Naval Research Grants No. N00014-89-J-1309, N00014-89-J-1344, and N00014-90-J-1691, Texas Instruments Incorporated, Texas Utilities Electric Inc., International Digital Modeling Corp., North Texas Research Institute, Combustion Engineering Inc., Westinghouse Electric Corp., Fluor-Daniels Inc., LTV Corporation, the State of Texas Higher Education Coordinating Board - Texas Advanced Technology Research Program, the Robert A. Welch Foundation, and the University of North Texas Organized Research Fund.

REFERENCES

1. Psaras, P.F. and Langford, H.D., ed., Advancing Materials Research, National Academy Press, Washington, D.C., 1987.
2. Duggan, J.L., McDaniel, F.D., Matteson, S., Golden, D.E., Anthony, J.M., Gnade, B., and Keenan, J.A., Nucl. Instrum. and Methods in Physics Research **B40/41**, 709 (1989).
3. Sokhi, R.S., Forcey, K.S., Ross, D.K., and Earwaker, L.G., Nucl. Instrum. and Methods in Physics Research **B40/41**, 780 (1989).
4. L'Ecuyer, J., Brassard, C., Cardinal, C., Chabbal, J., Deschenes, L., Labrie, J.P., Terreault, B., Martel, J.G., and St.-Jacques, R., J. Appl. Phys. **47**, 381 (1976).
5. Oliver, A., Miranda, J., Richards, J., and Cheang, J.C., Nucl. Instrum. and Methods in Physics Research **B40/41**, 627 (1989).
6. Chu, W.-K., Mayer, J.W., Nicolet, M.-A., Backscattering Spectrometry, Academic Press, NY (1978).
7. Jin, H.-S., Park, K.-H., Yapsir, A.S., Wang, G.-C., Lu, T.-M., Luo, L., Gibson, W.M., Yamada, I., and Takagi, T., Nucl. Instrum. and Methods in Physics Research **B40/41**, 817 (1989).
8. Anthony, J.M., Donahue, D.J., Jull, A.J.T., and Zabel, T.H., Nucl. Instrum. and Methods in Physics Research **B10/11**, 498 (1985).
9. Anthony, J.M., Matteson, S., McDaniel, F.D., and Duggan, J.L., Nucl. Instrum. and Methods in Physics Research **B40/41**, 731 (1989).
10. Dearnaley, G., Nucl. Instrum. and Methods in Physics Research **B40/41**, 571 (1989).
11. McDaniel, F.D., Duggan, J.L., Lapicki, G., and Miller, P.D., Nucl. Instrum. and Methods in Physics Research **B42**, 485 (1989).
12. Langley, R.A., Picraux, S.T., and Vook, F.L., J. Nucl. Mater. **53**, 257 (1974).
13. Hofsass, H.C., Parikh, N.R., Swanson, M.L., and Chu, W.-K., Nucl. Instrum. and Methods

in Physics Research **B45**, 151 (1990).

14. See for example the Proceedings of the Fourth International Conference on Particle Induced X-Ray Emission and Its Analytical Applications, Nucl. Instrum. and Methods in Physics Research **B22** (1987).
15. Maxwell, J.A., Campbell, J.L., and Teesdale, W.J., Nucl. Instrum. and Methods in Physics Research **B43**, 218 (1989).
16. Muller, R., Science **196**, 489 (1977).
17. Purser, K.H., Liebert, R.B., Litherland, A.E., Beukens, R.P., Gove, H.E., Bennett, C.L., Glover, M.R., and Sondheim, W.E., Rev. Appl. Phys. **12**, 1497 (1977).
18. Wilson, R.G., Stevie, F.A., and McGee, C.W., Secondary Ion Mass Spectrometry - A Practical Handbook for Depth Profiling and Bulk Impurity Analysis (John Wiley and Sons, New York, 1989) Table E9, App. E.17.
19. Purser, K.H., Litherland, A.E., and Rucklidge, J.C., Surface and Interface Analysis 1, 17 (1979).
20. Matteson, S., Duggan, J.L., Marble, D.K., McDaniel, F.D., Weathers, D.L., Wilson, D.K., Anthony, J.M., and Beavers, R.L., Fifth Inter. Conf. on Accelerator Mass Spectrometry, Paris, France, April 23-27, 1990 (To be published in Nucl. Instrum. and Methods in Physics Research).
21. Matteson, S., McDaniel, F.D., Duggan, J.L., Anthony, J.M., Bennett, T.J., and Beavers, R.L., Nucl. Instrum. and Methods in Physics Research **B40/41**, 759 (1989).
22. Gressett, J.D., Maxson, D.L., Matteson, S., McDaniel, F.D., Duggan, J.L., Mackey, H.J., and Anthony, J.M., Nucl. Instrum. and Methods in Physics Research **B40/41**, 755 (1989).
23. Destefani, J.D., Ion Implantation Update, Advanced Materials and Processes, Inc.-Metal Progress, 39 (1988).
24. Marble, D.K., Gressett, J.D., McDaniel, F.D., Duggan, J.L., Culwell, J.F., and Lapicki, G., Nucl. Instrum. and Methods in Physics Research **B40/41**, 98 (1989).
25. Proceedings of the Internatonal Conference on the Application of Accelerators in Research and Industry, Nucl. Instrum. and Methods in Physics Research **B40/41** (1989), **B24/25** (1987), **B10/11** (1985), etc.

FIGURE CAPTIONS

Fig. 1 - The Ion Beam Modification and Analysis Laboratory at the University of North Texas.

Fig. 2 - The University of North Texas Accelerator Mass Spectrometry facility showing the SNICS, ALPHATROSS, and CHIMERA ion sources, the 90° injection magnet, the 3 MV tandem accelerator, the 40° magnetic spectrometer, the 45° electrostatic analyzer, and the Si surface barrier detector. The SNICS and ALPHATROSS ion sources were purchased with the National Electrostatics Corporation (NEC) 9SDH tandem accelerator and the CHIMERA ion source is under construction and is the subject of a paper by Matteson et al. [20]. CHIMERA is an acronym for CHemical Identification by Mass and Energy Resolution and Analysis. Also shown are ion beam object slits (OS), image slits (IS), Faraday cups (FC), ion beam attenuators (Att), quadrupoles lenses (Q), and post stripper (PS). A single channel analyzer (SCA) and a current integrator (I) are also shown in the figure.

Fig. 3 - AMS mass scan of a float zone silicon sample for 3+ ions of energy 8069 keV. The large variation of isotopes present came from the float zone silicon and from the National Electrostatics Corporation SNICS ion source for atoms that could form negative ions.

Fig. 4 - AMS mass scan of ions that formed silicide molecules with ^{28}Si. By injecting molecules that form negative ions into the accelerator, additional isotopes may be observed for elements that do not readily form negative ions such as N, Mg, Ca, Mn, and Zn.

Fig. 5 - Calculated sensitivities as a function of mass number for the data displayed in figs. 3 and 4. SIMS relative sensitivity factors for Cs sputtering of Si [18] were used to calculate the sensitivities and do not correct for charge state variations.

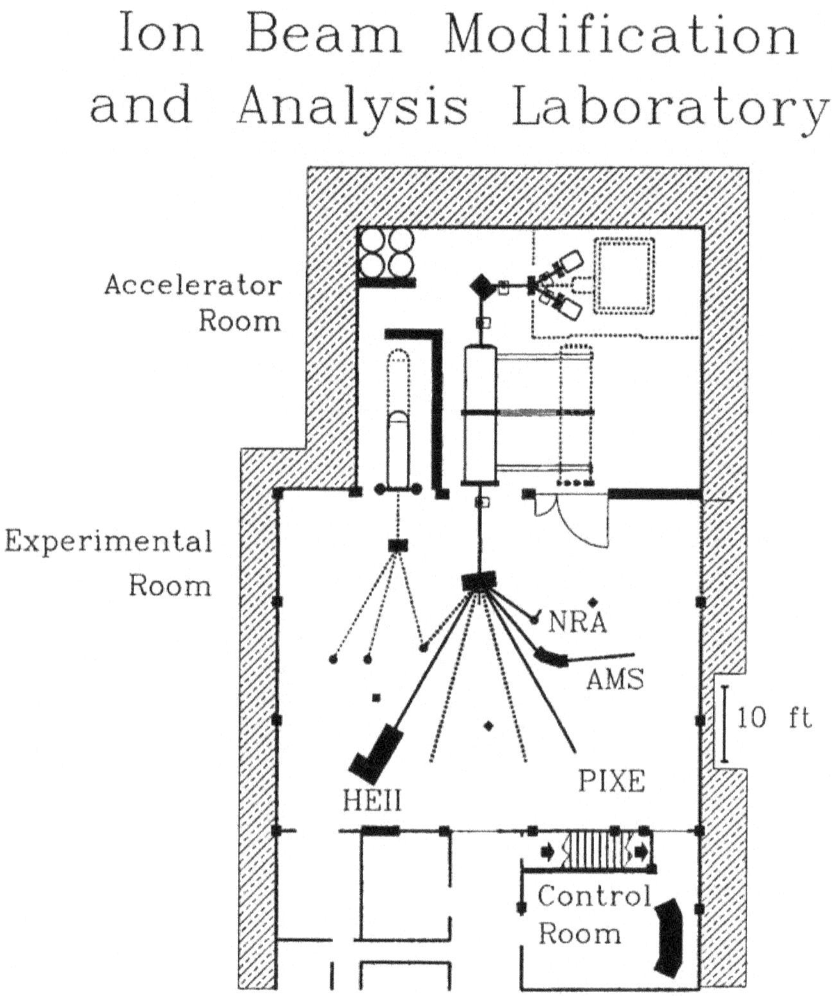
Ion Beam Modification
and Analysis Laboratory
Accelerator
Room
Experimental
Room
NRA
AMS
10 ft
PIXE
HEII
Control
Room

Fig. 1

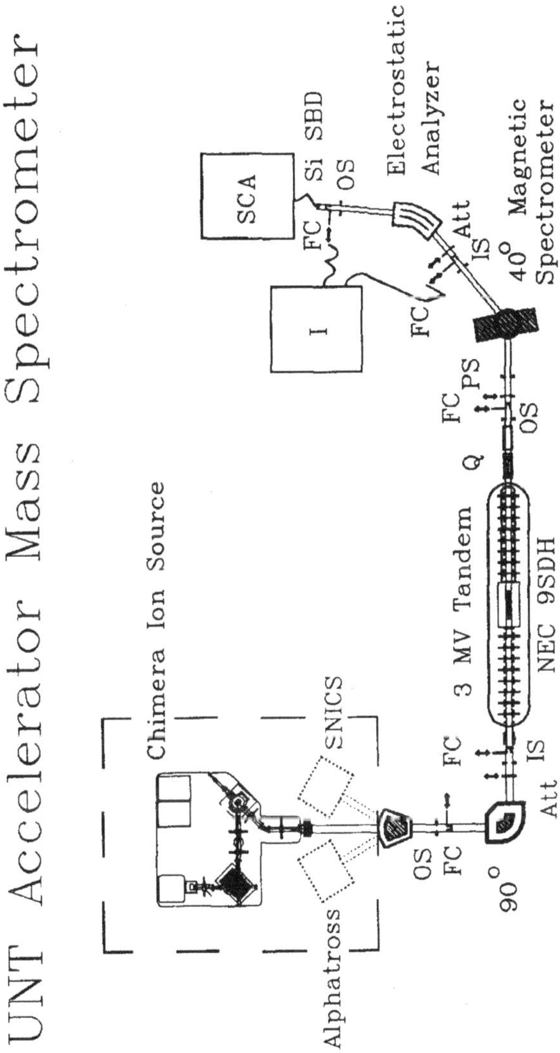
UNT Accelerator Mass Spectrometer
Chimera Ion Source
SNICS
Alphatross
OS
FC
90°
Att
FC
IS
3 MV Tandem
NEC 9SDH
Q
FC
PS
OS
FC
I
40°
Magnetic
Spectrometer
IS
Att
Electrostatic
Analyzer
FC
SCA
Si SBD
OS

Fig. 2

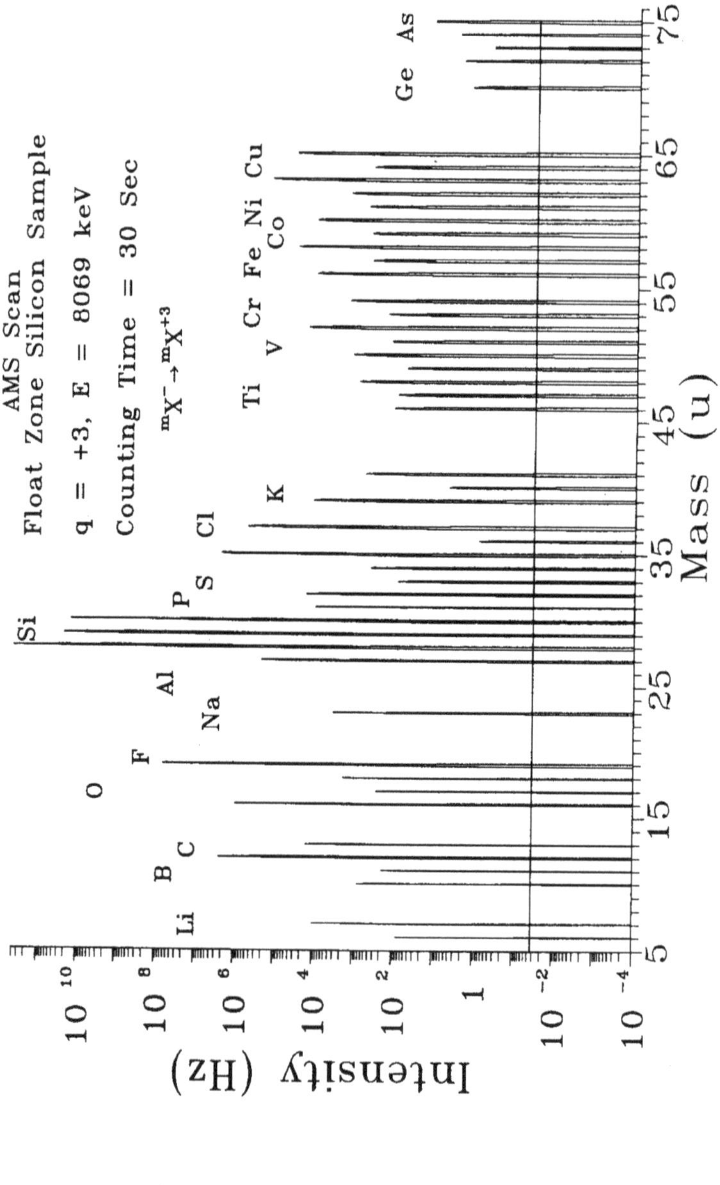
AMS Scan
Float Zone Silicon Sample
q = +3, E = 8069 keV
Counting Time = 30 Sec
Li B C O F Na Al Si P S Cl K Ti V Cr Fe Co Ni Cu Ge As
Intensity (Hz)
Mass (u)

Fig. 3

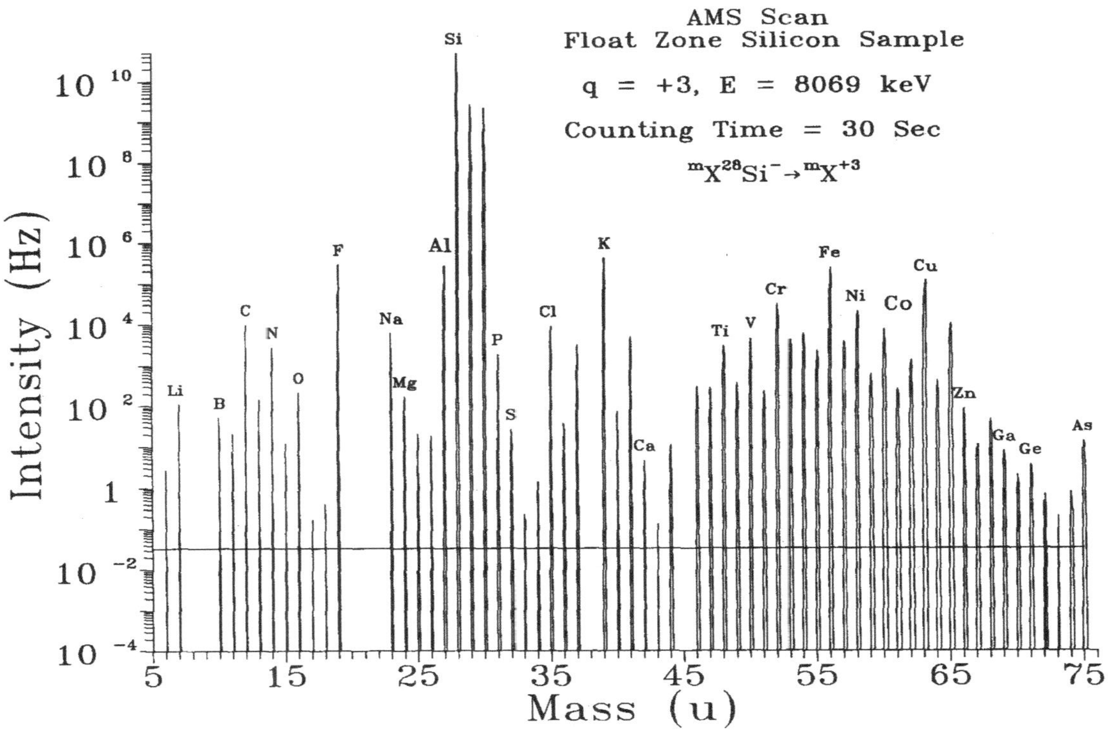
AMS Scan
Float Zone Silicon Sample
q = +3, E = 8069 keV
Counting Time = 30 Sec
$^{m}X^{28}Si^{-} \rightarrow {}^{m}X^{+3}$
Intensity (Hz)
Mass (u)
Li
B
C
N
O
F
Na
Mg
Al
Si
P
S
Cl
K
Ca
Ti
V
Cr
Fe
Ni
Co
Cu
Zn
Ga
Ge
As

Fig. 4

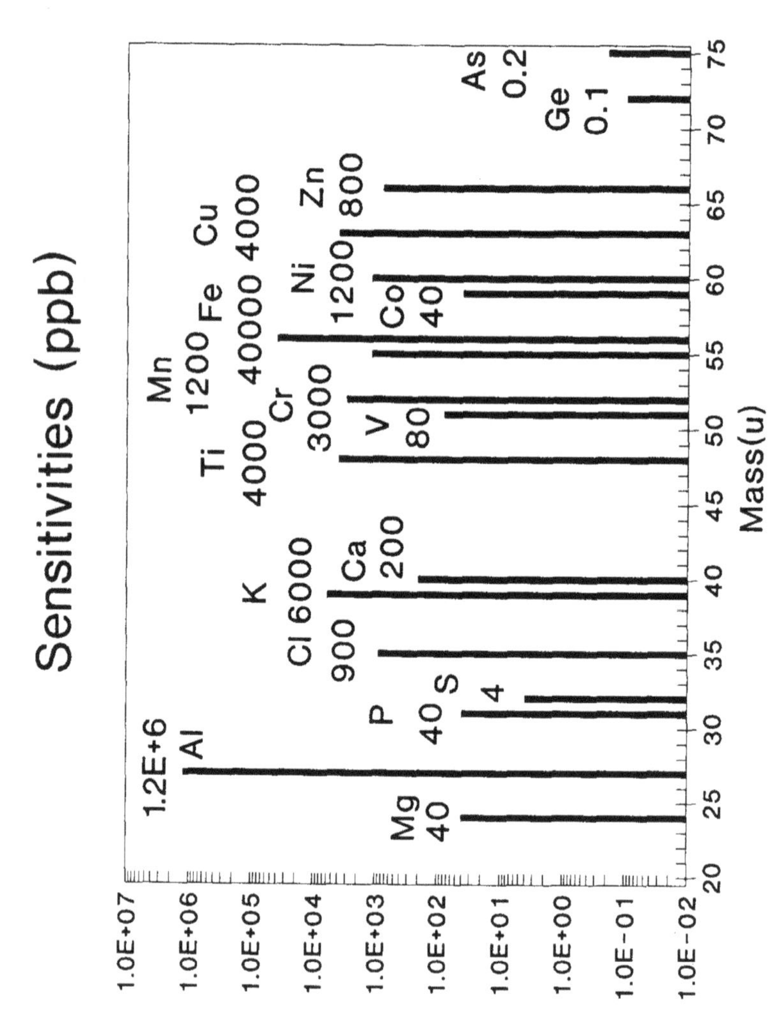
Sensitivities (ppb)
1.0E+07
1.0E+06
1.0E+05
1.0E+04
1.0E+03
1.0E+02
1.0E+01
1.0E+00
1.0E-01
1.0E-02
Mg 40
1.2E+6 Al
P 40
S 4
Cl 900
K 6000
Ca 200
Ti 4000
V 80
Cr 3000
Mn 1200
Fe 40000
Co 40
Ni 1200
Cu 4000
Zn 800
Ge 0.1
As 0.2
20
25
30
35
40
45
50
55
60
65
70
75
Mass(u)

Fig. 5

SEALED TUBES NEUTRON GENERATORS FOR DETECTION AND ANALYSIS

P. BACH

SODERN, 20 avenue Descartes
94450 Limeil Brévannes, FRANCE

ABSTRACT

Since many years, encouraging results have been obtained when using neutron analysis techniques, in a wide range of applications. Several neutron sources may be used ; nevertheless sealed tube neutron generators appear very attractive, because of their modulation and their high neutron output capabilities. Other characteristics such as transportability or lack of emission without electrical power make them adequate in a number of uses.

This paper reminds the main mechanisms in interaction between neutrons and atoms, and gives a description of some uses of sealed tube neutron generators. Some of them are using 14 MeV neutrons directly, the others lower energy neutrons. Main potential applications are : process control for bulk material or in metallurgy, in situ measurements mostly for oil and mineral logging, control of transuranic elements in waste drums, explosives detection hidden in luggages, neutron radiography. The future of these applications depends upon the availability of transportable equipments. SODERN is now able to put on the market different types of neutron generators, designed for intensive uses.

1 INTRODUCTION

Since many years, SODERN works in cooperation with the CEA (Commissariat à l'Energie Atomique, or French Atomic Energy Commission) for the development of neutron generators and their applications. The main features of the sealed tube high power neutron generators make them adequate to work in a wide range of applications, from detection of elements to analysis and neutron radiography.

Many physical mechanisms may be excited when neutrons interact with atoms. Most interesting of them involve gamma or neutron emission. At high energy (typ. some MeV), inelastic scattering of neutrons create in particular gamma rays at specific energies, which may be detected from outside of the sample and analyzed. This mechanism appears immediatly at the time of interaction. At high energy also, nuclear reactions may appear. Neutron activation processes are at the origin of new isotopes. If they are not stable, emission of particles or gamma rays may appear after the time of interaction.

At lower energies (typ. lower than 1 eV), neutrons may be captured by atoms. In a number of cases, characteristics gamma rays are emitted, immediately at the time of interaction. The use of selective detectors matched to these specific energies allows to determine the atomic content of the inspected material for these specific elements. Neutron activation processes at low energy appear also. If non stable isotopes are created, particles or gamma rays may appear after the time of interaction.

Neutron capture from heavy nuclei may gives also fission reactions, with emission of fast neutrons. The analysis of this neutron emission gives the number of initial heavy nuclei. This method of detection of fissile materials is more sensitive than the passive method, based on the self emission of the material.

For all this mechanisms, one of the better ways to obtain a high signal over noise ratio is to separate them with time, using a pulsed neutron generator. During the pulses, main contribution to the detected signals becomes from scattering and high energy neutron activation. Between pulses, main contribution becomes from lower energy neutron capture and activation. After irradiation, only activation products are detectable.

Neutron interrogation methods based on these mechanisms are very useful for detection and analysis of atoms concentrations in materials. If we consider now the neutron transmission properties of matter, neutron radiography may be also very useful for characterization of homogeneity of structures. As a non destructive testing method, neutron radiography gives a number of information such as corrosion detection, adhesive bounding defects,... The applications of neutron radiography for process controls are now possible, since high power on/off neutron generators appear on the market.

2 SEALED TUBES NEUTRON GENERATORS

Sealed tubes neutron generators are characterized by a very sharp neutron energy spectrum at 14 or at 2.5 MeV, by a high output (10^7 to more than 10^{11} n/s), by modulation possibilities, and by their installation possibilities with reduced safety rules compared to high strength isotopic sources. They are well adapted to all types of analysis, especially when physical mechanisms must be isolated ones from the others.

Sealed tubes neutron generators use the following nuclear reactions :

$$D + T \longrightarrow {}^4He\ (3.6\ MeV) + n\ (14.1\ MeV) \qquad (1)$$

$$D + D \longrightarrow {}^3He\ (0.8\ MeV) + n\ (2.45\ MeV) \qquad (2)$$

$$D + D \longrightarrow T\ (1\ MeV) + H\ (3\ MeV) \qquad (3)$$

The two fusion reactions (1) and (2) are very prolific even at low ion energies. On this basis, Sodern is able to propose two kinds of generators.

The first kind is based on the reaction (1) which is the more effective. In this configuration, a mixture of deuterium and tritium ions is accelerated at more than 100 kV up to a tritiated target. Monoenergetic neutrons at 14 MeV are emitted in a near isotopic manner. Neutron emission is controlled either in continuous mode or in pulsed mode, at frequencies up to 10 KHz. These generators are very interesting, for their high output and their high neutron energy. The lifetime of the sealed tubes is very extended. As an example for the TN 26 B Sodern tube, lifetime is of the order of a few thousand hours for a continuous emission at 10^8 n/s.

The second kind of generators is based on the reactions (2). In this configuration, no more radioactive material is used. The deuterium beam is accelerated at more than 100 kV up to a deuterated target. Neutrons between 2.45 and 3 MeV are emitted, especially in the direction of the beam : neutron emission is 2 to 2.5 times higher in the ion beam direction than in the perpendicular plane. These generators may be used when too high energy neutrons are not suitable. The maximum neutron output of these generators is limited to 10^{10} n/s, because cross section for reaction (2) is lower than for reaction (1).

These generators are convenient either in continuous or in pulsed mode. For example, if short pulses of 10 μs are used with a repetition rate of 5 kHz, then gamma rays from inelastic scattering of neutrons can be measured during the first 10 μs, and gamma rays from capture of thermal neutrons during the time delay from 10 μs to 200 μs. If the generator is switched off after a given number of pulses, it is also possible to measure the neutron activation contribution. Table 1 summarizes some characteristics of the Sodern tubes.

Table 1 : Neutron emission of some SODERN generators

Generator type	Tube type	High voltage power supply	Neutron output (n/s)	
			D-T tube (14 MeV)	D-D tube (1) (2.5 MeV)
GENIE46S(2)	TN 46 S	2 x 225 kV	10^{11} to 10^{12}	10^{9} to 10^{10}
GENIE 46	TN 46	225 kV	10^{10} to 10^{11}	4×10^{7} to 5×10^{8}
GENIE 36(2)	TN 36	160 kV	10^{8} to 10^{9}	4×10^{5} to 5×10^{6}
GENIE 26	TN 26 B	125 kV	10^{7} to 10^{8}	-

(1) in continuous mode - (2) not yet available

3 SOME APPLICATIONS OF SEALED TUBES n GENERATORS

3.1 Process control

The fluorine content of alumina used as a feed for electrolytic production of aluminium may be detected with an absolute accuracy of 0,1 %. On line elemental analyzers for bulk materials such as coal or cements may include a pulsed neutron generator[1], for an accurate determination of C, Ca, Al, Si, Fe, O,..

3.2. Oil, mineral, uranium logging

The use of prompt neutron gamma techniques is well established with 14 MeV neutron generators in the areas of geochemical exploration[2] (oil or gas logging) and environmental monitoring [3-4-5].

3.3 Hidden explosives detection

Neutrons may be used to detect hidden explosives[6-7] in luggages, or explosive materials in a number of special devices.

3.4 Detection of alpha emitters in nuclear waste

The more sensitive systems[8,9] are using induced fission from a pulsed 14 MeV neutron generator, working for example 20 µs with a repetition rate of 100 p.p.s. : detection limit is 3mg of ^{239} Pu in a 220 l drum for a 10 minutes measurement.

3.5 Neutron radiography

An european Eureka project, conducted by SODERN in cooperation with IABG in Germany, Sener in Spain, and with LTV in United States, is devoted to a mobile neutron radiography system using a more than 10^{11} n/s sealed tube neutron generator[10]. Compared to other equipments, its main advantages are the mobility on a truck (or the installation possibility anywhere in a restricted area), and the short exposure times to obtain good contrast pictures.

REFERENCES

1) Baron J.P. and al., Int. Symp. on Nuclear techniques in the exploration and exploitation of energy and mineral ressources, Vienna, Austria, 5-8 June 1990

2) Roscoe B.A., Gran J.A., SPE formation evaluation p.76 (March 1988)

3) Evans L.G. and al., IE^3-NS28 n° 2 p.1626 (April 1981)

4) Baron J.P., Cariou J., Cluzeau S., Conference on the application of accelerators in research and industry, Denton, Texas (November 86)

5) Humphreys D.R. and al., IE^3-NS 28 n° 2 p. 1691 (April 1981)

6) Shea P.M. and Lin V., Int. Conf. on neural network - Washington D.C., June 18 - 22, 1989

7) Grodzins L., invited paper/this conference

8) Smith G.W. and Rice L.G., IE^3-NS28 n°2 p.1637 (April 1981)

9) Beroud Y. and al., CEA-SGN paper for non destructive assay or radioactive waste meeting, Cadarache (France), Nov. 20, 1989

10) Dance W.E., Huriet J.R. and al., N.I.M. B 40-41 p. 1316 (1989)

"MINERALS PHYSICS .... AND ION PROBES"

J P F Sellschop
Schonland Research Centre for Nuclear Sciences
University of the Witwatersrand
P O WITS, 2050 Johannesburg
South Africa

ABSTRACT

The argument is advanced in this paper for focussing the tools and methods of physics, specifically those of nuclear physics, on the study of the minerals that abound in the Earth's crust. The study of specific mineral phases by harnessing a number of well-established physics techniques in a complementary way, inevitably brings a rich crop of data and the prospect of addressing basic and applied problems in a more quantitative way.

INTRODUCTION

The many facetted aspects that are embraced within the subject nuclear physics have, not surprisingly, led to a rich variety of applications, in an impressive array of diverse fields. It is the purpose of this contribution to attempt to highlight one of these and give it specific identification and recognition, namely that which we name "Minerals Physics". We do this for two reasons, the first of which is a fascination with the wealth of knowledge not generally accessible other than by the application of ion beams in the study of geological minerals, the second is a missionary endeavour to bring to the attention of physicists, and in particular physics students, that this field is not the preserve of the earth scientists alone, but offers rewarding scientific interest and indeed career opportunities to the physicist.

MINERALS PHYSICS

Minerals physics is the scientific study of the materials which make up the Earth and the planets and, thus, provides the critical data and theories to address a number of truly fundamental issues of central concern to all scientists and thinking, questing people.

In formulating these ideas, we have drawn heavily on the arguments developed by the Mineral Physics Committee of the American Geophysical Union ("Frontiers in Mineral Physics", 1988), on the statement on "Mineral Physics and Mineral Chemistry in Canada" (- a contribution to the Mineral Physics column of EOS, the weekly publication of the American Geophysical Union, 1989) and on correspondence with the Directors of a number of Institutes concerned with this interface between physics and the earth sciences.

Minerals physics addresses *inter alia* questions such as

- how did the Earth and the other planets form?
- how did minerals themselves form in the primordial solar nebulae?
- what detailed mechanisms prevailed in the accretion that led to the formation of the Earth?
- was the Earth formed by homogeneous or heterogeneous accretion?
- what were the mechanisms and the determining parameters that led to the differentiation of the Earth?
- the nature of the (seismic) 400 and 670 km discontinuities: are there compositional changes?
- what are the vertical and horizontal scales of mantle convection, and the dynamics thereof?

- what are the causes of the Earth's apparent upper mantle and inner core's anisotropy?

- how far do subducted slabs descend into the mantle and what is the extent of their perturbation on the mantle?

- what is the composition of the liquid outer core of the Earth?

- what reactions and thermal transport take place at the core-mantle boundary?

- how does the core dynamo work? and how and why does it periodically reverse?

- to what extent do the (microscopic) properties of materials at high pressures and temperatures determine some of the (macroscopic) phenomena identified above?

- do important boundaries in planets depend on the details of metallization of their constituent light elements such as hydrogen and nitrogen, as well as compounds such as water and hydrocarbons?

- how do fluids control the formation, migration and trapping of oil and gas and of a large class of minerals in the Earth's crust?

- what diagnostic information is contained in the fluid inclusions in minerals?

- to what extent are selected minerals, such as diamond, an undistorted window into the upper mantle of the Earth?

- hydrogeology or geohydrology ... the content and dynamics of water in the Earth's crust. Groundwater and the quantitative study thereof relates to ultimately the most vital of commodities to Man

- the mastery of understanding of fluid flow in the Earth's crust bears direct relevance into the burgeoning problem of ethically and technically acceptable strategies for the disposal of waste, nuclear and non-nuclear but toxic and hazardous

- the ultimate stability against failure through earthquakes or volcanism of occupied regions of the Earth's crust and the determinants thereof

- the stability against rockburst of deep excavations in pursuit of mineral exploitation

- what mechanisms determine the aggregation or concentration of minerals (of economic interest) in the Earth's crust?

- from the study of minerals-of-interest and of minerals cogenetic thereto, can the research tools of physics establish predictors for mineralized and barren deposits?

The list can evidently be expanded both in scope and in detail .... hopefully enough has been said to make convincingly the point of the centrality of physics, its tools and philosophical approach, to these questions both basic and of clear economic or practical relevance.

And what of the methods of physics in pursuit of these questions in the field of Minerals Physics? .... they are many!

Much stress has been given in the list above to the physics of the Earth's interior ... hence it is evident that high pressure physics exploiting the latest in presses and in the use of, e.g. diamond anvils is a central technique.

X-ray Crystallography enjoyed its origins in physics and should still be seen to reside as much in the domain of physics as of chemistry and geology. It is powerfully complemented by X-ray topography and by neutron diffraction.

The study of minerals still responds to optical microscopy, particularly if supplemented by e.g. hot stages and by cathodoluminescence.

The whole realm of modern optical spectroscopy needs no embellishment. Thermoluminescence plays a complementary role.

Modern techniques of transmission electron microscopy, scanning electron microscopy both with (electron induced) energy dispersive X-ray analysis, the electron microprobe and electron diffraction are rich in their information.

Many are the techniques for the study of the magnetic properties of minerals.

Powerful are the surface studies of minerals, in particular in modern ultra high vacuum chambers using electron excited techniques such as low energy electron diffraction, (scanning) Auger spectroscopy, as also X-ray induced electron emission.

This list is not exhaustive and to it must at least be added the "nuclear" contributions. These include Rutherford Backscattering, Elastic recoil analysis, nuclear reaction analysis (charged particle analysis and charged particle activation analysis), neutron activation analysis, proton induced X-ray analysis, accelerator mass spectrometry, Mössbauer analysis, time dependent perturbed angular correlations and distributions, ion channeling and dechanneling, combined with ion implantation and supplementary (on-line) techniques such as laser annealing.

By any account, it is my argument, both the questions to be addressed and the methods to be mounted, attest to the relevance of physics in the earth science field and fully justify the identification and accreditation of the field of Minerals Physics.

The evolving success of the field is due to not less than two factors...

- for the earth scientist new methodological approaches have been opened and a more quantitative, theoretically-linked basis exposed. As Seifert has pointed out (private communication, 900309) this has given the earth scientist a predictive capability that he was denied before, taking him into regimes of behaviour of the earth's matter under conditions not easily or at all accessible to experiments

- for the physicist it meant being introduced to materials that under normal circumstances he would not know of and if he did he would consider as much too complicated structurally and chemically! ... but now there is a mineralogical rationale which has diagnostic relevance and may even have commercial relevance! In addition he can take what Nature has offered and in a controlled, well-tailored way modify it revealing e.g. limits on physical conditions of the environment in which it had its genesis.

ION PROBES IN MINERALS PHYSICS

It is commonly assumed that nuclear techniques of analysis provide elemental information only. This is frequently the case but by no means is this always the case. To give emphasis to this important aspect of the nuclear probe, this section will be structured in three components, in increasing order of the sophistication of the information that can be extracted.

Simply Analytical

The ambition of any analytical method is to be multi-elemental in a simultaneous sense. This is but rarely realized but some nuclear methods can make this claim:

Rutherford backscattering (RBS)

is one such multi-elemental probe and is widely used in routine analysis of surface and near-surface assay. The energy dispersion disadvantages the method as one goes to higher masses, while the sensitivity improves as one goes to higher Z-values. It is particularly useful for analyzing higher Z-value surface layers on a lower Z-value substrate. Modern systems can be highly automated. The method is in principle absolute since it exploits the well-established electromagnetic interaction. The use of heavier ions and time-of-flight techniques can give significant enhancement to the analytical power of RBS.

Proton induced X-ray emission (PIXE)

is also in principle multi-elemental and has become a routine tool, with extensive automation of sample changers and data analysis. Its applicability is enhanced for thin samples. Its main (routine) competition comes from the electron microprobe, but it is significantly more sensitive.

Charged particle analysis (CPA)

is element-specific. It is so that enough is generally known about nuclear reactions that a reaction can be chosen which is specific if not unique to the element of interest in a defined matrix. The use of a wide variety of (heavy) ions enriches the prospects of this method. Nuclear cross sections are not generally well enough known, so that standards and reference materials become essential. It is a "prompt" method.

Charged particle activation analysis (CPAA)

is again element-specific and differs from CPA in that the emitted radiation that is observed is delayed. This has the implication that the activation process and the detection step can be separated not only in time but also in space. This has proved to be a very useful technique in minerals analysis and can be sophisticated by dual irradiation with different ions, by careful energy selection of the incident ions, by choice of the emitted radiation observed. It can be most closely compared with neutron activation analysis where the latter is truly a bulk technique.

Accelerator mass spectrometry (AMS)

is often known as "super-SIMS". This powerful technique exploits the sputtering negative ion source and uses a tandem accelerator for molecular dissociation. The MeV-energies involved permit single-atom detection and the combination of advantages leads to ultra-sensitive detection limits, but for one element at a time. In minerals physics it can be expected to play a very useful role, such as for example assaying the full suite of platinum group elements at low levels. With two-dimensional lateral rastering of the sample in the ion source, element imaging becomes possible.

Analytical integrity in these ion beam methods

The problem posed is not due to doubts about the nuclear method, but arises from a typical problem in earth science materials and that is the inhomogeneity of the distribution of the element of interest or the mineral of interest in the matrix. The implication is that stringent statistical tests are essential that take as input additional petrological parameters which have to be independently obtained.

Sensitivity amplification in ion beam analyses

If the element of interest occurs at low level in the geological matrix, or if there is the problem of inhomogeneity referred to above, the situation can be assisted by sample preconcentration. This is particularly relevant in ion beam analysis since the ion beam typically interrogates only a very small volume of the sample. While the (nuclear) physicist is understandably keen to maintain a purely instrumental technique, the demanding chemistry of sample preconcentration is in the field of minerals science often well worth the effort.

More than Simply Analytical

We now move to the next level of sophistication.

Two-dimensional analysis

This is a simple step and exploits the typically small diameter of the ion beam used and can be simply achieved by raster movements of the sample itself or by beam deflections, leading to imaging.

Analysis in the third dimension (depth)

Even simple RBS spectra are usually translated into depth information, but depth profiling at high resolution in an element-specific way is very effectively carried out using resonant nuclear reactions. One of the fine examples in mineral studies is of the depth profile analysis of hydrogen using either fluorine-19 or nitrogen-15 ion beams.

Microbeams

The technical skill now exists to produce ion beams of sub-micron diameter which now compete with the electron microprobe even more forcefully. The microbeam is scanned over the sample to give elemental images. The technique most commonly used is PIXE, supplemented by CPA. Scanning proton microprobes can be expected to make a significant impact on minerals studies.

Migration of species

Ion beam analysis, and in particular resonant nuclear reaction analysis, is used in a unique way to follow the spatial or depth migration of nuclear species in a matrix, as a function of time, or temperature, for example. This is singularly useful if combined with ion implantation to produce a well-defined dopant in a well-defined profile at a well-defined depth, and then to study in an entirely controlled way the dispersion behaviour as a function of parameters of choice. Such studies are key to the understanding and control of dopants in, for example, the mineral diamond in the production of semi-conductors.

Ion channeling

The channeling of well-collimated ions directed towards the major axes or planes of single crystals is a powerful ion beam tool in the study of minerals. It is uniquely powerful in the determination of the lattice location of elemental impurities. The changes in lattice location can be dynamically followed in their parameter dependence. The careful study of ion dechanneling, particularly in its energy dependence, permits one to delimit the types of defect structure responsible for the dechanneling.

Even more sophisticated than Just Analytical

Suffice it to name just a few examples that are relevant to minerals physics in this regard.

Bonding of surface layers

The surface layer of, for example, hydrogen is well-determined by a resonant nuclear reaction analysis. If the resonance width is sufficiently narrow and the ion beam energy width sufficiently small (and the accelerator sufficiently stable in energy) the Doppler spreading of the line width becomes observable. There will only be such an effect if there is true bonding of the hydrogen to the surface atoms, so that one is able to observe the vibrational modes of the bonded hydrogen. This can be followed in its orientation and temperature dependence for the next level of sophistication in the analysis and interpretation.

The hyperfine interaction

In particle-solid interaction studies, the hyperfine interaction can be exploited to excellent effect. A well-known case is that of the Mössbauer effect and its diagnostic use. Widely applied is the use of a suitable (implanted) radioactive species with a gamma ray cascade in

Time dependent perturbed angular correlations (TDPAC)

Frequently the beta particle initiating the gamma ray cascade can itself be detected so that the TDPAC measurements can be complemented by beta channeling studies of the lattice location of the radiating species.

With pulsed ion beams one can move to

Time dependent perturbed angular distributions (TDPAD)

giving a unique microscopic snapshot of the environment(s) in which the radiating nuclide finds itself. The frequencies associated with the coupling of the local environment electric field gradient to the quadrupole moment of the nuclear excited state, reveal the existence of specific chemical bonds. Even further information can be unravelled through the TDPAD study of species dependence, temperature dependence, dose dependence, orientation dependence.

A particular version of TDPAD is

Muon spin rotation

with the singular advantage of minimal radiation damage effects. The MuSR technique is further enriched in the information it gives in minerals studies through the formation of muonium, which is equivalent to a light form of hydrogen, which element plays a singularly important role in earth science and minerals physics research.

ACKNOWLEDGMENTS

The author is indebted to many persons and organizations for ongoing support and encouragement of many years standing in both nuclear physics and minerals physics, the confluence of which has been the subject of the paper.

Among these are some generations of research students, colleagues, the de Beers group of companies, and the Foundation for Research Development.

Nuclear Technologies for Finding Clandestine Explosives

Lee Grodzins
Massachusetts Institute of Technology
Cambridge, MA 02139

ABSTRACT

For many years explosives have been used by terrorists, malcontents and criminals. The terrorist acts are increasingly being viewed as instruments of foreign policy and governments throughout the world are developing countermeasures. This paper deals with nuclear-based technologies that have been proposed for finding clandestine explosives. We briefly review 13 techniques for utilizing the signatures of penetrating nuclear probes. An expanded description is given of the photon probes that make use of gamma ray resonances in the nuclei of bomb materials.

I. INTRODUCTION

Finding clandestine explosives is one of the most difficult of technological challenges. Consider the problem of finding a bomb in luggage checked by an overseas-going passenger. The typical suitcase will weigh 20 kg and have dimensions of 70 cm x 45 cm x 25 cm. A bomb, on the other hand, may weigh only 2% of the luggage; most experts believe that a half kilo of high-performance explosive is enough to destroy a jumbo jet flying at altitude. The deadly plastique can be shaped into almost any configuration, including thin sheets that might look like the lining of the suitcase. Moreover, the preferred bombs -- those with high yield that are still very stable against shock and heat -- are made of hydrogen, oxygen nitrogen and carbon -- H O N C; the same elements that make up most clothing and plastics. Add to the difficulties the rarity of what we are trying to find: so far there has only been about one bomb in every 10^9 bags checked on international flights each year. The last statistic sets the practical criteria for a useful security system: It must be reasonably rapid; the FAA calls for screening 10 bags per minute. It must not alarm often on innocuous bags that would then have to be hand-searched; one such search in a thousand bags seems to be a tolerable number. The system must not have a flaw that will allow the knowledgeable to beat it.

The task is formidable; ten years ago, it would doubtless have been considered hopeless. But advances in nuclear detectors, electronics and accelerator design have been rapid in the last decade and the problem is no longer insuperable. Practical solutions, of course, must not only be rapid, automated, fool-proof, and have excellent discriminating powers, but they must fit into existing airlines with minimal inconvenience and must operate at a cost tolerated by the paying passenger. Such "practicalities" are not all that restrictive, however. International airlines are large buildings with high ceilings and wide corridors designed to accommodate the transportation of thousands of passengers each hour. And these passengers are willing to pay for a guaranteed safe trip. If we assume a surcharge of $5 per flight (now being charged by some airlines) then it is easy to show that the capital costs of a security system that handles 1000 passengers per day (2,000 pieces of checked luggage) can be as high as $3 million dollars and the operating costs can be as much as $500,000 dollars a year.

The paper begins with a review of the properties of explosives relevant to nuclear detection schemes. We then summarize the basic physics of the techniques that are being considered. The last section considers one of those methods -- gamma ray, "resonant" probes -- in some detail.

II. ON DENSITIES AND ELEMENTAL CONCENTRATIONS

1. Density:

1. Densities of Luggage: Checked luggage may weight up to 70 pounds; carry-on can weight 75 pounds. Dimensions may total 143 cm. for checked bags (Example: 20 cm x 80 cm x 43 cm) and 112 cm for carry-on (Example: 20 x 40 x 52). In fact, since some airlines do not

enforce the regulations and, in any case, larger bags can be checked for a surcharge one must be prepared for larger and heavier containers, including huge duffel bags and large wooden or cardboard boxes.

TABLE 1

NOMINAL <u>MAXIMUM</u> DENSITIES

	Average density	Average Projected Density	
		Thin Dimension	Thick Dimension
Checked	0.5 gm/cm^3	10 gm/cm^2	40 gm/cm^2
Carry-On	0.8 gm/cm^3	15 gm/cm^2	40 gm/cm^2

Luggage must be among the most non-uniform of all containers. One must be prepared for local densities up to 10 gm/cm^3 (The density of copper, for example, is 8.9 gm/cm^3.) One must be prepared for local projected densities up to 20 to 30 gm/cm^2 or more, even through the short dimension of "average" bags.

2. Densities of Nitrogen. Every cubic centimeter of air contains about 1 mg of nitrogen. A cubic meter of air -- the nominal cavity size of an X-ray machine for checked bags -- contains a kilogram of nitrogen, enough to make at least 4 kilograms of high performance explosive.

To first approximation, almost every piece of luggage checked for overseas contains enough nitrogen to make a kilogram bomb. Total nitrogen content is not a useful measure of the presence of an explosive.

2. Elemental Composition of Explosives and Common Materials:

The following series of 8 graphs shows the densities and atomic weight fractions of explosives and common materials that are useful to analyze and evaluate nuclear detection schemes.

1. The density of clothing has been taken as 0.2; that is, about one-fifth the material density. The number comes from personal tests of packing of sweaters, etc.. I expect that the densities of clothes in most suitcases is less than 0.2 gm/cm^3, but some materials, such as leather, will have densities around 1.

2. Distributions are listed in atomic percent, since this is most useful for most nuclear probes. In any case, the difference between atomic and weight percent is not significant except for hydrogen, which is generally a high atomic percent of explosives, clothing and many plastics, but is almost always a small weight percent of the materials. (Example: 66% of the atoms of water is hydrogen, 11% of the weight of water is hydrogen.)

3. Explosives are unique in their high concentrations of both nitrogen and oxygen. All explosives fall within a confined area in the phase space of oxygen and nitrogen, and no innocent material falls in that space.

4. Other correlations, carbon plus oxygen for example, are not unique but can be excellent discriminants if the distribution of the correlated concentrations can be well-measured.

5. Hydrogen is not a strong discriminant for detecting explosives, for two reasons: First, explosives contain less hydrogen, by weight or by atom, than most common materials. Second, those materials that falsely alarm nitrogen detectors -- wool, melamine, etc -- invariably contain more hydrogen than explosives do. Thus if the nitrogen content is not a discriminant then the hydrogen content will not be either, unless one can measure the hydrogen density accurately

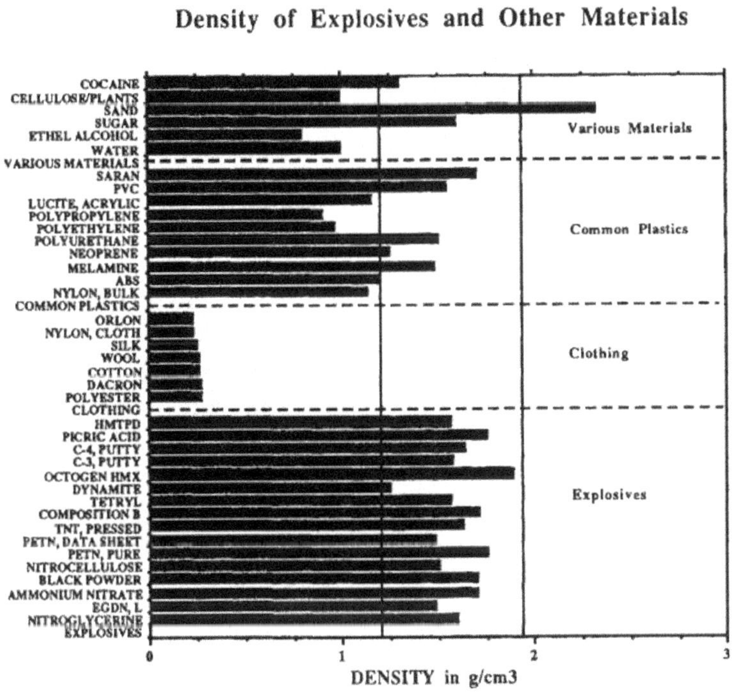

Figure 1. Many common materials have densities similar to explosives. And we emphasize that this figure is far from complete: there are explosives that fall outside the range indicated by the vertical bars; not all clothing has low density (leather has a density close to 1); and there are thousands of other materials that passengers pack in luggage..

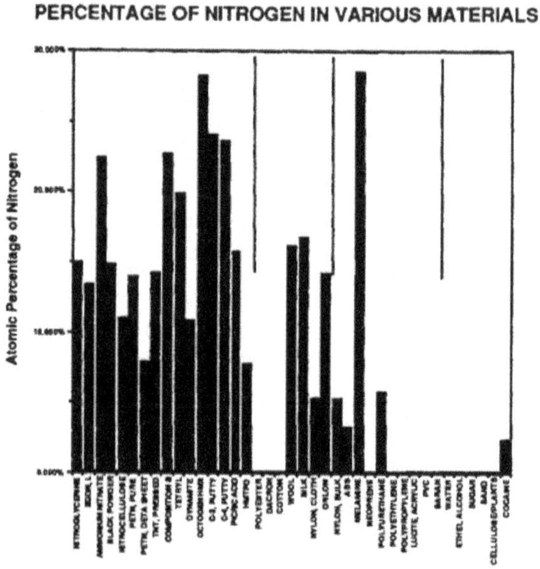

Figure 2. The atomic percentage of nitrogen in explosives. Explosives are N-rich but so are many common materials. Total nitrogen is not a particularly useful discriminant.

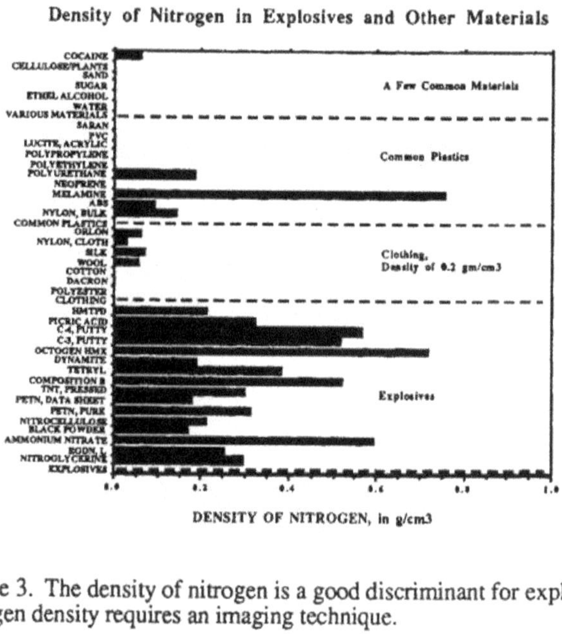

Figure 3. The density of nitrogen is a good discriminant for explosives. To measure the nitrogen density requires an imaging technique.

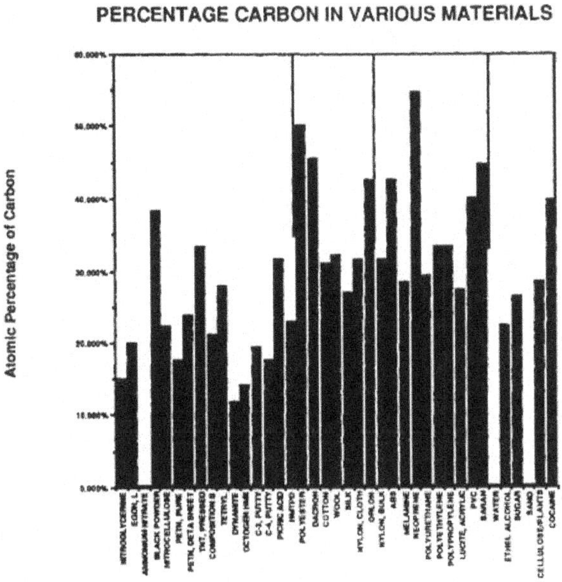

Figure 4 Atomic percentage of carbon in materials. Explosives are carbon poor.

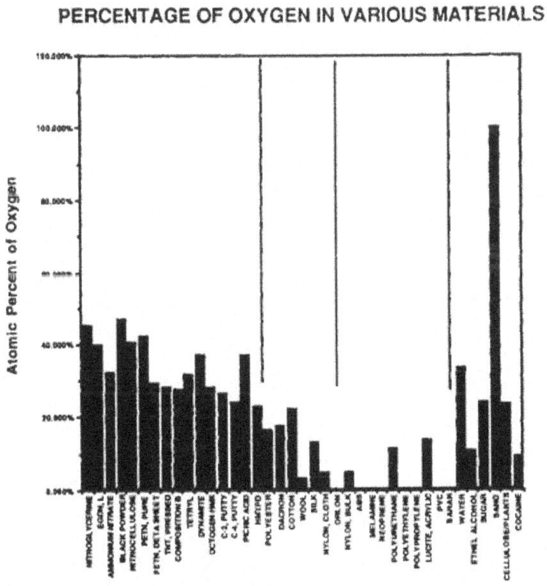

Figure 5. Atomic percentage of oxygen in materials. Explosives are oxygen rich.

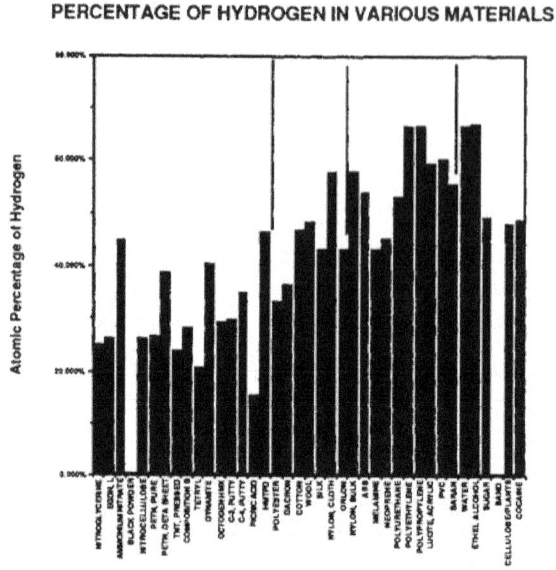

Figure 6. Atomic percentage of hydrogen in materials. Explosives are hydrogen poor.

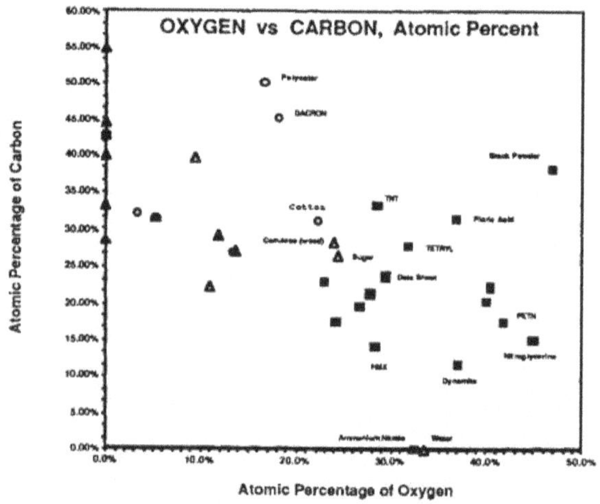

Figure 7. Correlation between oxygen and carbon in materials. Black squares are explosives. All explosives cluster in this graph, though some non-explosives have similar correlations.

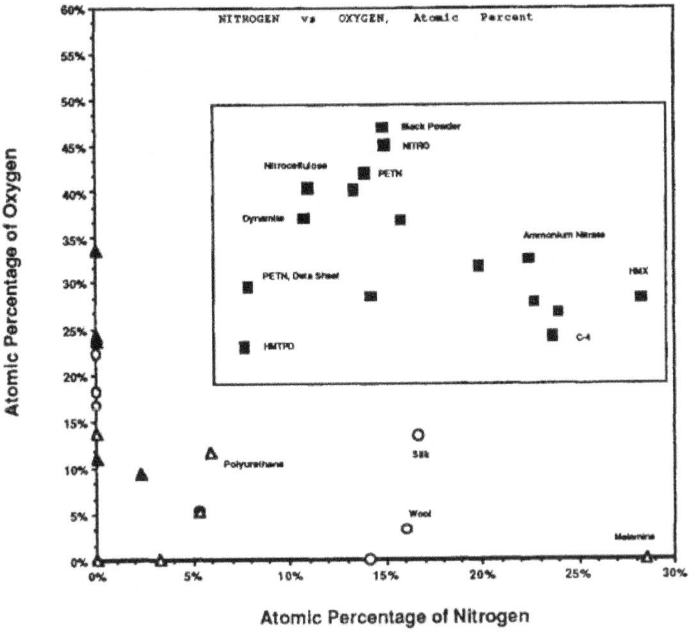

Figure 8. Correlation between nitrogen and oxygen in various materials. Black squares are explosives. Open triangles are plastics. Open circles are clothing. Black triangles are miscellaneousl. All explosives fall within the inner rectangle. No non-explosive material falls within the rectangle.

III. NUCLEAR TECHNIQUES FOR FINDING EXPLOSIVES

The summary is divided according to the type of nuclear probe. Most of the techniques irradiate the bag with neutrons. Almost all techniques detect gamma rays as the signature of the explosive. Some of the techniques have been tried in laboratory mock-ups on "representative" luggage Some have been proposed to the Federal Aviation Administration on the basis of the effectiveness of the technique in other applications. Some are still untried.

The techniques are listed and summarized to impress with their variety and to stimulate interest. Neither the merits nor the pitfalls are discussed here. Nor do I mean to compare, by intention or implication, one technique with another; the adversary will find the flaws, the advocate will find the virtues.

TABLE 2

NUCLEAR TECHNIQUES

Method	Incoming Radiation	Outgoing Radiation	Element Measured
Thermal Neutron Analysis	Thermal Neutrons	Characteristic High Energy γ	N
Fast Neutron Analysis	Fast (High-Energy) n	Characteristic High Energy γ	C,N,O
Pulsed Neutron Beam	Monoenergetic Fast n	Energetic Characteristic γ	C,N,O
Associated Particle	Monoenergetic Fast n	Energetic Characteristic γ	C,N,O
Pulsed FNA/TNA	Fast then slow n	Characteristic High Energy γ	C,N,O
γ-ray Resonance Absorpt.	Unique High-Energy γ	Unique High-Energy γ	N,(O?, C?
γ-Ray Resonance Absorpt.	Continuum High-Energy γs	Unique High Energy γ	C,(N?,O?)
Nitrogen 13 Production	Continuum High-Energy γs	Annihilation 511 kev γ	N
Nitrogen 12 Production	Continuum High-Energy γs	Annihilation 511 kev γ	N
Boron 12 Production	Continuum High-Energy γs	Bremsstrahlung γ	N
Pulsed n Backscatter	Monoenergetic Fast n	High-Energy n Spectrum	C,N,O
Pulsed n Absorption	Continuum High-Energy n	High-Energy n Spectrum	C,N,O
Nitrogen 13 Production	Continuum High Energy γs	Prompt n	N

Each of the methods has merit, but they are not equally effective. For example, some rely on annihilation or bremsstrahlung signatures that are not unique to the H,O,N C elements. Neither are they equally practical. For example, some require beams of nuclear particles that require extensive shielding and that produce unwanted residual radioactivity; others, to be practical, require advances in accelerator or detector or target design.

Each of the techniques are briefly described to bring out the underlying physics. They are taken up in the order of Table 2 with the exception that the resonance gamma ray techniques are treated separately, and in greater detail, in the latter part of the paper.

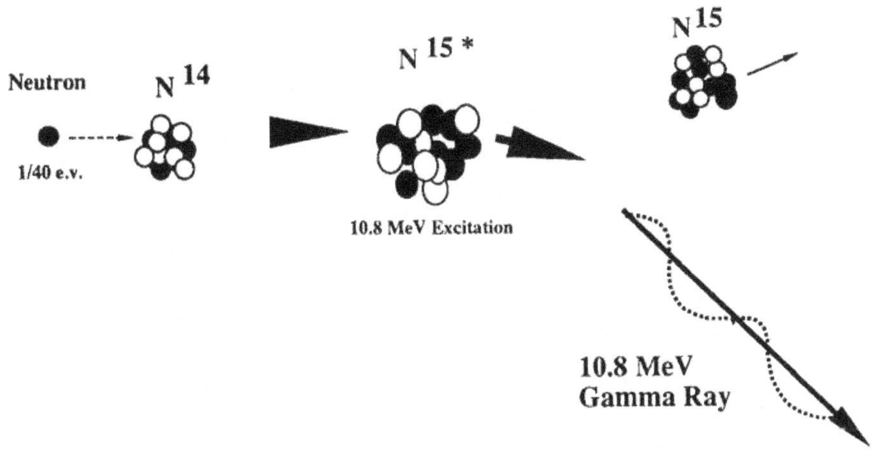

The Capture of Very Low Energy Neutrons Results in the Emission of a Unique Gamma

Figure 9. Cartoon showing the capture of thermal neutrons by nitrogen, leading to the emission of a unique 10.8 Mev gamma ray.

1. Thermal Neutron Analysis, TNA. Figs 9 and 10.. Luggage moves through a bath of slow neutrons which get captured by the nuclei of the elements that make up the luggage material. Gamma rays, characteristic of the elements, are emitted. The capture of slow neutrons by nitrogen is a weak process with an cross section of only 10 millibarns but it results in the highest energy gamma ray produced by any readily-observable thermal neutron capture.

$$^{14}N + n = ^{15}N + 10.8 \text{ Mev } \gamma.$$

The unique 10.8 Mev signals from a hodoscope of gamma-ray detectors are tomographically analyzed to give the spatial distribution of nitrogen. TNA is intrinsically matched to finding bulk quantities of nitrogen. Spatial resolution is limited. TNA can detect hydrogen and many other elements but it cannot detect oxygen or carbon.

Backgrounds are especially troublesome with this neutron-immersion technique. The neutrons from the source have Mev of energy and unwanted gamma rays are produced during the slowing down process. Once slowed, the neutrons interact mainly with innocuous materials producing gamma rays that can interfere with the wanted signal. It requires a tour de force of technology to sort the 10.8 Mev gamma ray from the copious unwanted gamma rays of lower energy.

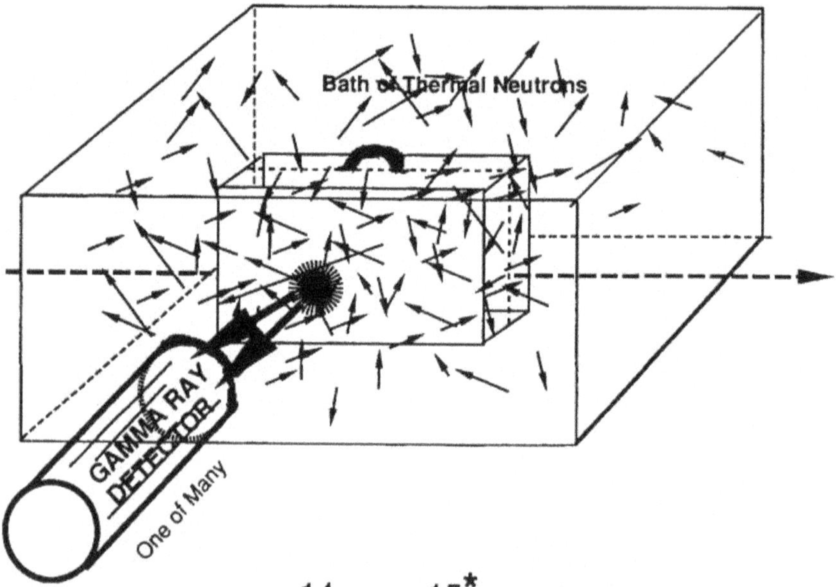

$$n + N^{14} \rightarrow N^{15^*}_{(excited)} \rightarrow N^{15} + 10.8\ \text{MeV}\ \gamma$$

Figure 10. Schematic of Thermal Neutron Analysis chamber.

2. Fast Neutron Analysis, FNA. Figs. 11 and 12. Luggage moves through a bath of fast neutrons. The fast neutrons interact with the nuclei producing gamma rays that are characteristic of the elements. The interaction strength in nitrogen is spread among several levels, illustrated in the list below.

Carbon: ^{12}C + 14.8 Mev n = ^{12}C + 10.4 Mev n + 4.4 Mev γ.
Oxygen: ^{16}O + 14.8 Mev n = ^{16}O + 8.67 Mev n + 6.13 Mev γ
Nitro.: ^{14}N + 14.8 Mev n = ^{14}N + 12.5 Mev n + 2.3 Mev γ.
10.8 Mev n + 3.95 Mev γ
10.8 Mev n + 1.63 Mev γ
9.9 Mev n + 4.92 Mev γ
9.7 Mev n + 5.11 Mev γ
9.7 Mev n + 2.79 Mev γ
8.9 Mev n + 5.84 Mev γ

The neutron energy must be at least as great as that of the produced gamma ray. In practice that means using at least 8 Mev neutrons to observed oxygen. 14.8 Mev neutrons result in greater backgrounds that neutrons of lower energy. On the other hand, such neutrons are readily produced by small, relatively inexpensive accelerators.

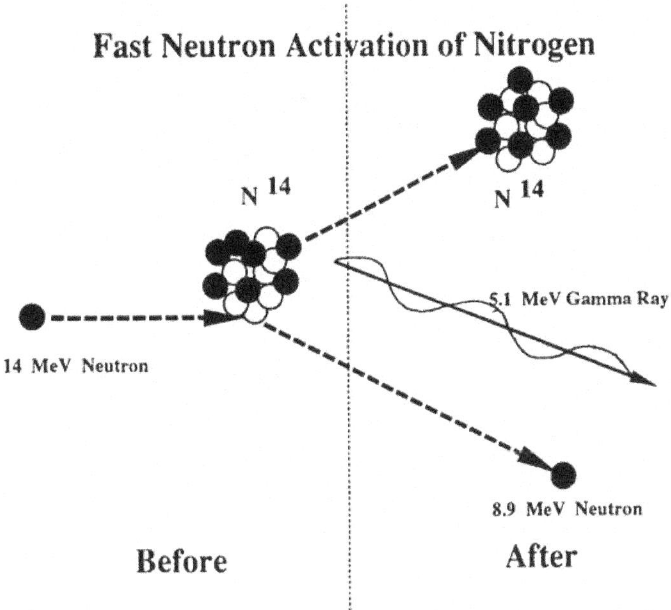

Figure 11. Cartoon showing the excitation of nitrogen as a result of being struck by a fast neutron. The excited nitrogen nuclei deexcite by emitting unique gamma rays.

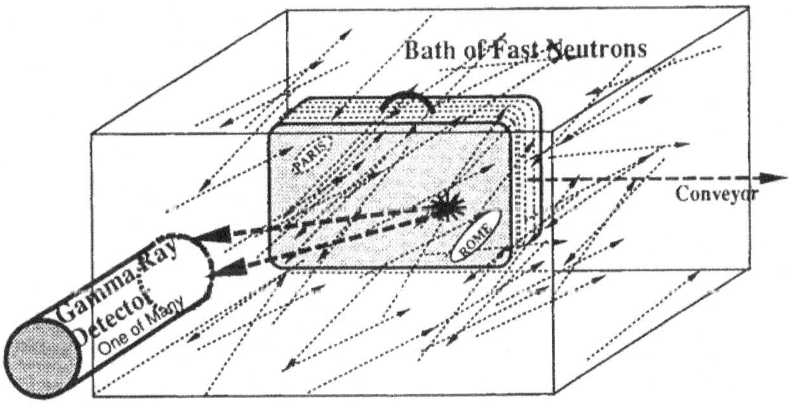

$$14\ \text{MeV n} + \text{O(16)} \longrightarrow \text{O(16)}^* + 7.9\ \text{MeV n};\quad \text{O(16)}^* \rightarrow \text{O(16)} + 6.1\ \text{MeV}\ \gamma$$

$$14\ \text{MeV n} + \text{C(12)} \longrightarrow \text{C(12)}^* + 9.6\ \text{MeV n};\quad \text{C(12)}^* \rightarrow \text{C(12)} + 4.4\ \text{MeV}\ \gamma$$

$$14\ \text{MeV n} + \text{N(14)} \longrightarrow \text{N(14)}^* + 8.9\ \text{MeV n};\quad \text{N(14)}^* \rightarrow \text{N(14)} + 5.1\ \text{MeV}\ \gamma$$

Figure 12. Schematic of Fast Neutron Analysis chamber.

The signals from a hodoscope of gamma-ray detectors (one is shown in Figure 12) are tomographically analyzed to give the spatial distribution of the nitrogen, carbon and oxygen. The method is a bulk-explosive detector similar to the successful TNA system (#1).

The presence of oxygen and carbon is revealed by strongly produced, unique gamma rays of 6 Mev and 4.4 Mev respectively; the cross sections for producing these gamma rays are about 200 millibarns. Nitrogen is more difficult to identify since the characteristic gamma rays are not so strongly produced and might be masked by other gamma radiations.

3. Pulsed, Fast-Neutron-Beam Analysis, PFNA. Fig. 13. Pulses of fast neutrons are collimated and sent through the luggage. The characteristic gamma rays from O, N, and C are sorted according to the position of the neutron beam. A three-dimensional image can be obtained if the gamma rays are correlated in time with the neutron pulse that produced them. A timing resolution of 1 nanosecond gives a depth resolution of 5 cm, for 14 Mev neutrons. Lower energy neutrons will have better depth resolution.

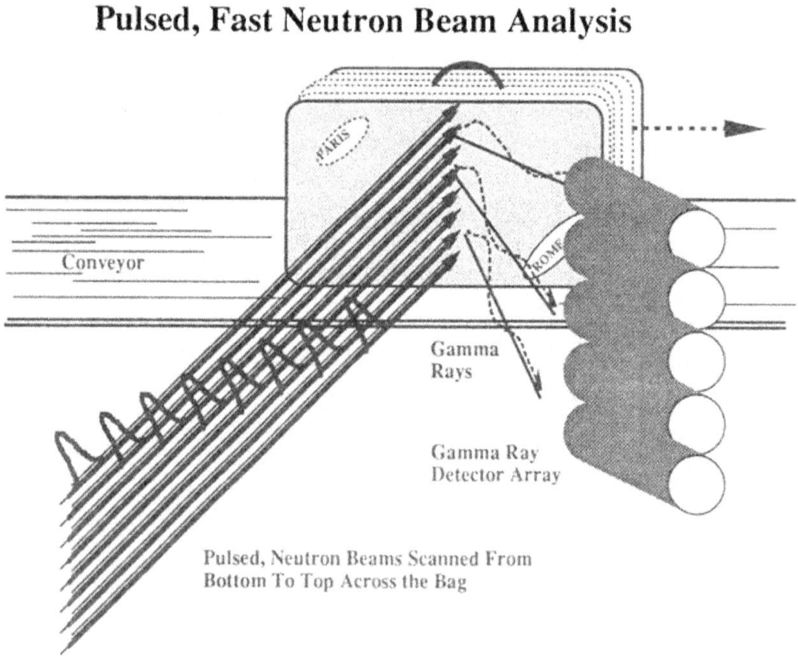

Figure 13. Schematic of Pulsed, Fast Neutron Analysis.

PFNA is an imaging technique. If the spatial resolution of PFNA can be made equal or smaller than that of the explosive, then the gamma ray lines observed from that voxel will contain only gamma rays emitted from O, N and C. Quantitative measures of the three constituents would give an unambiguous signature of the explosive.

Compact cyclotrons with natural pulse widths of the order of a nanosecond or less appear to be well-matched to the pulsed, fast-neutron, gamma technique.

4. Associated Particle Production This method has been worked on for decades. It has many features in common with method 3; a timed and directed neutron excites a nucleus, which is identified by its deexcitation gamma rays. The difference is that the timing and direction of the neutrons are determined by the kinematics of reaction of 100-300 kev deuterons on tritons.

$$0.2 \text{ Mev } {}^2\text{H} + {}^3\text{H} = 3 \text{ Mev } \alpha + 14.8 \text{ Mev n.}$$

The alpha particle and the neutron are emitted back-to-back in this exoergic reaction. The direction and timing of the alpha particle gives the direction and timing of the neutron. The distribution of characteristic gamma rays, measured in coincidence with the timed and positioned alpha particles, results in three-dimensional maps, one each for the C,N and O in the bag. In principle, this technique (and method 3) can yield a 3-dimensional map of every element that can be excited strongly by inelastically scattered neutrons. And most elements have such strong (n,n') cross sections.

The X,Y position of the alpha particles establishes the X,Y position of the nucleus that emitted the gamma ray. The energy of the gamma ray specifies the element. If the timing of the gamma ray with respect to the alpha particle can be determined to 1 nanosecond, then the depth (Z) can be determined to about 5 cm. The spatial resolution of this, and method 3, can, in principle, be less than the minimum threat explosives. In principle, this imaging technique (and method 3) should be well-adapted to finding sheet explosives.

5. Pulsed Fast/Slow Neutron Analysis, PFSNA. This method combines methods 1 and 2. Fast neutrons, without directionality, are produced in short bursts that last a few microseconds and are spaced about one millisecond apart. The gamma rays produced during the initial neutron burst come from the inelastic scattering of the fast neutrons; O and C stand out. The neutrons slow down between the beam bursts,and become thermalized. The presence of nitrogen is then signaled by the 10.8 Mev gamma ray. This method combines the effectiveness of fast neutrons for finding oxygen and carbon with the effectiveness of slow neutrons for finding nitrogen and hydrogen.

A variant of this method utilizes two independent sources each operating in its own cavity. A low-energy neutron beam is created to produce the thermal neutrons and the d,t reaction is used for the fast neutrons. The pulsing of the neutron-producing systems has several possible advantages: 1) Improved signal to noise in each system. 2) Increased information. 3) Longer-lifetimes for the neutron sources.

The next three techniques use gamma ray probes to cause nuclear reactions in luggage. The signatures are, by themselves, not unique to the explosive materials. We have reserved a separate section, IV, for gamma ray probes that produce fluorescent gamma rays whose energies characterize the nuclear species being excited.

6. Nitrogen 13 Production, Gamma Detection, NPG. This method, schematized in Fig 14 is being proposed to look for nitrogen by measuring the amount of the 10 minute

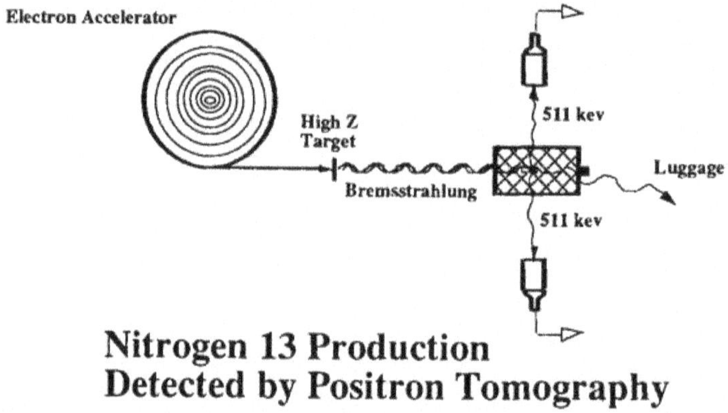

Figure 14. Schematic of the nitrogen 13 production method and analysis.

isotope, ^{13}N, formed when high energy photons are absorbed by ^{14}N.

$$^{14}N + \gamma = {}^{13}N \text{ (10 mins)} + n.$$

The ^{13}N is identified by the back-to-back annihilation radiation emitted when it decays to carbon.

The luggage is made radioactive, but the radioactivity of N^{13} is turned to an advantage by using one accelerator to rapidly irradiate many bags that are analyzed off-line and in parallel by positron emission tomography techniques.

This method must deal with fact that annihilation radiation is not a unique signature and other elements have much larger γ,n (and γ,p) cross sections that be difficult to deal with.

7. Nitrogen 12 Production, NPA This method, proposed and patented by Luis Alvarez, uses a short pulse of very high energy γs on nitrogen to make ^{12}N. ^{12}N decays in only 11 milliseconds, emitting a very high energy positron. 10 milliseconds after the production pulse has ended, the positron decay is detected by positron emission tomography.

$$>30 \text{ Mev } \gamma + {}^{14}N = {}^{12}N + 2p$$
$$^{12}N = {}^{12}C + 15.3 \text{ Mev } e^+$$

The method has clever features. The initiating beam is naturally collimated to about 1°. The short lifetime of positron annihilation is a unique identifier of nitrogen, and the short half life lends itself to scanning. Positron tomography gives good three-dimensional imaging.

Bremsstrahlung Photons In - Bremsstrahlung Photons Out.

Electron Accelerator

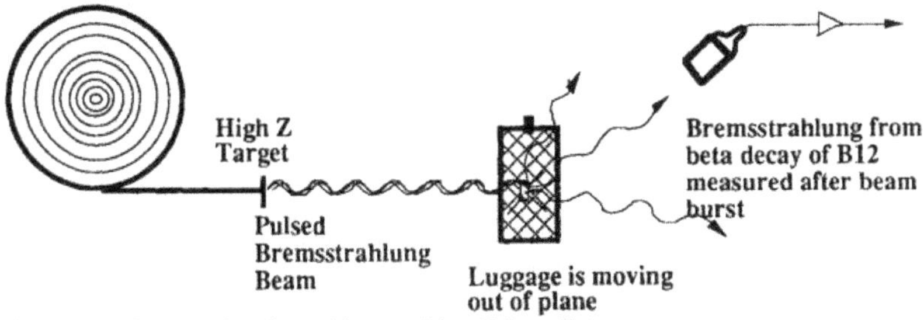

Figure 15. Schematic of the Alvarez-Plus-Others Scheme.

The scheme requires very high energy photons and has a very low production probability. The scheme is not operationally much different from method 10, which developed from it.

8. Boron 12 Production, BPA. Fig. 15. When measurements were made of the probability for making ^{12}N (Method 7) by the γ,2n process, it was found that the boron 12 production via the γ,2p process was some 100 times larger. ^{12}B also has a short lifetime, decaying in 20 milliseconds with the emission of a 13 Mev, high energy electron. The detection of the delayed bremsstrahlung, emitted by the high energy electrons, is being proposed as the signature of nitrogen and hence the explosive. One complication is that ^{12}B is also made from the γ,^{13}C (1% natural abundance) reaction. The basic method is shown in Figure 15.

9. Pulsed-Neutron Backscattering. PNB. Fig. 16. A pulsed beam of monoenergetic neutrons is scanned across the luggage. The intensities of the time-of-flight groups of elastically, backscattered neutrons yield the mass distributions inside the bag. The time-of-flight spectrum simultaneously measures the position of the target nuclei.

The probability for elastic scattering of neutrons is large, typically about 5 barns at resonances that are generally unique for each element. The signals from individual elements can be optimized by choosing the right neutron energy.

To effectively search luggage for explosives one should scan sequentially with energies optimized first for nitrogen, then for oxygen, and, if necessary, for carbon. In principle, this technique results in a three-dimensional map of the elemental distributions in the luggage. To obtain such images rapidly will probably require advances in the technologies of accelerator design and neutron time-of-flight counting.

Resonance Scattering of Neutrons

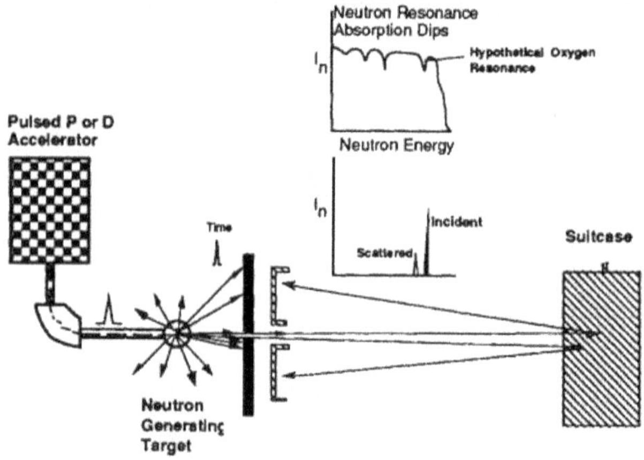

$$E\,(\text{backscattered neutron}) = [(A-1)/(A+1)]^2\; E\,(\text{Incident neutron})$$

$$v\,(\text{backscattered Neutron}) = [(A-1)/(A+1)]\; v\,(\text{Incident Neutron})$$

Figure 16. Schematic of the Resonance Backscatter method using pulsed monoenergetic neutrons.

10. Pulsed-Neutron Absorption. PNA. Fig 17. This is the absorption complement to the scattering technique, Method 11 above. A broad energy spectrum of pulsed neutrons is created. The elements in the path of the beam absorb those neutrons whose energies correspond to the characteristic neutron resonances of the elements. The dips in the intensity spectrum of the neutrons that pass through the bag, measured as a function of the position of the beam, results in a picture of the elemental distribution in the bag. The method has been described by Professor J. Overley of the University of Oregon[1]; It has not yet been applied to the airport security problem.

11. Nitrogen 13 Production, Neutrons. N13n. This method begins like that of method 8 above; high energy photons transmute nitrogen into ^{13}N. The ^{13}N is identified, however, by the neutron that is emitted in the direct reaction. The uniqueness of the neutron as a signature for identifying nitrogen is not clear.

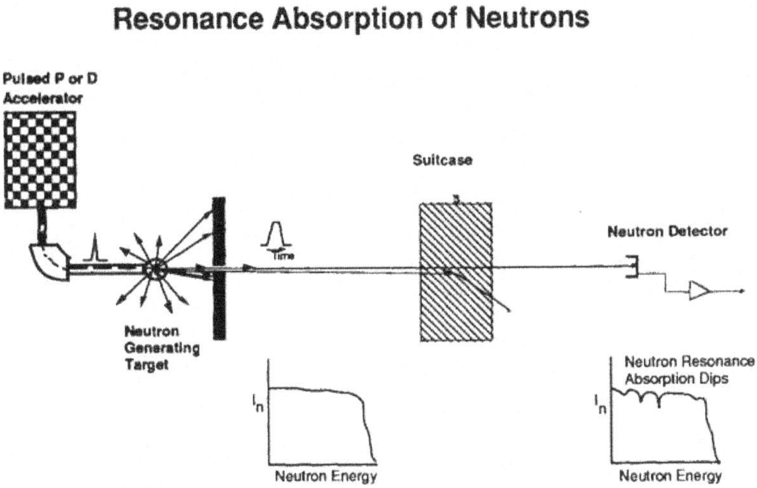

Figure 17. Schematic of the Resonance Neutron Absorption method using pulsed neutrons that have a flat energy spectrum.

IV. Resonant Fluorescence Methods

The study of the interaction of photons with atomic nuclei is one of the fundamental branches of nuclear physics. Resonances can occur when the energy of the photon matches that of one of the nuclear states. The importance of resonance excitation for the study of nuclear states was pointed out long ago by Schiff.[2] An excellent general reference is F. Metzger, Resonance Fluorescence in Nuclei.[3]

The probability for excitation is given by the Breit-Wigner formula:

$$\sigma_o(E\gamma) = (\lambda^2/4\pi)\frac{\Gamma\Gamma_o}{[E\gamma-E_o]^2 + \Gamma^2/4} \qquad 1)$$

where σ_o is the cross section in cm^2, λ is the wave length of the photon in cm., Γ and Γ_o are, respectively, the total width of the state and the partial width for deexcitation to the ground state, E_o is the energy of the eigenstate and $E\gamma$ is the energy of the photon; spin terms have been omitted.

The maximum cross section, σ_o, occurs when the photon energy is exactly on resonance ($E\gamma - E_o = 0$).

$$\sigma_o(E\gamma = E_o) = \lambda^2/\pi\frac{\Gamma\gamma}{\Gamma} = \frac{5 \times 10^3}{E_o{}^2}\frac{\Gamma\gamma}{\Gamma} \qquad (2)$$

where σ_o is in barns (10^{-24} cm^2) and E_o is in MeV, and spin dependent terms have been omitted. The ratio $\Gamma\gamma/\Gamma$ measures the strength with which the nuclear excited state is connected with the ground state. If the excited state only decays to the ground state then $\Gamma_\gamma/\Gamma = 1$. In that case a 10 Mev state can have an absorption cross section of order 50 barns, which is 100 times the electronic stopping cross section for 10 Mev photons in light materials; the mean free path in bomb material will be of the order of 1 centimeter.

Low-lying nuclear states are typically very narrow; e.g., less than 1 ev wide. To photo-excite them one needs photons that have the precise energy of the resonance plus extra energies needed to account for the energy shifts that are the inevitable consequences of interactions. There are three general sources of photons that have been used for the resonant excitation of high-energy nuclear states: A. Bremsstrahlung from an X-ray source. B. A broad-energy source of gamma rays produced in a nuclear interaction. C. Resonant gamma-rays produced in a nuclear reaction.

Method A. Fig. 18. The bremsstrahlung probe method has been used by F.D. Seward[4] to study the 15.1 Mev state in ^{12}C. The bremsstrahlung photons, created by a 20 Mev linear accelerator, are naturally focussed into a cone of half-width angle, $\theta = \frac{mc^2}{E_e} = 1.4^\circ$. The fluorescent detector was at 90° to the incident photon beam.

Seward was able to observe the resonantly excited 15.1 Mev state well above the background but objections may be raised to the placement of a high-energy X-ray machine in an airport environment.

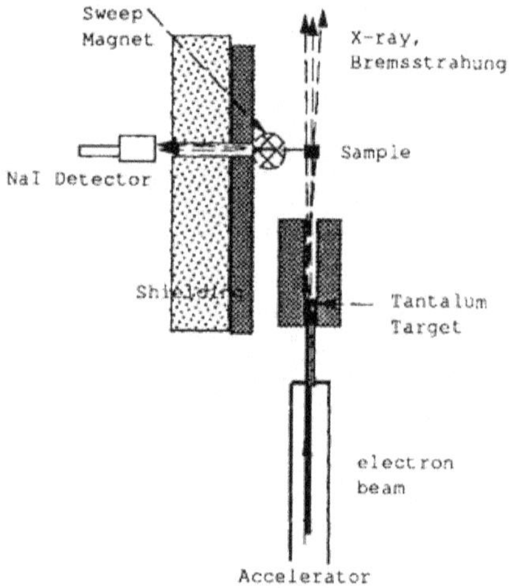

Figure 18. Seward's geometry.

Method B. The 15.1 Mev state of ^{12}C has also been studied extensively using the gamma rays generated by the $^7Li(p,\gamma)^8Be$ reaction.[5] The $^7Li(p,\gamma)^8Be$ reaction is exoergic by 17.25 Mev and a low-energy proton beam (about 470 kev) generates gamma rays in the range from 14 to 16 Mev with an efficiency of about 10^{-8} photons per proton.

Method C. The third resonance-fluorescence method uses the resonance gamma ray itself as the resonant probe. The method is especially practical for nitrogen where it has been exploited for bomb detectors by groups in Israel and LASL. The physics and basic technology is well-documented in the literature.[6]

The state in 14N at 9.172 Mev can be excited by the direct, resonant capture of protons of 1.7476 Mev.

$$1.7476 \text{ Mev protons} + {}^{13}C \rightarrow {}^{14}N^* \text{ (9.172 Mev)}$$

The state has a width of only 140 electron volts. If the energy of the proton is too high (e.g. 1.749 Mev) or too low (e.g. 1.745 Mev) the resonance will be missed and the state will not be excited.

The excited nucleus is created with a kinetic energy of 125 kev as a result of the momentum imparted by the 1.7476 Mev protons. Gamma rays emitted from the moving $^{14}N^*$, will be Doppler shifted according to the well-known formula:

$$E\gamma = E\gamma^{0}(1 + \frac{v}{c} \cos \theta), \tag{3}$$

where $E\gamma^{0}$ is the energy of the emitted gamma ray and θ is the angle between the directions of gamma ray emission and the recoil. Finally, one must take into account the kinematics of the emission and absorption of the 9.172 Mev gamma ray. In these processes, the nucleus recoils taking up an energy, E_R, given by:

$$E_R = \frac{E\gamma^2}{2M_o c^2}. \tag{4}$$

The consequences of these kinematics is that 9.172 Mev photons emitted at 80.2° with respect to the proton beam have precisely the correct energy to cause resonance absorption in nitrogen. Figures 19 and 20 sketch out the process described in the next paragraphs.

A beam of 1.7476 Mev protons strikes a thin target of ^{13}C creating ^{14}N in the resonant state at 9.172 Mev. The state deexcites emitting gamma rays whose energy are monotonic functions of the angle of emission. Those gamma rays emitted in a 0.5° wide cone centered at about 80.2°, are in "resonance." Gamma rays emitted in all other angles are not.

The 9.172 Mev gamma ray excites the ^{14}N in the luggage. One can, in principle, detect the absorption by measuring the presence of deexcitation gamma rays; that is, by resonance fluorescence. Unfortunately, 95% of the time the 9.172 Mev state decays through the inverse reaction that produced it. It is, therefore, more effective to measure the resonance absorption by the direct measurement of the reduction in the photon flux. The Israelis measure the absorption with a resonant detector, sensitive only to resonant gamma rays. The LASL group relies on highly-efficient, non-resonance gamma ray detectors, shown in Figure 19, to measure the reduction of the signal strength when the resonant gamma rays pass through nitrogenous materials.

Method C has a number of inherent merits for nitrogen: The low-energy accelerator requires little shielding and produces no radioactivity. The large cross section makes the method sensitive to small amounts of nitrogen. The high-energy gamma rays have exceptional penetrating power so that the technique can be used for quite large containers. The technique produces both a normal X-ray radiography of the projected density distribution and a nitrogen radiograph of the projected nitrogen distribution in the bag. The major disadvantage of the technique is the low gamma-ray production probability. But this limitation does not appear to be fundamental.

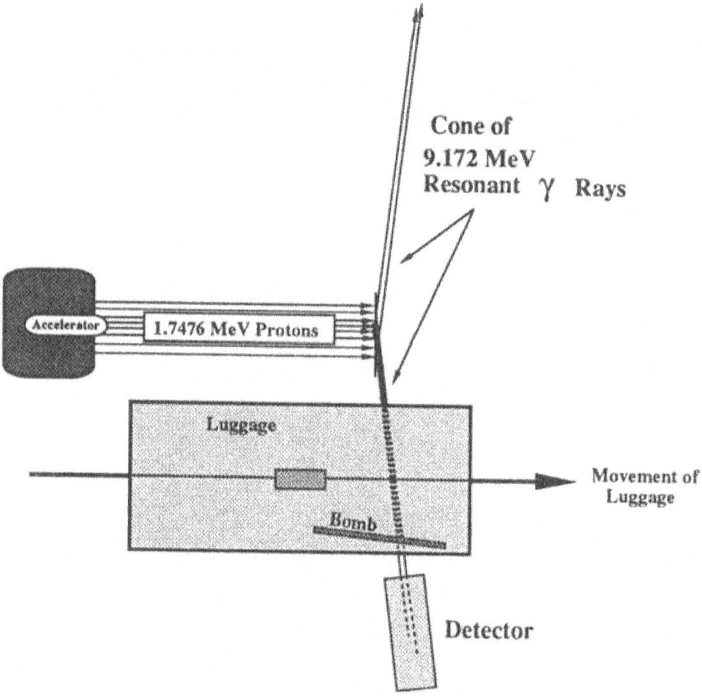

Figure 19 Schematic of the RAA method showing the special γs, produced by 1.74 Mev protons , passing through the bag to the detector.

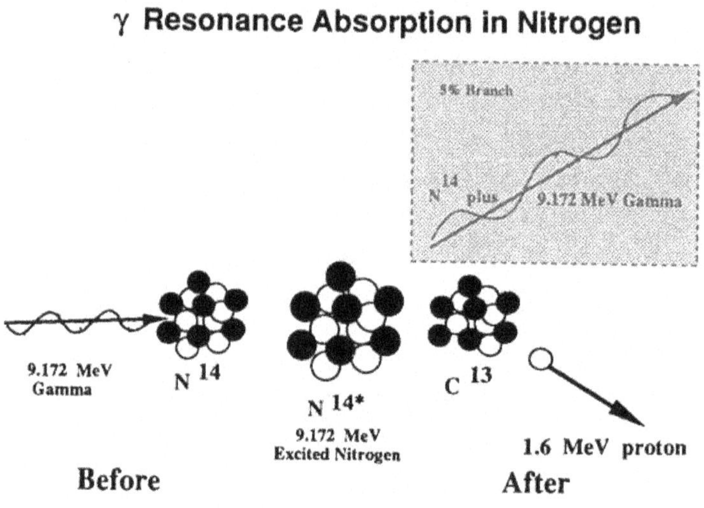

Figure 20 Cartoon showing the interaction of 9.172 Mev γs with ^{14}N.

We close this section by noting that there are other variations of the three resonant techniques described above. For example, the 6.917 and 7.117 Mev states of ^{16}O have been studied extensively by resonance scattering and absorption of these gamma rays produced with large cross sections in the reaction, $^{19}F(p,\alpha)^{16}O$.[7] The 6.9 and 7.1 Mev gamma rays produced in this reaction are strongly Doppler shifted so that only a small fraction have the correct energy for resonance absorption. However, those photons that have that correct energy have absorption cross sections of 150 to 250 barns; that is, mean free paths of only a few millimeters of explosives. To exploit this favorable absorption cross section requires the development of an efficient resonant absorption detector.

SUMMARY

In the past few years we have seen rapid progress in the technological means for securing airplanes against terrorist's bombs. At this time, the most powerful of the techniques use nuclear probes. This paper summarizes the basic physics of each with emphasize on the resonant gamma ray techniques. The deliberate omission of any critical examination of either the merits or shortcomings of any of the techniques stems from our belief that, in many cases, it may be too early to see all the faults or the virtues; clever antagonists will find subtle faults, clever protagonists will uncover new virtues. We note only that some of the most promising technologies are ready for prototyping, after which they can be beta-tested for final evaluation. I am confident that several of the nuclear techniques will pass their tests successively and that in a relatively few years we will have the technological means for rapidly and reliably detecting small and thin explosives in airplane luggage.

The author is pleased to thank the Federal Aviation Administration for their support of this work.

REFERENCES

1. Overly, J. Nucl. Inst. and Methods B24/25, 1058-1062 (1987) and references cited therein.
2. Schiff, L.I. Phys. Rev. 70, 761 (1946).
3. Metzger, F. Resonance Fluorescence in Nuclei. Progress in Nuclear Physics Vol 7, 54 (1959).
4. Seward, F.D. Phys. Rev. 125, 335-340 (1962). [See also, Garwin, E.L. Gamma-Ray Excitation of the 15.1 MeV Level in C^{12}, Phys. Rev. 114, 143 (1959).]
5. Hanna, S.S. and Segel, R.E. Fluorescence of the 15.1 Mev level of ^{12}C Proc. Roy. Soc. (London) A 259, 267 (1961) and Schmid H. and Scholz, W. Resonanzstreuung am 15.1 Mev-Niveau in C^{12}, Zeits fur Phys 175, 430 (1963).
6. Hanna, S.S. and Meyer-Schutzmeister, L. Resonant Absorption by the 9.17 Mev Level in N^{14}. Phys. Rev. 115, 986 (1959). Biesiot, W. and Smith, Ph.B. Phys. Rev. C24, 808 (1981).
7. Swann, C.P. Nucl. Phys. A150, 300 (1970) and references therein.

Radiometry of Heavy Mineral Sands as a Tracer of Coastal Erosion

M. B. Greenfield
Div. of Nat. Sci., ICU, Mitaka, Tokyo 181, Japan
and Phys. Dept., Fla. A. and M. Uni., Talla., Fl., USA

R. J. de Meijer, H. M. E. Lesscher and L. W. Put
KVI, Rijksuniveristeit, Groningen, TheNetherlands

J. F. Donoghue
Dept. of Geology, Fla. St. Uni., Talla., Florida, USA

and
J. Wiersma
Rijkswaterstaat, Rijkswijk, The Netherlands

Abstract: Changes in the natural radiation level concurrent with changes in elevation have been measured on the beaches of the Northwest Gulf of Mexico, Florida, USA, The North Sea Coast of The Netherlands, and preliminarily the Pacific Coast of Japan. A negative correlation between the log changes of the gamma radiation levels and changes in beach elevations has been observed. This is due to the absorption of γ activity by quartz sand deposited over the radiogenic heavy mineral sand. It is suggested that radiometric mapping of coastal regions on a large scale may be a sensitive, practical and efficient way to monitor the extent and degree of coastal changes.

Radiometry of Heavy Mineral Sands as a Tracer of Shore Erosion

I. INTRODUCTION

The γ-radioactivity originating from trace amounts of uranium, thorium and their radiogenic daughters found in heavy mineral sands may be used as a sensitive method for both tracing and quantifying sand transport processes. Areal distributions of this activity observed on beaches changes according to the amount of wind-driven, low-activity sand covering the underlying radioactive placer deposits. The absorption of γ–radioactivity explains the observed negative correlation [Gre89].

The γ-radioactivity from radiogenic isotopes occurring in minerals such as rutile, zircon and monazite, is a good locator and quantitative indicator of heavy mineral deposits. Monitoring the changes in γ-radioactivity as a means of tracing sand transport mechanisms has been reported on the beaches of the Northwest Gulf of Mexico, USA [Don88], on the North Sea Coast of the Netherlands [Gre89,Les89]. Observations of these γ-radioactive placer deposits on Little St. George Island, USA and Texel, The Netherlands using grid sizes on the order of hundreds of square meters have indicated local concentrations on the order of tens of meters, were built and then destroyed by the action of wind and waves within days or weeks.

The natural logarithm of the changes in γ-radioactivity, due primarily to wind-driven sand transport, has been shown to be negatively correlated with changes in elevation. This may be explained by a relatively simple model based upon the absorption of gamma-radiation by inactive quartz sand as it accumulates or erodes away from existing deposits [Gre89] and summarized below.

In addition to 1000 m^2 grids, a larger grid of 120mX200m on Texel, NL, was also monitored for several months, not only to observe the morphology of the transported sand, but also to determine the most efficient way to continue such observations on land, and in future the sea bottom, and perhaps measurements from the air. Using a linear grid spacing of 4m and then successively increasing that resolution, the optimal resolution, balancing accuracy and efficiency, was estimated to be about 10-20m.

All of the above locations were selected for the study of sand transport phenomena because of their dynamic nature as well as the locally occurring, high levels of natural radioactivity. This activity gives rise to γ-radiation count rates on the order of from 10 to 100 counts per sec. in portable, field monitors. Employing small, hand-held monitors enables easy, *in situ* monitoring of the spatial extent and movement of the activity and associated heavy mineral sands. Although the rapid migration of heavy mineral sands is a well known phenomenon [Don89], it is difficult to quantify such processes. A data base of measurements of γ-radioactivity on the barrier islands and neighboring shoals of the Gulf of Mexico, North Sea and Pacific Coast of Japan, is enabling the establishment of more definitive and quantitative relationships between changes in activity and the movement of those sands (see [Gre90] and references therein.

Since the existing activity in naturally occurring placer deposits of heavy minerals is employed, the presence of heavy mineral placer deposits might seem to be a limiting feature to the global applicability of this technique. On the other hand, it is known that such concentrations of heavy minerals tend to occur in high energy regions of the coastal zone, precisely where it is most interesting and relevant to make such measurements.

II. EXPERIMENTAL METHODS AND OBJECTIVES

A. Methods

1. Gamma-ray detection: Field measurements were obtained with a Scintrex model GIS-5, portable, solid-state, integrating spectrometer. This spectrometer contains a 4.1 x 4.1 x 5.0 cm^3 thallium activated, sodium iodide - NaI(Tl) - crystal. The scintillated light is proportional to the absorbed gamma-ray energy which is then converted to an electrical impulse by an attached photo-multiplier tube. Since the output pulse height is proportional to the energy of the absorbed photon, determination of various gamma rays is facilitated by adjusting a threshold level. Naturally occurring radioactive isotopes of potassium, uranium and thorium and their radioactive decay products, found in detrital heavy minerals, produce the majority of the gamma radiation observed in the field.

2. Background: Estimates of background radiation from cosmic rays were obtained by taking readings over water with a depth of greater than 5 meters (nearly all of the gamma radiation from the underlying sediments are absorbed in a few meters of water). The activity from the water itself is at least two orders of magnitude less than that from the sediment. Although specific activities from potassium, uranium and thorium and their associated radioactive decay products may be measured in the field, the total counts, inclusive of all gamma-ray energies above a threshold of about 0.05 MeV, are usually indicative of the presence of heavy minerals and were therefore used consistently in these measurements.

3. Detection geometry: The total gamma-ray activity at a distance of approximately one meter (hip-height) was integrated over a ten second interval for each grid point. For a thin surface of uniform activity and effectively infinite horizontal extent, it can be shown that the count rate observed in the detector is roughly independent of the distance above that surface activity. In cases where the activity varies horizontally over distances similar to the distance between the detector and the active surface,

the activity recorded varies with the height that the detector is held above the surface.

4. Determination of detector height: To determine the optimal detector height above the ground for typical gamma-source distributions observed in the field, the observed activity was measured as a function of the height that the detector was held above the ground. In Table 1, the number of integrated counts observed in three typical situations was recorded for four distinct detector geometries - on the ground, 1 m above the ground, 4.5 m above the ground and averaged over a circle of radius 6 m around the original data point. In the region where the count rate was relatively flat (i.e. a uniform planar distribution of radiogenic heavy minerals), the counts recorded were independent of detection geometry within statistical errors of the measurement. In a second region where the count rate was higher and more variable over horizontal position, the count rate varied as a function of height as expected. In a third region the variation of count rate with horizontal position was about an order of magnitude, but once again the detector position of 1 m above the surface gave a reasonable estimate of the measurements averaged over a circle of 6 m radius. In conclusion we see that the activity observed with the detector held 1 m above the ground gives a reasonable estimate of the averaged activity for that grid point and is adopted throughout the measurements described herein.

Table 1: γ count rates for different heights and locations.

Height of detector	Flat activity	High activity	Variable activity
ground	1391	7912	4675
1m	1350	7512	5352
4.5m	1330	6323	4963
1m averaged circle of 6m	----	7923	5747

5. Elevation measurements: Elevations were measured at the same time employing transit and stadia rods. The details of both radiometric and elevation measurements have been described in detail elsewhere [Gre89].

B. Objectives

The general objectives in these measurements were to: 1. observe the presence of gamma radioactivity associated with highly concentrated placer deposits of heavy mineral sands in regions where there was significant energy from wind and water; 2. to attempt to quantify the changes in that observed activity and correlate them with the corresponding changes in beach elevation due to sand transport phenomena; and 3. to investigate the feasibility of applying these techniques to the large scale monitoring of shore erosion. In order to accomplish these general objectives the following has been attempted:

- mapping of gamma radioactivity at several sites on the order of 1000 m^2 along the Florida Gulf, Dutch North Sea, and Japan Pacific Sea Coasts;
- monitoring (monthly) the difference in activity on these grids over time (about six months to a year);
- monitoring the changes in beach elevation during the same periods;
- attempting to quantify changes in radioactivity and correlate them with changes in beach elevation;
- ascertaining the optimum resolution, grid spacing, overall grid dimensions and general feasibility for large scale measurements (on grids of about 20,000 m^2);
- mapping beach profiles up to and over these grids;
- taking sand samples of about 1 kg at each location for laboratory γ-ray spectroscopy and mineralogical analysis (details of sample analysis are found in [Gre89]).

Measurements attempting to quantify shore erosion through radiometric means described herein have taken place over a period of approximately three years on three continents beginning in 1988 in Florida, 1989 in the Netherlands and continuing through June of 1990 in Japan.

III. QUALITATIVE RESULTS

A. Small Scale (1000 m^2) Grids and Features Observed

The gamma-radioactivity and corresponding beach elevations were simultaneously measured with 2 m linear spacing. Sites were selected which contained placer deposits or significant concentrations of heavy minerals. Such deposits may be visually identified by their darker appearance. Grids of approximately 1000 m^2 inclusive of such deposits were established, designated with beach poles at each corner and measured at approximately monthly intervals. Three such sites on the North Sea Coast of the island of Texel are shown in Figure 1.

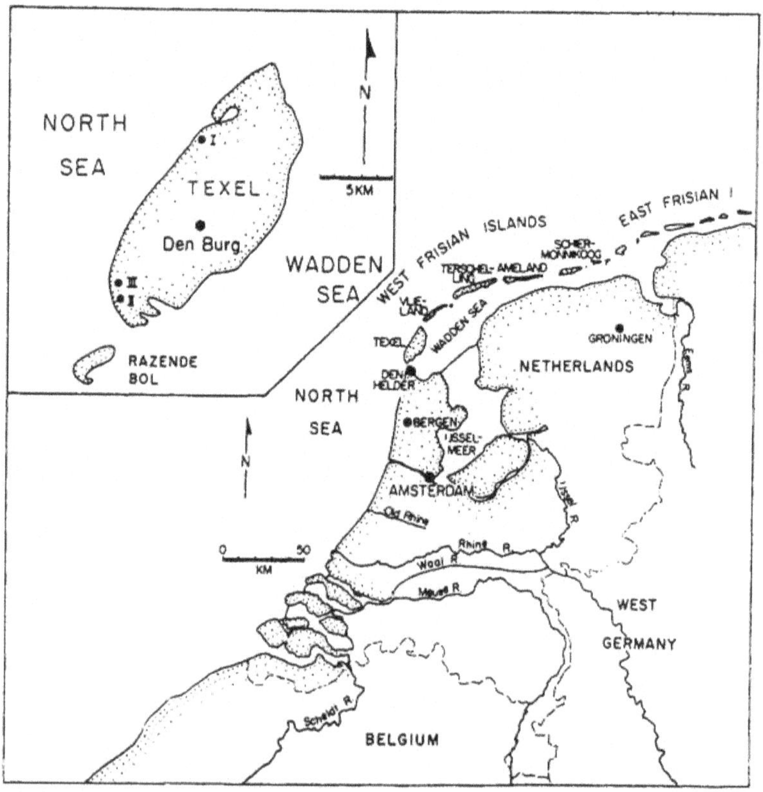

Fig. 1. Regional map of The Netherlands from the Belgian to German borders with inset of Texel showing measurement sites.

Figure 2 shows both elevation and radiometry grids measured over a three month period at site I. The trends shown are consistent for the two other sites on Texel as well as earlier measurements on the Gulf of Mexico Coast of Little St. George Island.

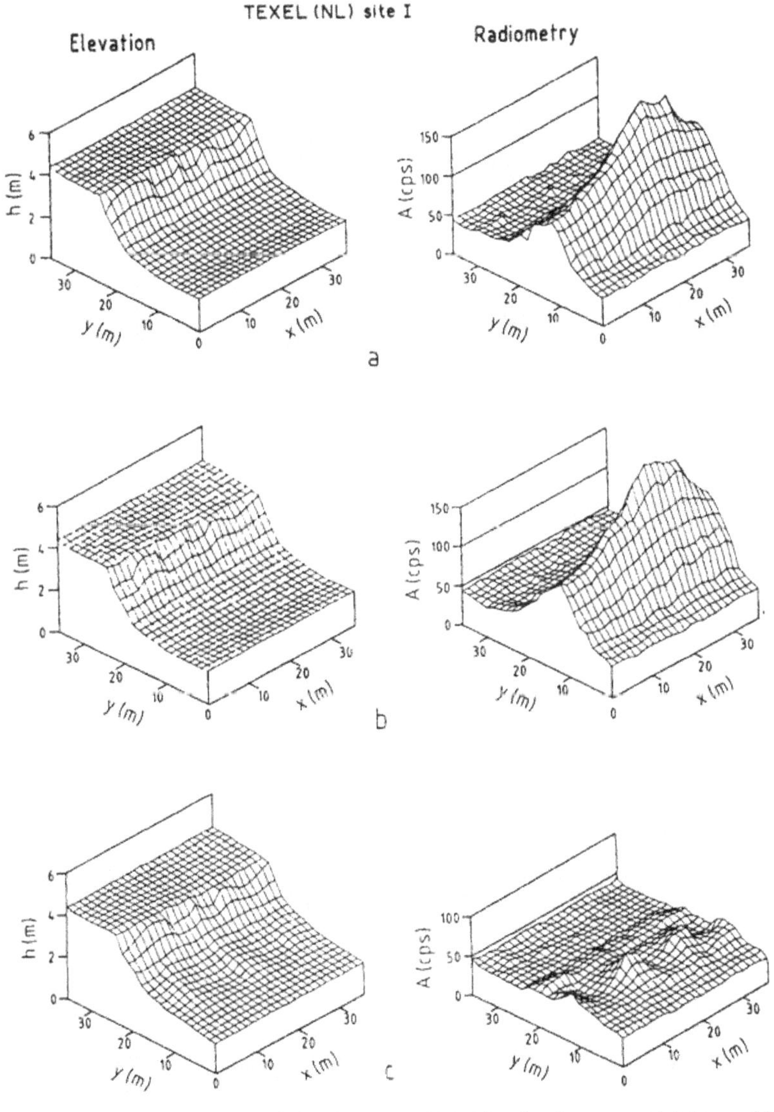

Fig. 2. Three-dimensional representation of elevation (above N.A.P.) and radiometry grids taken on 8 June (a), 29 June (b) and 27 July 1988 (c) for site I on the island of Texel, The Netherlands. The orientation of the grid is such that the x-axis is parallel to the shore.

The left side of Figure 2, designated "Elevation", indicates a 2m berm which is the remnant of an artificial renourishment program and which serves as a baffle for wave and wind action. Erosion selectively concentrated heavier grains at the base of the berm creating a heavy mineral concentrate and enhanced radioactivity. This was clearly observed in the earlier radiometry measurements labeled (a and b). In the later measurement aeolian activity had covered much of the area seaward of the berm with a 20 cm thick layer of sand. This was evident not only in the elevation measurements, but by digging down 20 cm below the surface and observing that the underlying sand indeed had the dark laminae characteristic of heavy mineral concentrations.

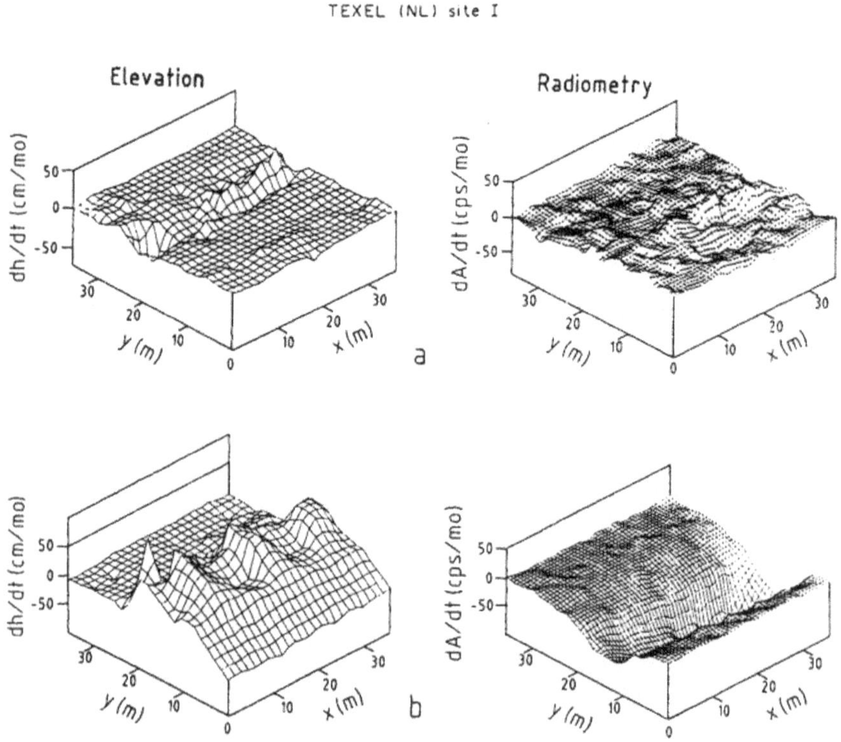

Fig. 3. Difference grids showing changes in elevation and radiometry for site I (differences between successive pairs of dates in Fig. 2). For the orientation see Fig. 2.

Changes in elevation, barely perceptible in the elevation grid, show up as negatively correlated and amplified changes in the radiometry grids. In the period between the first two measurements, with relatively light winds, the winnowing of quartz and the subsequent concentration of heavy minerals occurred leading to a slight increase in count rate. In the period between the second and third measurements, winds reported up to 60 km/h, transported sand over the existing peak. The new sand drifted across the site producing dunelets with approximately 8 m wavelength and amplitude of nearly 0.5 m. In areas where the new sand accumulated the observed count rate decreased. This effect is illustrated with even greater sensitivity in the difference grids (see Figure 3) showing both changes in elevation and corresponding negatively correlated changes in radioactivity. The new sand is mostly quartz, which covers the former surface enriched in heavy minerals and thereby absorbs the γ radiation emitted by those minerals. The changes in activity negatively correlated to changes in elevation may be used as a highly sensitive monitor of aeolian sand transport.

These preliminary results are remarkable in the degree to which the change in elevation corresponds to a decrease in count rate. As will be shown, strong reduction of the radioactivity is correlated with the covering of existing heavy mineral concentrations by sand of lower activity. Similar surveys in Florida [Don89] and preliminary observations in Japan [Gre90] reveal locally concentrated deposits of heavy minerals which are presently being monitored for this now familiar pattern.

B. Large Scale (>20,000m^2) grids

Large scale grids were monitored for longer periods and with varying grid space resolution. The purpose of these measurements was not only to observe, classify and monitor the radiometric features present and improve our understanding of the sand transport mechanisms involved, but also to determine the most efficient way that these features might be monitored on land and perhaps in the future along the nearshore and by airborne detection systems. We selected a site, adjacent to Site II on the southwest shore of Texel (see Figure 1) for investigation of the changing features produced by wind and water driven sand transport. This site was appropriate for many reasons. Firstly, this portion of the coast of Texel and its nearshore is a region of longstanding interest to coastal engineers, geologists and sedimentologists (see [Sha89] and references therein). Wiersma and Van Alphen [Wie88] noted a net landward (eastward) transport through the Texel inlet to the Wadden Sea. In a study of sand transport patterns in the ebb-tidal delta off Texel inlet, Sha determined: "the sand is transported in both directions parallel to the coast by tidal currents, predominantly northwards and landwards. Some sand is deposited where a weak rotational tidal current pattern is produced north of the inlet by the interruption of tidal currents parallel to the coast by the inlet" [Sha89]. The 120m x 200m beach site which we chose to monitor is occasionally awash during storm surges, subject to predominantly south-southwest winds and most certainly the long-term recipient of much of the above mentioned transported sand. In addition this site, being somewhat removed from the more heavily used recreational portions of the beach was less prone to disturbances from human activity. Although the perturbing influences of human activity, including our measurements, may not entirely be ignored, they are at least minimized at this site.

Recently de Meijer et al. [deM90] determined that the heavy mineral content of sediments deposited along the shores of the northwest coast of North Holland and the West Frisian Islands seem to consist of two distinct radiometric groupings. The heavy minerals found to the south of den Helder seem to have roughly half the specific activity of those found to the north of den Helder (see Figure 1). This difference, which may be due to the different provenance of the sediments, makes this area of even greater interest.

Monthly visits to Texel from February through the end of July 1989 were carried out in order to take radiometry and elevation data with a 20m grid spacing and to establish whether that 20m grid spacing leads to loss of essential information compared with the 2m spacing on small grids.

The first goal was accomplished by measuring a grid of 77 points with a grid spacing of 20 m and a 10 second integration time for each monthly measuring period. The data obtained were analyzed using graphics software [GG87], for the purpose of observing dominant features and how they changed over time (for details see [Gre90]). The second goal was accomplished by measuring the same area with l58l points, a linear grid spacing of 4 m and a 10 second integration time. In Figure 4 a radiometry grid from Texel, NL, with 4 m grid space resolution is shown.

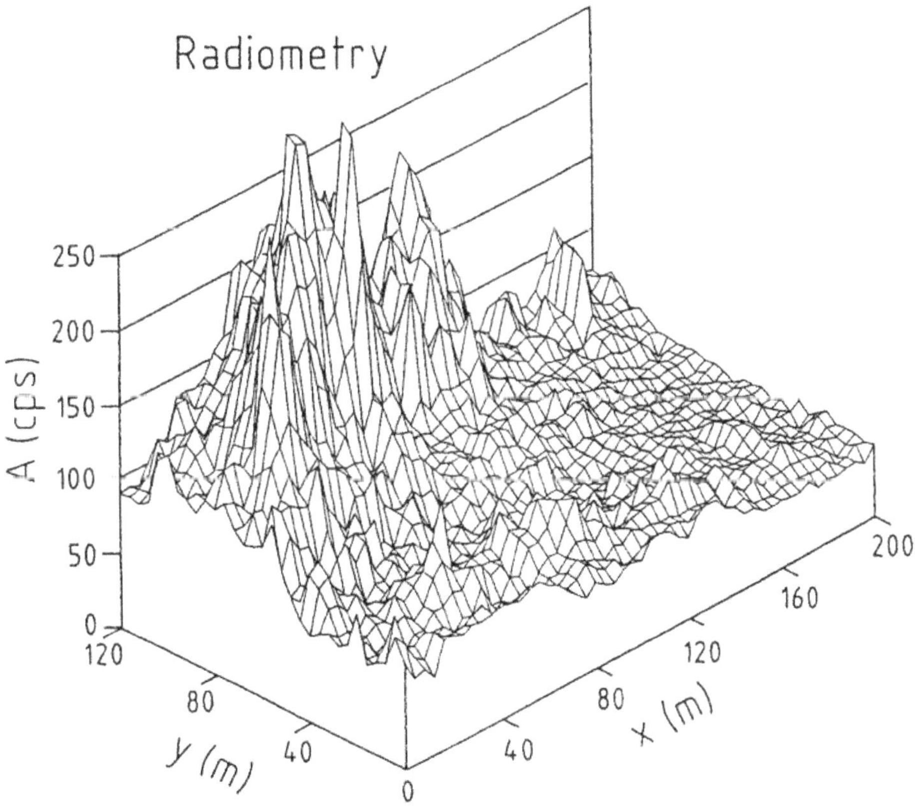

Fig. 4. 120m x 200m radiometry grid with 4 meter grid spacing.

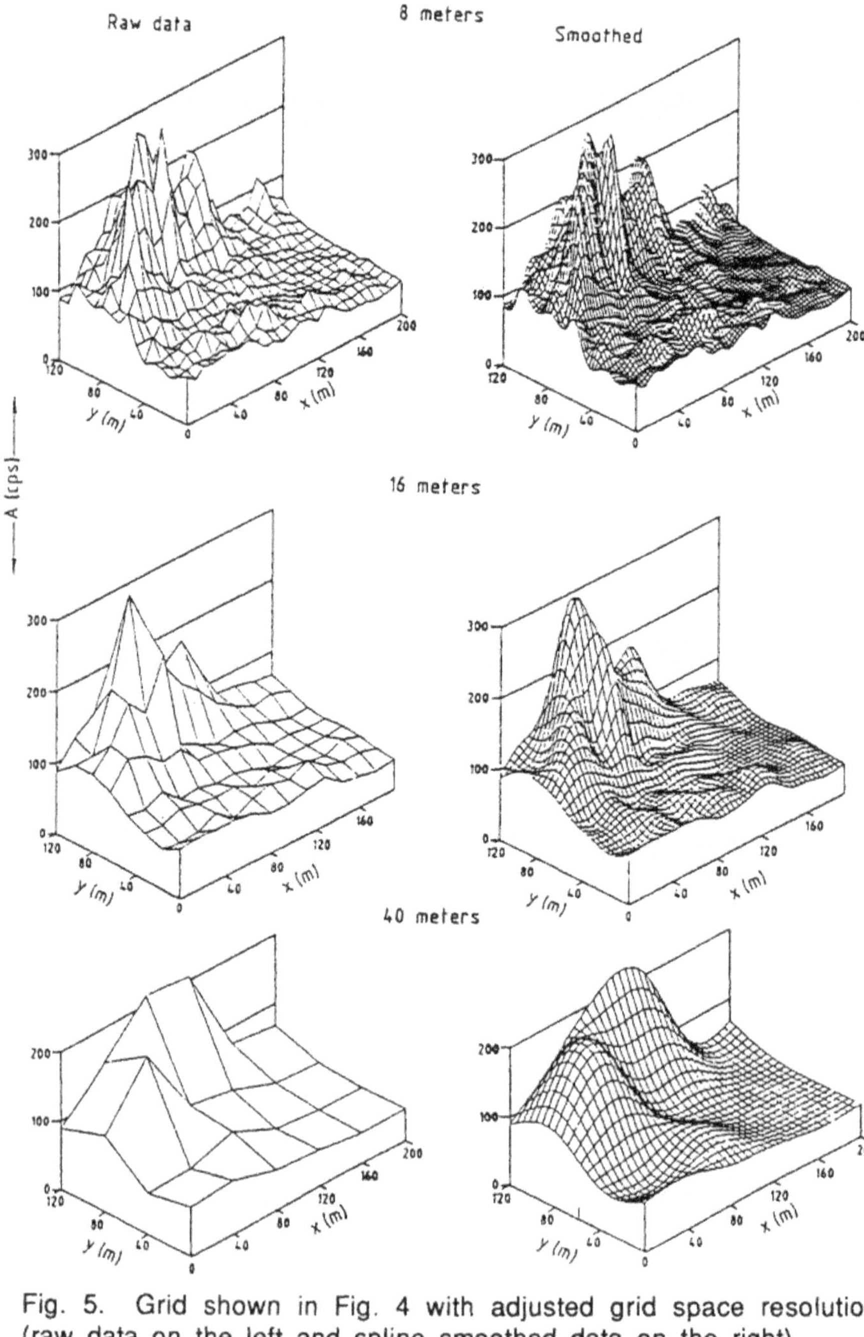

Fig. 5. Grid shown in Fig. 4 with adjusted grid space resolution (raw data on the left and spline smoothed data on the right).

Features with characteristic wave lengths of from 10-50 meters may be seen. A finer structure on the order of 4-8 meters, consistent with the previous small scale measurements is also evident. In Figure 5 the same data with successively poorer resolution (two, four and ten times the linear grid spacing corresponding to 1/4, 1/16 and 1/100 of the resolution) are shown. The raw data are shown on the left hand side while the same data smoothed with a spline smoothing algorithm [GG87] is shown on the right hand side. The dominant features in Figure 5 are reasonably represented with grid spacing 2 times that of the original data, but become successively washed out and distorted for poorer resolution (4 and 10 times the original grid spacing).

These results are of interest in planning for subsequent undersea or aerial measurements. We are planning to look at larger tracts and at the near shore in future measurements employing techniques previously described in the literature [Tho83], [Mil73], and [Sum70]. Underwater and aerial measurements may not be feasible with grid spacing of less than 10m due to inherent errors.

From our preliminary measurements we conclude that some features may be located with a linear resolution on the order of 40m. A clear representation of the shape and magnitude of the dominant features requires a resolution of from 8 to 20 m. The implications for large scale measurements are that a 40 m linear spacing, which is feasible over large surface areas may be used to locate dominant features. After such a preliminary survey, finer grid spacing with better resolution might be employed. It still remains uncertain whether sea-floor and aerial measurements will be able to obtain such resolution.

IV. QUANTITATIVE ANALYSIS

A. Absorption of Radiation in Sand

A simplified absorption model has been established [Gre89], wherein the measured intensity of radiation is exponentially reduced by an absorber of a known thickness and density. The relationship between the activity emitted, I_o, and that actually observed, I, above a surface layer of homogeneous sand covered by an absorbing layer of homogeneous sand of thickness d and specific activity, I_1, may be given by (for further details see [Gre89]):

$$(I - I_1)/(I_o - I_1) = e^{-\mu d} \qquad (1)$$

where μ is the effective linear absorption coefficient for γ radiation in sand. The results of an experiment in which a flat surface layer of radiogenic heavy mineral sand was covered with successive layers of lower activity (primarily quartz) sand are shown in Table 2.

Table 2. Change in elevation, count rate and deduced absorption coefficient for an experiment in which "light sand", with I_1 = 1169 counts in 30 s, was added in successive layers

d (cm)	I (/30 s)	$e^{-\mu d}$	μ (cm^{-1})	Remarks
0	2270			$= I_0$
3	2087	0.834 ± 0.059	0.061 ± 0.024	Uncompacted
7	1803	0.576 ± 0.049	0.079 ± 0.013	Uncompacted
11	1649	0.436 ± 0.045	0.075 ± 0.010	Uncompacted
14	1346	0.161 ± 0.038	0.131 ± 0.018	Sand packed with spade
18	1287	0.107 ± 0.036	0.124 ± 0.023	Sand packed with spade

From this experiment it may be concluded that, within the uncertainties of the experiment, the effective absorption coefficient, μ, for a given type of sand is reasonably constant. As expected, μ for the uncompacted sand is less than for compacted sand, with the lower value expected to be an upper limit for wind-transported sand. In conclusion the average change in count rate due to the absorption from wind-blown, uncompacted, quartz sand may be described by equation (1) with μ < 0.07 cm^{-1}.

B. Measured Elevation-Radiometry Anti-Correlation

Taking the above model into account, it might be expected that the addition of new, less active, sand would provide a corresponding exponential decrease in the observed radioactivity. Figure 6 shows the correlation between the

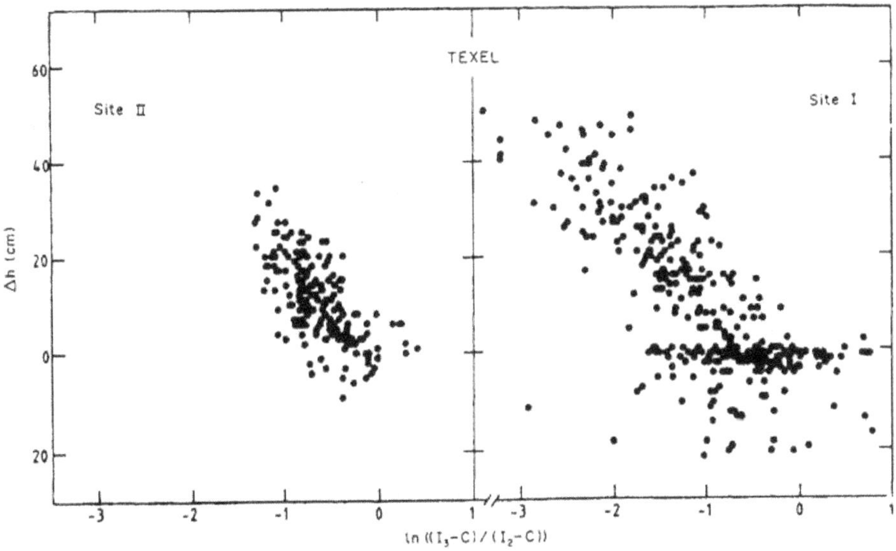

Fig. 6. Correlation plot of the change in elevation versus the logarithm of the background adjusted count rate.

changes in height, Δh, and the natural logarithm of the change in radiometry for each grid point from sites I and II. The value of C = 34 cps was the base line count rate, just below the lowest radiometry readings of each grid, that was subtracted from each intensity. The point-by-point count rates measured for the third and second measuring periods, I_3 and I_2, respectively were used since the observed differences were greatest between those two periods. From Figure 6 there is a negative correlation (Pearson correlation coefficient, r = -0.7 [SPSS75]) between the change in elevation and the change in the effective count rate, and hence equation (1) provides a reasonable quantitative description of the data. Part of the scatter in the data is because the change in elevation measured at a precise grid point is correlated with a change in radiometry representing the average over a larger area. The high density of points at about Δh = -3cm in site I are from the top of the "artificial" berm. For this flat, hard-packed, plateau, removal of light components by the wind has been artificially limited to a uniform depth.

The slope of the regression line yields $\mu = 0.05\ cm^{-1}$. This value agrees reasonably well with the limiting value of $\mu < 0.07\ cm^{-1}$ obtained for uncompacted sand in the above absorption experiment as well as with the value obtained from similar measurements preformed on Little St. George Island, Fla., USA [Don89].

V. DISCUSSION AND CONCLUSIONS

Radiometric mapping of beaches with grid spacings of 2m, 4m, and 20m has provided information on wind-driven sand transport. From a comparison of the changes in elevation and natural radiation levels it was found that wind velocities up to about 50 Km/h transport mainly lower-activity sand. This selective removal of the light sand fraction leads to the concentration of heavy minerals while deposition tends to cover the heavy minerals and absorb the radiation emitted. Thus wind driven sand transport leads to an inverse correlation between changes in elevation and changes in the observed radiation levels.

A model of exponential γ ray absorption with a linear absorption coefficient of $\mu = 0.05\text{-}0.07\ cm^{-1}$ for wind transported, uncompacted sand quantitatively describes this anti-correlation. The change in radiometry is a sensitive, quantitative indicator of sand-transport. If one may assume that this model is valid for more extreme transport conditions such as storms and flooding, airborne or vehicular γ-ray surveillance of beaches might allow a fast method for monitoring short term changes in beach profiles. Such measurements may be confined to regions where there are significant deposits containing radiogenic minerals. It does seem, however, that when such heavy minerals are present that they tend to be concentrated precisely where the coastline is eroding or rapidly changing. At these sites, radiometric techniques provide the opportunity to observe coastal changes with appropriate sensitivity and resolution in a relatively simple manner and within a reasonable period of time.

From observations of a 120m x 200m area of the southwest beach on Texel with a linear resolution of 4m and the subsequent artificial spoiling of that resolution with computer graphics, we concluded that although the location of the dominant radiometric features is feasible with a resolution on the order of 10-20m or more, the precise evolution and theoretical modelling of these large scale features may necessitate a resolution of 4-8 m. We continued to monitor this same area with 20 m resolution and with such resolution could observe the evolution of large scale sediment transport [Gre89]. It therefore seems quite feasible to monitor the changes of radiometry and anti-correlated movement of sediments at sea or from the air employing these techniques. Such investigations are in progress.

VI. ACKNOWLEDGEMNENTS

We gratefully acknowledge the assistance of the Department of Natural Resources of the State of Florida, especially Woodard Miley and the staff of the Apalachicola National Estuarine Reserve for assisting in all aspects of the Florida field work. We acknowledge the assistance of Rijkswaterstaat Dienkstkring Texel, especially C.W. Smit and D. Maas for their assistance in the elevation measurements on Texel. We are especially grateful to Ms. Yuko Yamamoto and Mr. Masumichi Seta for assisting with preliminary measurements in Japan. One of us (MBG) gratefully acknowledges the administration, faculty and staff of the Kernfysisch Versneller Instituut for hosting him during research visits to The Netherlands and the entire Division of Natural Sciences as well as the Administration and Staff of the International Christian University for providing support for preliminary investigations in Japan and the presentation of this paper.

VII. REFERENCES

[deM89] de Meijer R. J., L. W. Put, R. D. Schuiling, J. Wiersma and J. H. de Reus, Proc. KNGMG Symp. Coastal Lowlands Geol. Geotech. (l987), pp 355-361.

[deM90] deMeijer R. J., H. M. E. Lesscher, R. D. Schuiling and M. Elburg, private communication; publication in progress.

[Don89] Donoghue, J. F. and M. B. Greenfield, J. Coastal Research, Vol. 6 no 3 (1989).

[GG87] Golden Graphics Version 3.II, Golden Software Inc., P.O. Box 281, Golden Colorado, USA 80402 (1987).

[Gre89] Greenfield M. B., R. J. de Meijer, L. W. Put, J. Wiersma, J. F. Donoghue, Nucl. Geophys. 3 (1989) 231-244.

[Gre90] Greenfield, M. B., H. M. E. Lesscher, R. J. de Meijer and L. W. Put, KVI Internal Report Z-05, and to be submitted to Nuclear Geophysics (1990).

[Les89] Lesscher, H. M. E.,"Blown with the wind", KVI Internal Report 157i (1989).

[Mil73] Miller, J. M., P. D. Roberts, G. D. Symons, N. H. Merrill and M. R. Wormald, "A towed sea-bed gamma-ray spectrometer for continental shelf surveys", IAEA, Vienna, (1973) 465.

[Mil73] Miller, J. M., G. D. Symons, Nature 242 (1973) 184.

[Sha89] Sha, L. P., Mar. Geol., 86 (1989) 137.

[SPSS75] Statistical Package for the Social Sciences, Version 9.0 (Ed. N. H. Nie), McGraw-Hill, (New York - 1975).

[Sum70] Summerhayes, C. P., B. H. Hazelhoff Roelfzema, J. S. Tooms and D. B. Smith, Economic Geology, 65 (1970) 36.

[Tho83] Thomas, B. W., C. G. Clayton, V. C. Ranaasinghe and I. M. Blair, Mineral Exploration of the sea-bed by towed sea-bed spectrometers, Perg. Press, (London-1983) 437.

[Wie88] Wiersma, J. and Van Alphen, J. S. L. J. ,Tide-Influenced Sedimentary Environments and Facies, Riedel, Dordrecht (1988) 102.

The Neutron Spectrum of the Hiroshima A - Bomb and the Dosimetry System 1986

W. Rühm, K. Kato *, G. Korschinek, H. Morinaga, A. Urban, L. Zerle and E. Nolte

Faculty of Physics, Technical University of Munich, FRG

* Suzugamine women's junior college, Inokuti, nishi-ku, Hiroshima, Japan

Abstract

Five radioactivities of ^{36}Cl, ^{41}Ca, ^{60}Co, ^{152}Eu and ^{154}Eu produced by the Hiroshima A - bomb in a granite stone 107 m from the hypocenter were analysed. ^{36}Cl and ^{41}Ca have been measured by accelerator mass spectrometry at the Munich accelerator laboratory. The other isotopes have been detected with γ spectroscopy at the Hiroshima University. Because of different energy dependences of the neutron capture cross sections of these isotopes, it was possible to deduce the neutron spectrum. This spectrum is much harder than the one used in the standard dosimetry system DS86. Within this study, the resonance integral for the reaction $^{40}\mathrm{Ca}(\mathrm{n},\gamma)^{41}\mathrm{Ca}$ was determined to be (0.22 ± 0.02) b.

1. Introduction

In the tentative 1965 dose evaluation T65D, the rates of leucemias in dependence of the radiation dose were different by a factor of about 4 for Hiroshima and Nagasaki. This discrepancy was considered to be due to large relative biological effectivenesses for neutrons and the stronger neutron fluence in Hiroshima. This descrepancy was removed in the new dosimetry system 1986, DS86 [1]. But still, this neutron spectrum consisting of a Maxwell - Boltzmann distribution for thermal energies and of a 1/E dependence for non - thermal energies at a distance of 600 m from the explosion was never confirmed. Because of resonances in the neutron capture cross sections and different energy dependences of these cross sections it is possible to deduce spectrum and fluence of neutrons from the Hiroshima A - bomb by analyzing several bomb produced radioactivities.

In the present work we describe AMS measurements of ^{41}Ca in a gravestone from Hiroshima and of the resonance integral for the reaction $^{40}\mathrm{Ca}(\mathrm{n}, \gamma)^{41}\mathrm{Ca}$. These data together with measurements on ^{36}Cl, ^{60}Co, ^{152}Eu and ^{154}Eu are compared with the neutron spectrum used in DS86. By means of least squares fits, neutron spectra were obtained which can reproduce the measured data.

2. Existing measurements

All samples described in this work are taken from a gravestone made of granite located 107 m from the hypocenter of the explosion in Hiroshima (fig. 1). In table 1 all isotopes measured in the first two centimeters of the granite stone are shown. The measured isotopic ratios are normalized to the corresponding thermal cross section. The ^{36}Cl and ^{152}Eu values are part of depth profiles measured with AMS and γ spectroscopy [2]. The ^{41}Ca measurement is described below.

isotope	measured fluences [1/cm^2]	method	reference
^{36}Cl	$(3.6 \pm 0.4) \cdot 10^{12}$	AMS	[2]
^{41}Ca	$(10.6 \pm 2) \cdot 10^{12}$	AMS	this work
^{60}Co	$(9.7 \pm 1.4) \cdot 10^{12}$	γ spectr.	[3]
^{152}Eu	$(6.8 \pm 0.7) \cdot 10^{12}$	γ spectr.	[2]
^{154}Eu	$(17.4 \pm 3.3) \cdot 10^{12}$	γ spectr.	[3]

Tabl. 1: Measured isotopes from the first 2 cm from the top of the granite stone

Fig. 2 [4, 5, 6] and fig. 3 [7, 8] show the (n,γ) cross sections responsible for the production of the isotopes mentioned above.

3. AMS measurements of ^{41}Ca

Hiroshima sample

AMS measurements of ^{41}Ca were carried out at the Munich accelerator laboratory. ^{41}CaF and ^{41}KF as pilot beam were injected into the MP tandem. After the tandem and the two IH type postaccelerators ^{41}Ca and ^{41}K were stripped completely at an energy of 5.75 MeV/n. Both ions were seperated by a 90° analysing magnet and ^{41}Ca was identified by a $\Delta E/E$ ionization chamber. The system in detail is described in [9].

During the runs a detection limit which is defined by one count per hour in the detector of $3 \cdot 10^{-13}$ has been reached.

For the Hiroshima sample a ratio of $^{41}\mathrm{Ca}/^{40}\mathrm{Ca} = (4.4 \pm 0.8) \cdot 10^{-12}$ was obtained.

Resonance integral for the reaction $^{40}\mathrm{Ca}(\mathrm{n}, \gamma)^{41}\mathrm{Ca}$

The (n,γ) cross section for ^{40}Ca shows resonances in the energy region above 3 keV. Below 3 keV no measurements exist and a 1/v shape is assumed. To exclude resonances in this region the resonance integral defined by $\mathrm{I}_\gamma = \int_{0.5\mathrm{eV}}^{2\mathrm{MeV}} \sigma(\mathrm{E}) \frac{\mathrm{dE}}{\mathrm{E}}$ was measured by AMS. For this purpose, $^{40}\mathrm{CaCO}_3$ and gold foils as flux monitors covered by 1 mm thick cadmium were irradiated at the Nuclear Research Reactor Munich. After 2400 s with an epithermal flux of $3.7 \cdot 10^{11} \frac{1}{\mathrm{cm}^2\mathrm{s}}$ measured by the gold foils the samples were chemically converted to CaF_2. The isotopic ratio of $^{41}Ca/^{40}Ca = (1.47 \pm 0.12) \cdot 10^{-10}$ was obtained. This corresponds to a resonance integral of $I_\gamma = (0.22 \pm 0.02)$ b which is consistent with the assumed 1/v shape of the cross section.

4. Determination of the Neutron Spectrum

With the different (n,γ) cross sections of the 5 isotopes and the measured concentrations at **one** location, we are now able to test whether the neutron spectrum used in DS86 is able to reproduce these data. The DS86 spectrum seen in fig.4 consists of a Maxwell - Boltzmann part at 300 K below 0.125 eV. The part above 0.125 eV was fitted from [1]. Fig. 5 compares theoretical ratios we achieved by calculating the integral

$$\frac{1}{\sigma_{\mathrm{th}}} \int \sigma_{\mathrm{i}}(\mathrm{E}) \cdot \frac{\mathrm{d}\Phi \mathrm{t}}{\mathrm{dE}} \cdot \mathrm{dE} \qquad \mathrm{i} = 35, 40, 59, 151, 153$$

using the neutron spectrum of fig. 4 and the cross sections of fig. 2 and fig. 3 with the measured data. If the isotopes were activated only by thermal neutrons this integral normalized to the cross section for thermal neutron activation would yield the same result for all isotopes.

It appears that especially the ^{36}Cl activation cannot be explained by the neutron spectrum of DS86.

A hypothetical neutron spectrum which can reproduce the measured data must fulfill the five following equations:

$$(\frac{^{i+1}N}{^{i}N})_{exp} = \int \sigma_i(E) \cdot \frac{d\Phi t}{dE} \cdot dE \qquad i = 35, 40, 59, 151, 153$$

Our first approach was to divide the DS86 spectrum of fig. 4 into five energy groups. These intervals are defined by the main resonances of the cross sections. For example a region between 10 eV and 1000 eV was used due to the main resonance of ^{59}Co (see fig. 3). The amplitudes of the spectrum in each region serve as fit parameters. Fig. 6 shows a spectrum resulting from this procedure. The measured ratios are reproduced completely with a negative amplitude of the Maxwellian part between 0 eV and 0.125 eV. However this part yields only small contributions to each activation. In fig. 6 an amplitude which contributes on the order of 1 % to each activation is shown.

This model spectrum suggests a spectrum which has less thermal neutrons but more neutrons with higher energies.

To find a neutron spectrum with less thermal neutrons the temperature in the Maxwell-Boltzmann part of the spectrum was used as a parameter. The amplitudes of the Maxwellian part and a part proportional to E^{-b} and b served as fit parameters. Fig. 7 shows the resulting neutron spectrum.

In a third model the neutron spectrum is constant within an energy region of 1.3 meV to 1 eV. Between 1 eV and 3 MeV a functional dependency of E^{-b} is assumed.

Parameters are the amplitudes of the two regions and b. Fig. 8 shows the spectrum and fig. 9 the corresponding fluences. For b, a value of 0.5 was obtained.

It turns out that these 3 different approaches require less thermal neutrons but more neutrons with higher energies to explain the measured data.

5. Comparison with DS86

The measurements of ^{41}Ca, ^{60}Co, ^{152}Eu and ^{154}Eu in the granite stone can be explained by the neutron spectrum used in DS86. However it cannot be explained together with the activity of ^{36}Cl. Due to the high quality of the granite a contamination of the first centimeters from the surface by rainwater with less ^{36}Cl content is very unlikely but has to be checked.

Provided all measured ratios are correct and there exists no mechanism which can diminish the ^{36}Cl/^{35}Cl ratio the neutron spectrum which has activated the five isotopes mentioned above is much harder than the neutron spectrum used in the Dosimetry System 1986.

This work was supported by the German Federal Ministry for Research and Technology (BMFT).

References

[1] William C. Roesch, editor. *US - Japan - Joint Reassessment of Atomic Bomb Radiation Dosimetry in Hiroshima and Nagasaki, Final Report.* Radiation Effects Research Foundation, 1987.

[2] K. Kato, M. Habara, T. Aoyama, Y. Yoshizawa, U. Biebel, G. Haberstock, J. Heinzl, G. Korschinek, H. Morinaga, and E. Nolte. Measurements of neutron fluence from the Hiroshima atomic bomb. *Journal of Radiation Research*, 29:261–266, 1988.

[3] K. Kato. private communication.

[4] M. Hatchya. JENDL-2. 1980.

[5] Livermore R. J. Howerton. ENDL-84/V. 1983.

[6] BNL S. F. Mughabghab. ENDF/B-5. 1977.

[7] Livermore R. J. Howerton. ENDL84/V. 1983.

[8] BNL S. F. Mughabghab. ENDF/B-5. 1978.

[9] D. Elmore H. E. Gove, A. E. Litherland, editor. *Proceedings of the Fourth international symposium on Accelerator Mass Spectrometry*, Canada, april 27 - 30 1987.

Fig.1. Location of the granite stone.

Fig.2. (n,γ) cross sections for Cl [5], ^{40}Ca [4] and ^{151}Eu [6].

Fig.3. (n,γ) cross sections for ^{59}Co [7] and ^{153}Eu [8].

Fig.4. Fitted neutron spectrum used in DS86.

Fig.5. Comparison of measured (•) and calculated (○) neutron fluences using the neutron spectrum of DS86.

Fig.6. Model neutron spectrum using five energy groups and five amplitudes as fit parameters (see text) compared with the spectrum of DS86.

Fig.7. Model neutron spectrum using two energy groups compared with the spectrum of DS86; the temperature of the Maxwellian part, the two amplitudes and the slope of the second part serve as fit parameters (see text).

Fig.8. Model neutron spectrum using two energy groups compared with the spectrum of DS86; the two amplitudes and the slope of the second part serve as fit parameters (see text).

Fig.9. Calculated fluences (○) for the spectrum used in fig. 8 compared with the measured data (•).

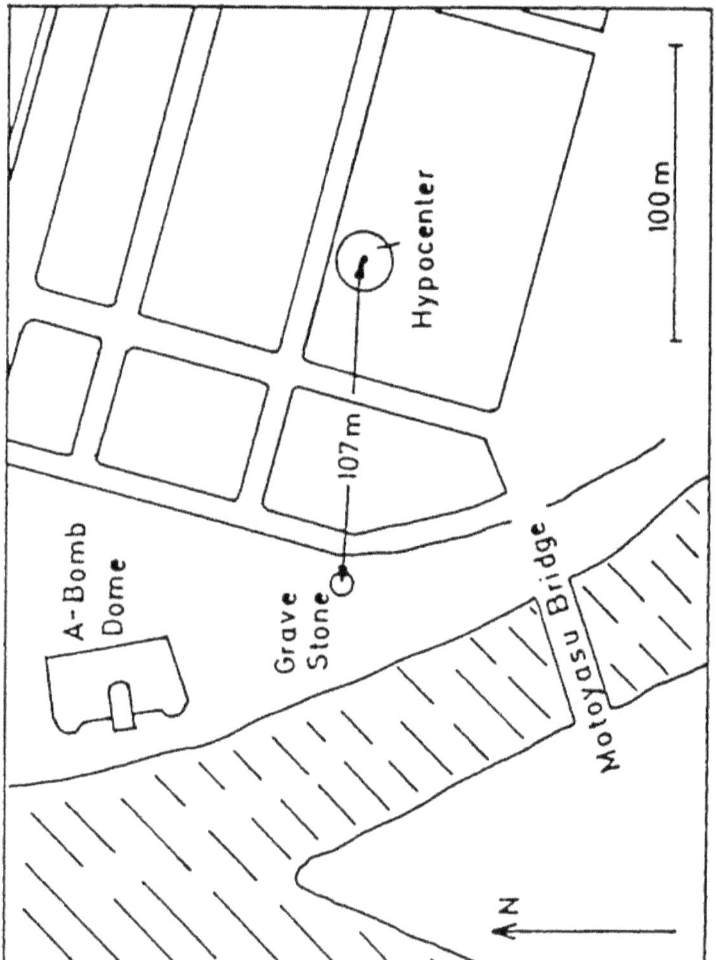
A-Bomb Dome
Grave Stone
107 m
Hypocenter
100 m
Motoyasu Bridge
N

Fig.1.

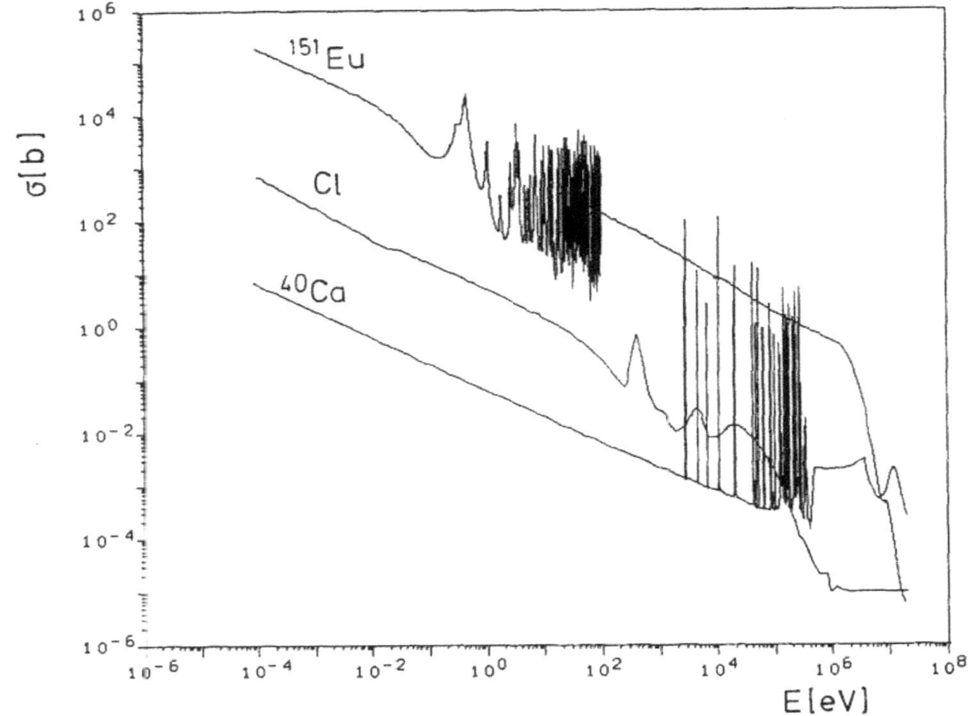

σ[b]
10^6
10^4
10^2
10^0
10^-2
10^-4
10^-6
151Eu
Cl
40Ca
10^-6
10^-4
10^-2
10^0
10^2
10^4
10^6
10^8
E[eV]

Fig.2.

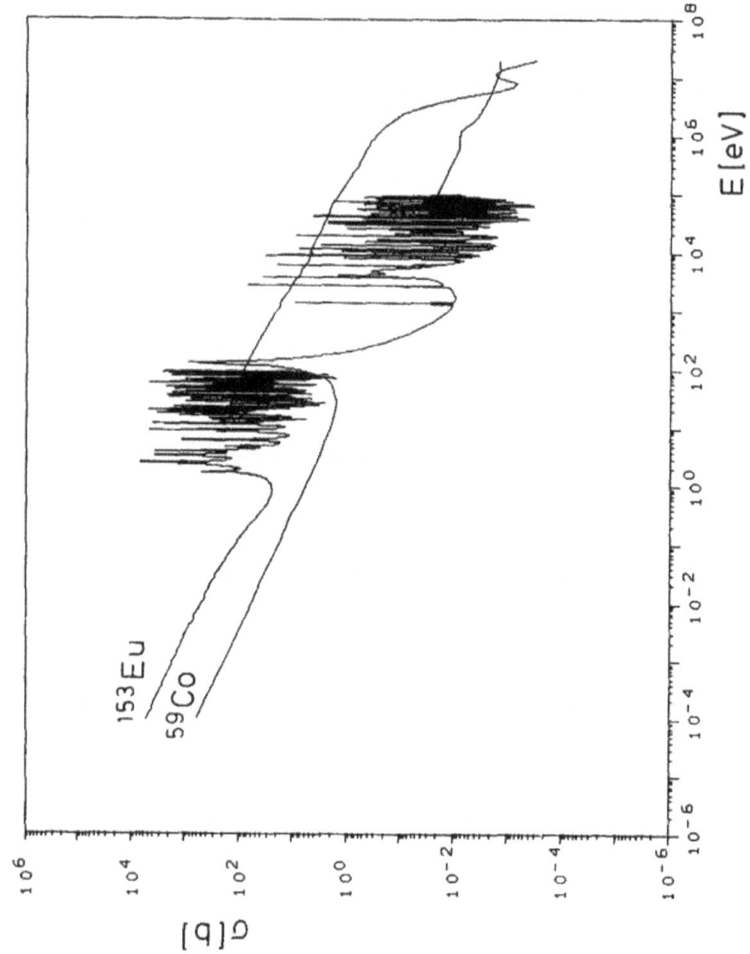

Fig.3.

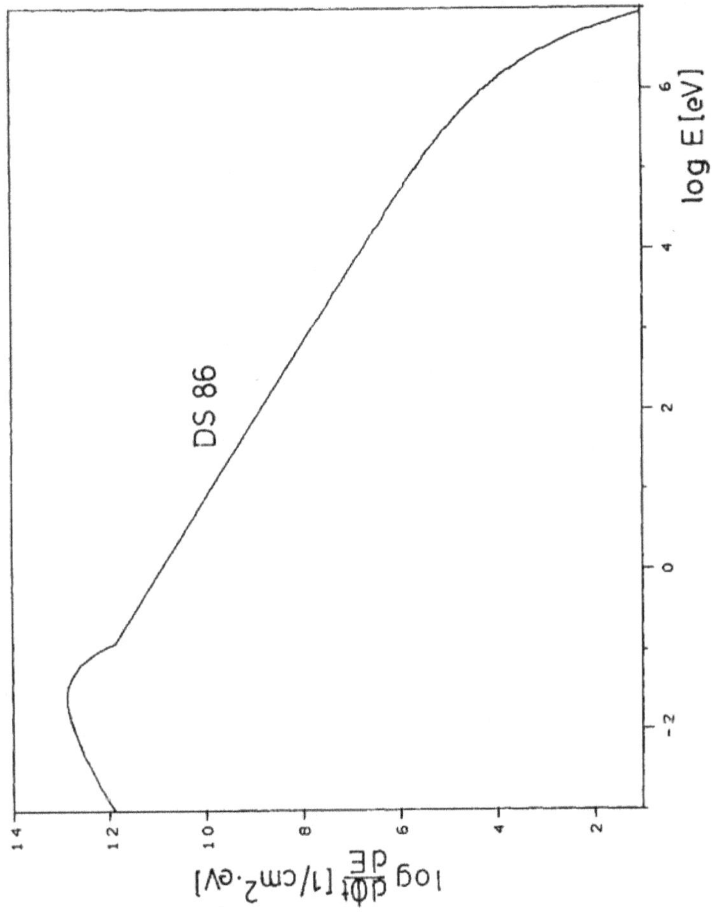

Fig.4.

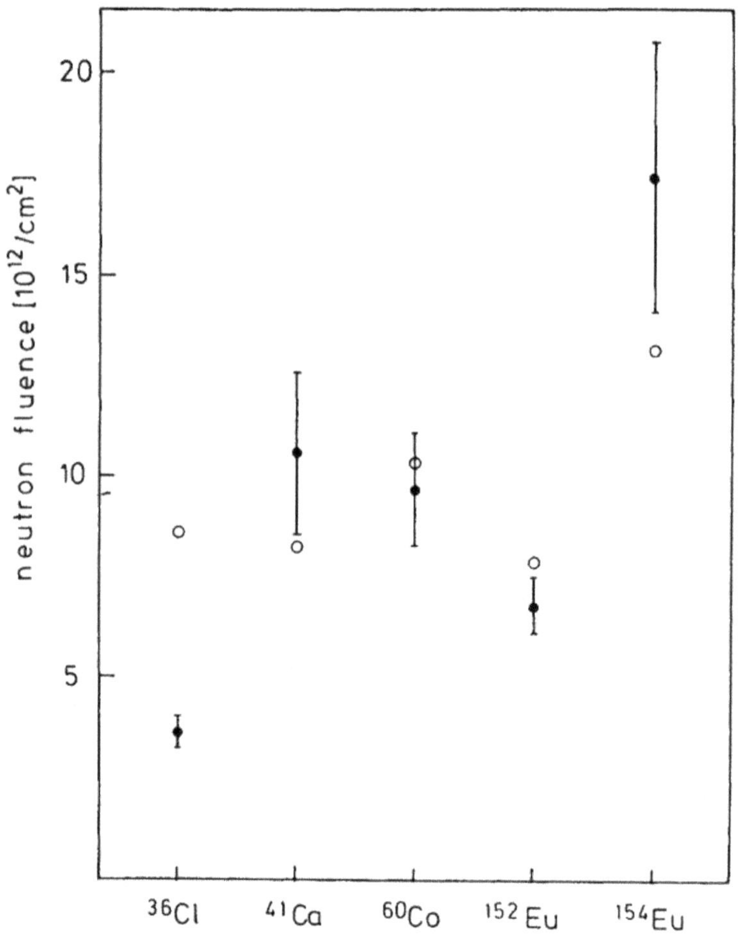

neutron fluence [10^12/cm^2]
20
15
10
5
36Cl
41Ca
60Co
152Eu
154Eu

Fig.5.

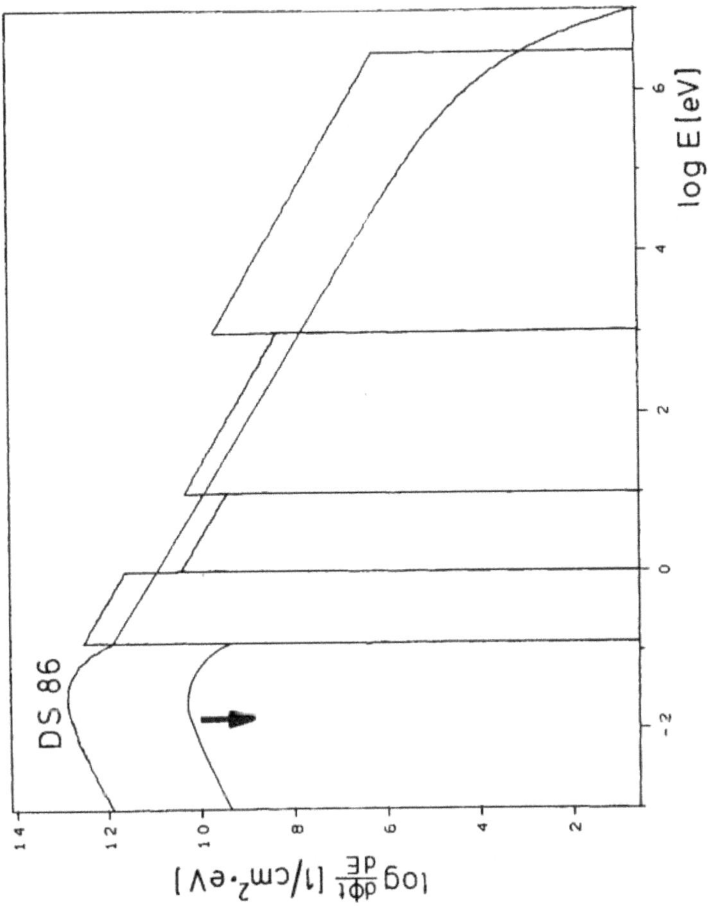

Fig.6.

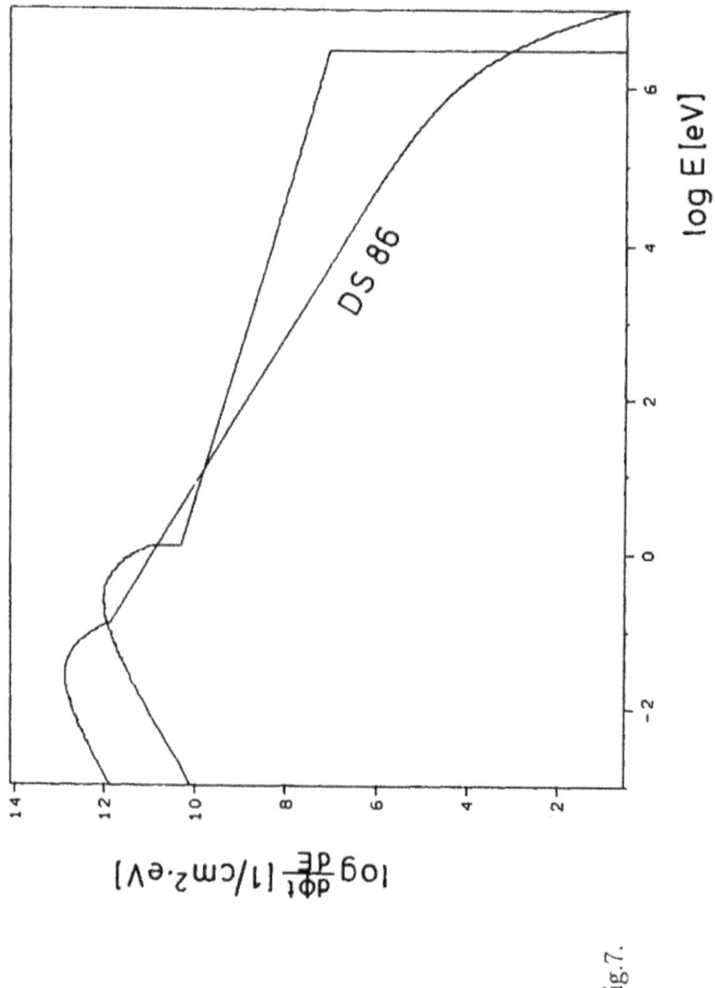

Fig.7.

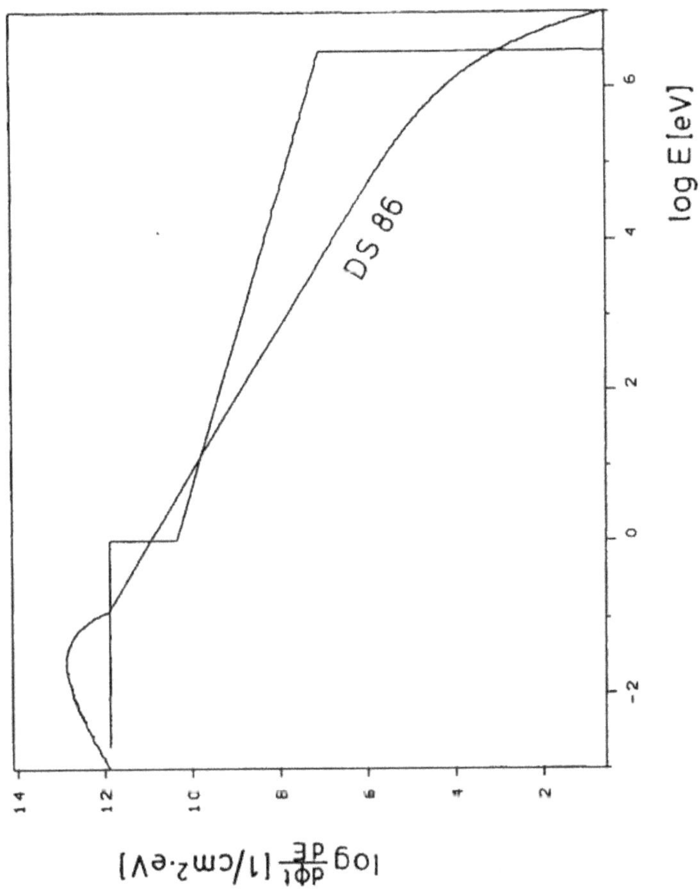

Fig.8.

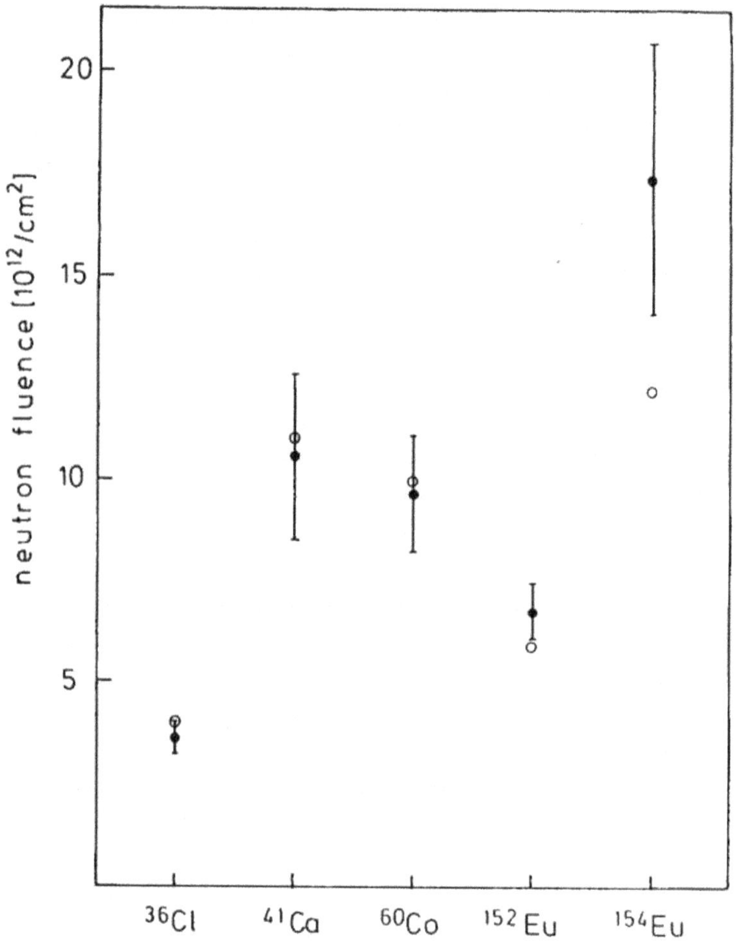
neutron fluence [10^{12}/cm^2]
20
15
10
5
^{36}Cl
^{41}Ca
^{60}Co
^{152}Eu
^{154}Eu

Fig.9.

A Universal Method of Producing Ultra-Fine Microbeam

Akira Isoya

Institute of Reserch and Development, Tokai University
Hiratsuka, Kanagawa-Pref. Japan
and
ULVAC CORPORATION, Chigasaki, Kanagawa-Pref. Japan

Abstract

A possible universal way of producing a ultra-fine ion beam focus of 0.1μm order is discussed, being based on the principle of conservation of normalized emittance in the process of acceleration of the beam. The ion-optical system under consideration is composed of an accel-lens of two stage and an einzel-lens. Contraction of the beam of a factor of 10^{-3} is mainly achieved by the former lens and the latter is used for converting a virtual image produced by the former to a real one. The spherical aberration of the accel-lens can be made small enough to produce a focus of 0.1μm, in spite of a very large contraction factor. The color aberration does not take place since the same high voltage is used in common for the both lenses.

§ 1 Introduction

The micro-beam technique is one of the most promising way of application of the electrostatic accelerator technology. This can be a very powerful tool for surveying micro-structures on the solid surface or cellular tissues when it is combined with the PIXE or RBS techniques. Micro-beam apparatus being used at the present time are made to squeeze the ion beam from the existing particle accelerators by using quadrupole magnetic lens. The minimum size of the focused beam usually obtained is about 1 μm. Limitation of the size comes in principle from the spherical aberration of the ion lens and the chromatic aberration due to energy spread of the primary ion beam. In order to reduce the beam size to a level of 0.1 μm, fractional energy spread should be reduced down to 10^{-5}. This is far from the present status of usual electrostatic accelerator technology. Further, precision of machining of quadrupole magnets comes to the limit.

I am considering that a possible way to overcome these difficulties and reach a level of less than 0.1μm is to utilize the contraction effect of beam width, associated with the acceleration process itself, which is enabled by the conservation of normalized emittance.

§ 2 Realizable Magnification with the accel lens

Fig.1 is a diagram showing the relation of the magnification of the accel-lens M and the ratio of beam divergence angles at the image and object points β/α, which are computed for various accel-lens systems in the course of the present work. Here $N(=V_f/V_i)$ is the ratio of the final to initial beam energies. Fig.1 shows that for small values of N, M decreases along a smooth curve as the increase of N, but at some N value M stops to decrease, being associated with sudden decrease of β/α. This value of N is about 100 commonly for all cases investigated. Beyond this point the curves for different lenses are quite strongly dispersed. Decrease of M seems to be recovered for much higher N values ($>10^5$). We will see that such an effect is caused by the strong lens action at the entrance of the accel-tube.

In Fig.1 we see that points, corresponding to the same N value along the various curves for different lenses, are aligned accurately on a straight line with the slope of 45°. This is a very good demonstration of the conservation of normalized emittance.

For realizing a micro-beam of 0.1μm, the magnification M of less than 10^{-2} is required. According to the above formalism the corresponding N value has to be larger than 10^4, assuming $\beta/\alpha=1$. But the diagram of Fig.1 indicates that M values of order of 10^{-2} are hard to be obtained with this N value. Then it is necessary to use two accel-lenses with N of 100 in cascade. The important problem is that how small spherical aberration can be obtained with the accel-

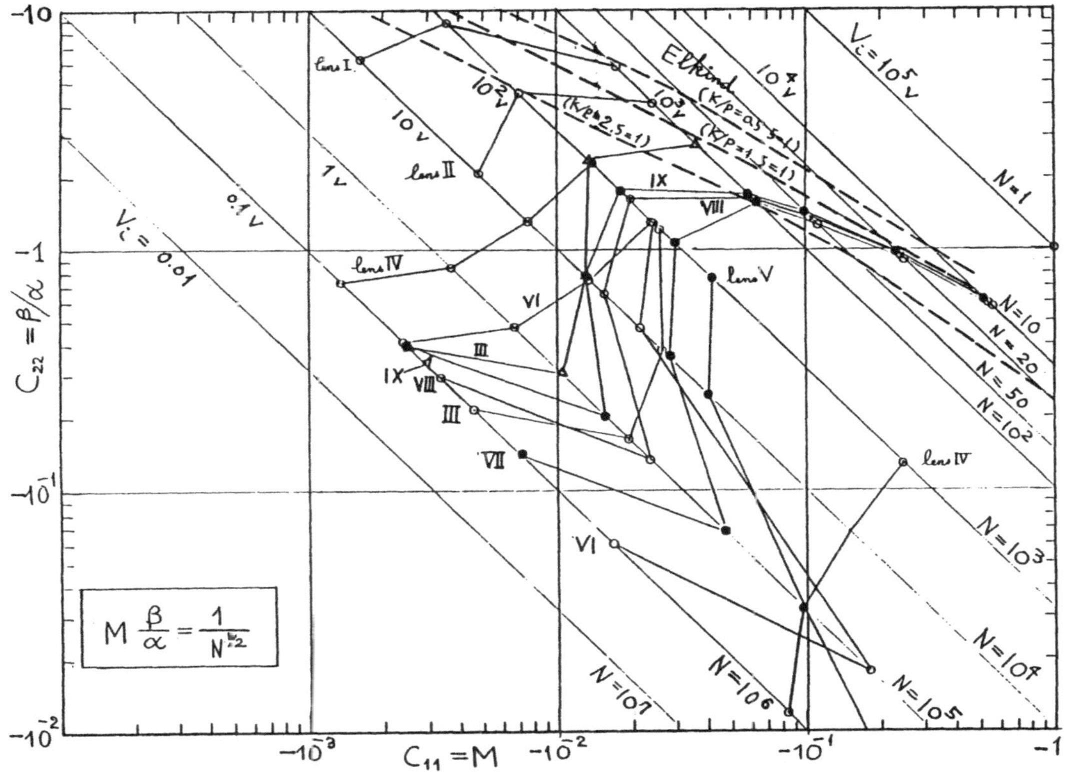

Fig.1 Relation of the magnification M and the ratio of beam divergence angles at the image and object points β/α .

lens. This point is a main subject of the present work. The conclusion is that the accel-lens is much superior to the usual ion lenses without acceleration.

§ 3 Features of ion optics of the accel-lens system

A formula which describes the ion optics in long accelerating tube has been presented by Elkind[1] in MIT in 1953. Since it was written in a convenient form for the actual use, it has been widely used for the analysis of the ion beam optics of the Van de Graaff accelerators in many laboratories in the world.

Fig.2 shows the form of this formula. Its derivation was made with a simple assumption. The lens action of the whole tube is divided into three parts: aperture lenses at the entrance and exit portions of the tube, and uniform accelerating field in the middle portion. In this model the incident beam of some energy is deflected by a strong entrance aperture lens with negligible change of its energy. The focal length of the entrance aperture lens is given by the following formula, in the approximation of constant deviation of the path from the tube axis,

$$f^{ent} = -\frac{4\zeta\ V_i}{E} \quad . \tag{1}$$

Here E is the strength of uniform accelerating field, V_i is the initial energy of ion beam, and ζ is the correction factor for the finite thickness of the lens, depending on the diameter of aperture. This parameter can be determined only by comparison with the computed trajectory. If the lens action is strong enough for the incident beam energy a small virtual image is formed just behind the entrance aperture, and the accelerating field makes this image shift along the tube axis without change of its magnification. The exit aperture lens farther modifies the image, but its effect is much weaker than that of the entrance aperture lens, so may be ignored.

A remarkable feature of the ion optics of the accelerating tube is that, if $N(=V_f/V_i)$ is large enough, a vertual image is formed a little outside of the tube entrance, as shown in Fig.2. The magnification M of this image is reduced smaller and smaller as the N is increased, with keeping the position L/K of the

---

[1] M.M.Elkind, R.S.I.24,129(1953)

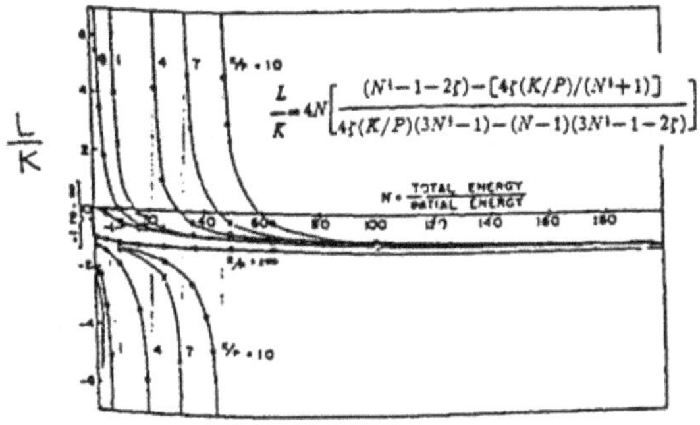

FIG. 3. L/K as a function of N for a three-element tube (or source system). $\zeta = 1$.

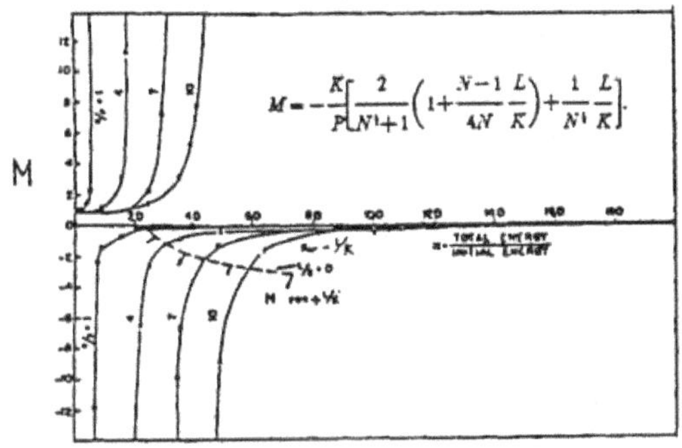

FIG. 5. Magnification as a function of N for a three-element tube (or source system). $\zeta = 1$.

$$f^{ent} = \frac{-4\zeta V_i}{E}$$

$$\frac{1}{P_1} + \frac{1}{L_1} = \frac{1}{f^{ent}}$$

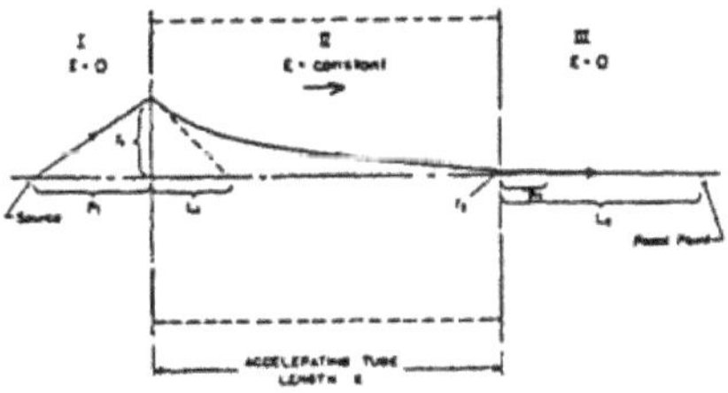

FIG. 1. Three-element tube focusing trajectory.

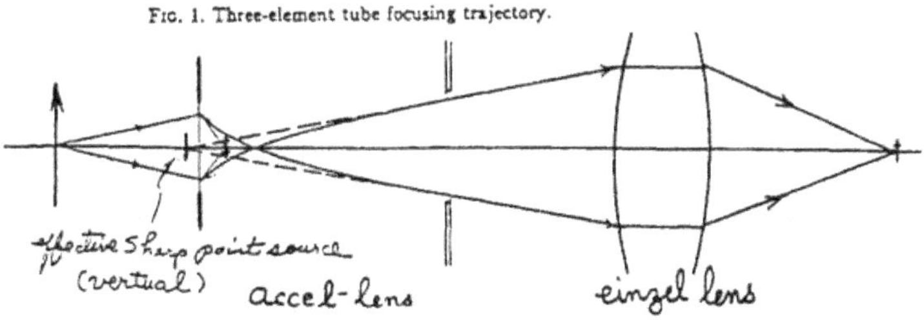

Fig.2 Elkind's formula

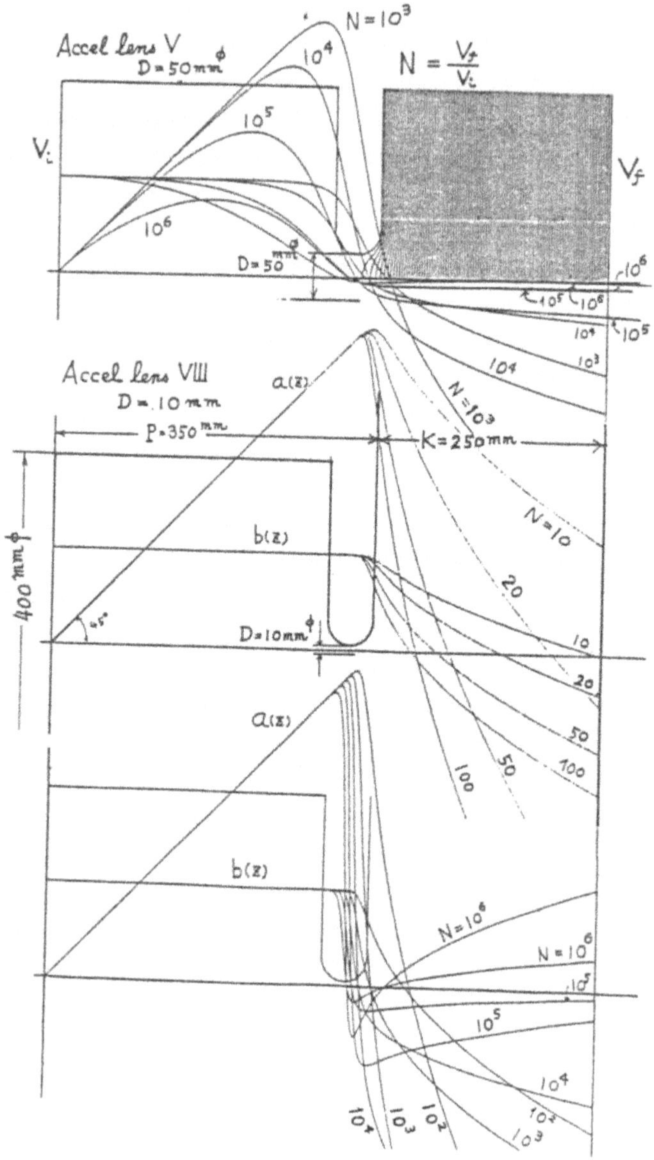

Fig.3 Arrangement of the electrodes of the accel-lenses used for the computor calculation and first order trajectories for several values of N.

image very constant. We take advantage of this feature in the ion optics of the micro-beam system. Since a v rtual image is produced in this lens system, some sort of focusing lens is necessary after the accel-lens to form a real image. We plan to use an einzel lens which is fed with the same high voltage as for the accel-lens. In this way difficulty of chromatic aberration can be avoided.

Trajectories in the accel-lens of the electrode arrangements, as shown in Figs.3 and 4, have been calculated for several different values of N. The size of diameter of the entrance aperture D was changed in several ways.

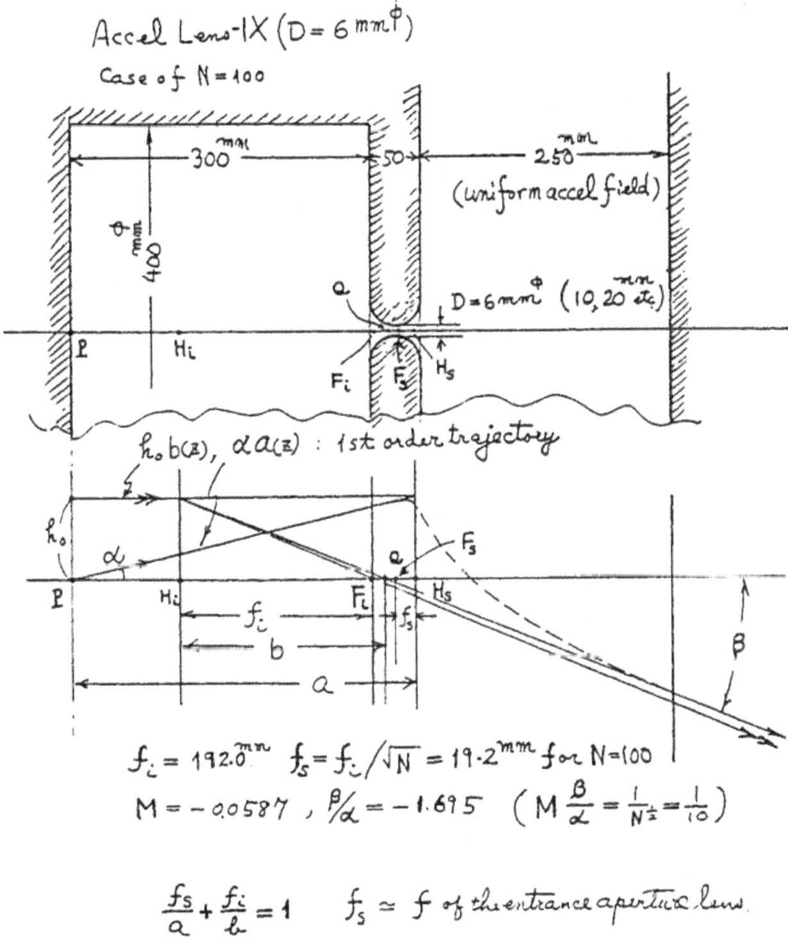

Fig.4 Ion optics of the accel-lens

Some of the computed trajectories a(z) and b(z) are shown in Fig.3*. These are two independent solutions of the first order trajectory equation

$$x'' + \frac{F'}{2F}\, x' + \frac{F''}{4F}\, x = 0$$

with the initial conditions shown in Fig.5.

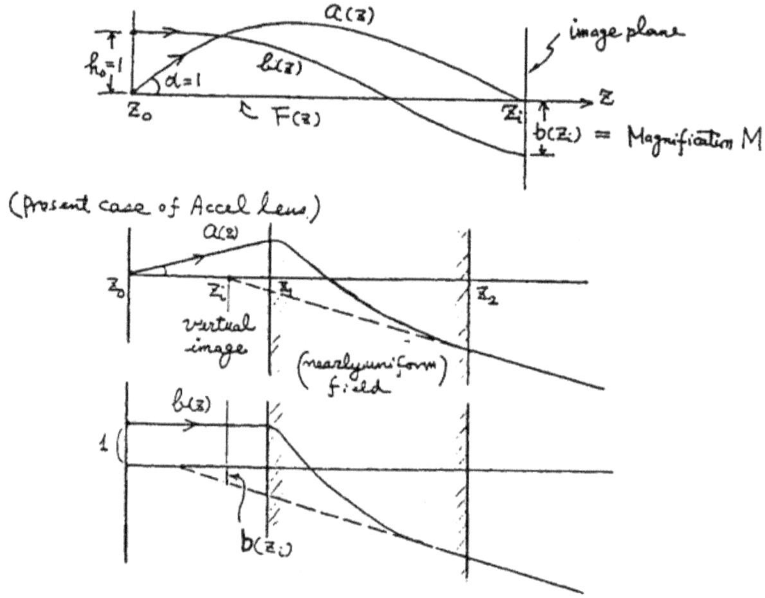

Fig.5 First-order trajectories

Fig.4 shows features of the ion optics of the accel-lens, obtained from this calculation. It is represented by a thick lens formula, in which both principal planes are situated in front of the entrance aperture, being crossed of each other. In particular, the principal plane Hi for the image space is very far from the tube entrance. This feature makes the optical property of the accel-lens quite different from those of usual lenses.

* We owe the computer calculation for the ion lens and its aberrations to Dr.T.Katagawa in ULVAC Corp.

From the lens property of a thick lens, the magnification of the accel-lens is expressed by the following formula,

$$M = \frac{-1}{\dfrac{P}{f_s} - 1} , \tag{2}$$

where f_s is the object-space focal length of the whole lens system and P is the object distance. With the assumption made by Elkind, this f_s and the location of the corresponding focal point F_s coincide with the focal length f_s^{ent} and the corresponding focal point F_s^{ent} for the entrance aperture lens.
This is because the rays passing through the latter focal point, being deflected by the entrance aperture lens, enter the accelerating field region in parallel to the tube axis and hence to the direction of the accel-field.
Therefore we have

$$M = \frac{-1}{\dfrac{P}{f^{ent}} - 1} = \frac{-1}{\dfrac{P(N-1)}{4\zeta K} - 1} , \tag{3}$$

putting
$$f^{ent} = \frac{-4\zeta V_i}{E} = \frac{-4\zeta V_i}{(V_i - V_f)/K} = \frac{4\zeta K}{N-1} . \tag{4}$$

The value of f^{ent} becomes very small as the N value is increased. Thus we see that the M value of the accel-lens can be made very small, compared to those of usual ion lenses without acceleration (einzel lens or magnetic lens), for which object-and image-space focal lengths are equal so that these focal lengths are necessarily limited to the order of the dimension of the ion lens (cf.Fig.7).

M is dependent on the ratio E/V_i or $(N-1)/K$, so M can not be decreased if N is increased by keeping E constant. For large N values, M does not necessarily decrease in inverse proportion to N, since the correction factor ζ also increases. And finally this simple model fails when the strength of protruding field from the entrance aperture becomes too strong relatively to the incident energy. Fig.1 includes the plots of the Elkind's formula. It is shown that validity of the formula is limited to the N-values up to about 100.

§ 4 Spherical aberration of the accel-lens

For the complete description of the third order deviation of the ion trajectory from the first order one, we need 12 aberration parameters, but in the present case, using a very small source point on the axis, spherical aberration is predominant. The deviation at the image point δW_i is given by the following formula*,

$$\delta W_i = \alpha^3 b(z_i) \int_{z_0}^{z_i} \left[a^4 \left\{ \frac{5}{64} \left(\frac{F''}{F} \right)^2 - \frac{3}{128} \frac{F''F^2}{F^3} \right\} + a^3 a' \frac{13}{64} \frac{F''F'}{F^2} + a^2 a'^2 \frac{1}{4} \frac{F''}{F} + aa'^3 \frac{1}{2} \frac{F'}{F} \right] \left(\frac{F}{F_0} \right)^{1/2} dz \quad (5)$$

$$= d_6 \alpha^3 \qquad (d_6 : \text{spherical aberration coefficient}).$$

Here $F(z)$ is the electrostatic potential on the tube axis ; the prime indicates the differentiation with respect to z, and $a(z)$ and $b(z)$ are the two independent solutions of the first order trajectory equation with the initial conditions as shown in Fig.5, and α is the emitting angle of the actual rays from the source point. $b(z_i)$ is a value of $b(z)$ at image point z_i. Since $b(z)$ is set to be 1 at the object point, $b(z_i)$ is equal to the magnification M.

In Fig.6 values of δW_i computed for several accel-lenses, mentioned in the last sections, are plotted as a function of M. In this figure we see that behaviours of the relation between δW_i and M are smooth for N values less than about 1000, but very rough for larger N values. This fact reflects a situation mentioned in the last section as for the relation between β/α and M, i.e. when the effect of the entrance aperture lens extends far from the aperture, the trajectory makes an oscillating motion before entering the main field, and then the focus point moves drastically between $+\infty$ and $-\infty$. Fig.6 indicates that stable condition may be available again for very large values of $N(\geqq 10^5)$. We can not judge at this moment whether such a condition at very high voltage is suitable for the micro-beam production apparatus or not. So at present we confine ourselves to a moderate range of smaller N values, where the prediction of Elkind's formula is valid.

---

* For derivation of this formula see, for example, W. Glaser, Handbuch der physik 33 (1956) 123.

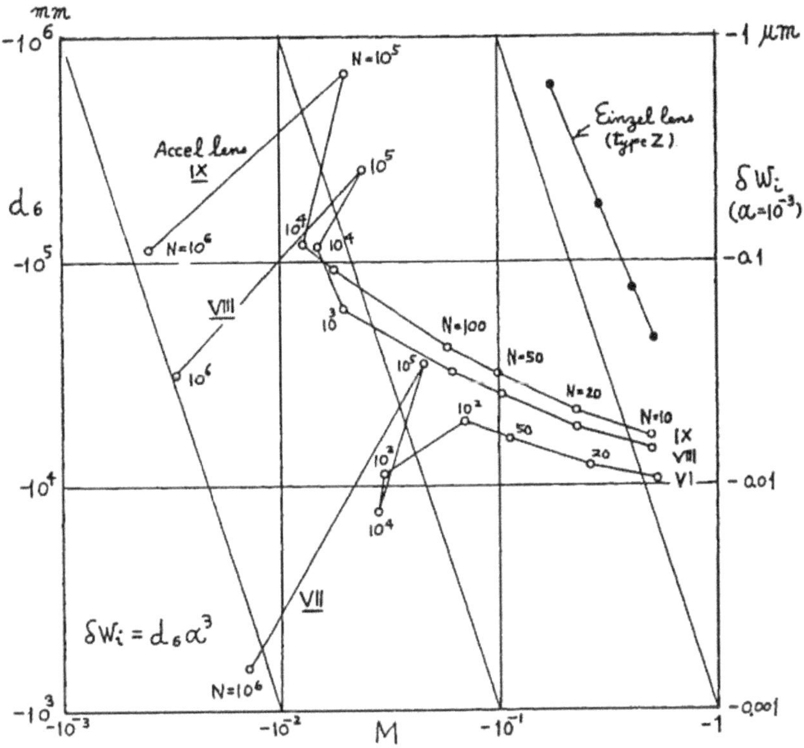

Fig.6 Spherical aberration δW_i for several accel-lenses as a function of magnification M

Fig.6 indicates that δW_i less than $0.1\mu m$ can be obtained with M value less than 0.1, for $N = 10^2 \sim 10^3$ and $\alpha = 10^{-3}$. This magnitude of spherical aberration is small enough to produce a microbeam less than $0.1\mu m$ in diameter. In particular, two accel-lenses in cascade, as mentioned in section 2 is considered to be just suitable for this purpose.

§ 5 Considerations on the magnitude of the spherical aberration of the accel-lens and its simple expression

In the last section we have seen that the magnification of the accel-lens can be much smaller than for usual non-accel lenses, owing to the fact that the object-space focal length f_s for the former is much smaller than those for the latters. By the same reason the spherical aberration of the accel lens is shown to be much smaller than for the non-accel lens. From formula 2, which is applicable to both types of lens, we can argue that for getting same values of M for both lenses the ratio P to f_s should be taken same. Therefore the object distance P for the accel-lens can be taken considerably smaller compared to the non-accel lens, owing to the much smaller value of the focal length f_s. This means that deviation of the beam path from the tube axis for the accel-lens is considerably smaller. Thus we can expect much smaller spherical aberration for the accel-lens.

According to the model by Elkind, the lens action of the accel-lens is mostly determined by the entrance aperture lens and the succeeding accel-field has no effect on the magnification. If the same reasoning is applied to the spherical aberration, it is determined by the entrance aperture lens. Since in the aperture lens section change of the beam energy due to acceleration is assumed negligibly small, its spherical aberation can be evaluated with the same formula as used for the einzel lens[2] ;

$$\delta W_i \fallingdotseq S q \left(\frac{r}{f} \right)^3 . \tag{6}$$

Here r is the deviation of outmost incident rays from the axis, q is the image point distance from the principal plane, and S is a dimensionless constant, specifying the change of the focal length for outmost rays Δf from f for paraxial rays,[3]

$$\Delta f = S \frac{r^2}{f} . \tag{7}$$

Therefore the spherical aberration for the present case can be expressed as follows;

$$\delta W_i \fallingdotseq \delta W_i^{ent} = S q^{ent} \left(\frac{r}{f^{ent}} \right)^3 . \tag{8}$$

---

[2] G.Liebmann, Proc. Phys. Soc. 62 (1949) 213.

[3] G.H.Bachman and S.Ramo, JAP 14 (1943) 8.

Using the relations $r = \alpha P$, $M = -q^{ent} / P$, and $1/P + 1/q^{ent} = 1/f^{ent}$, it can be rewritten in the following way,

$$\left.\begin{aligned} \delta W_i &= -S\,\alpha^3 f^{ent}\, M \left(1 - \frac{1}{M}\right)^4 \\ &\fallingdotseq -S\,\alpha^3 \frac{f^{ent}}{M^3} \qquad \text{for } M \ll 1 \end{aligned}\right\} \tag{9}$$

Table 1 indicates the accel-lens parameters ζ and S, which are obtained by applying formulas (3), (4) and (5) to the computar-calculated M and δW_i values, shown in section 4. Magnitude of S thus obtained is found to be one order less compared to those usually obtained for the einzel lens (cf.section 6).

Thus we see from formula (9) that smallness of spherical aberration of the accel-lens is assured not only by this small value of f^{ent} but also by the smallness of S factor. The latter fact can be realized by investigating the functional form of formula (5). In the entrance aperture portion, distance of the trajectory from the axis is kept approximately constant so that integration on the terms including $a'(z)$ vanishes. So the main contribution to the integration comes from the first term with the factor a^4. But this term is consisted of two terms which take opposite signs of each other owing to the special field configuration at the entrance portion so that strong cancellation occurs, resulting in a very small value of δW_i.

Table I Characteristic constants of accel-lens, ζ and S, determined from the lens performance, M and δW_i

	N	M	$d_s \left(= \frac{\delta W_i}{\alpha^3} \right)$	f^{ent}	ζ	S
lens VII	100	-0.069	-1.91×10^4 mm	22.6 mm	2.24	-0.21
D = 20mm ϕ	50	-0.111	-1.62×10^4	35.0	1.72	-0.42
	20	-0.244	-1.25×10^4	68.6	1.30	-1.11
	10	-0.531	-1.06×10^4	121.4	1.09	-2.4
lens VII	100	-0.062	-3.31×10^4	20.4	2.02	-0.30
D = 10mm ϕ	50	-0.102	-2.58×10^4	32.4	1.59	-0.57
(cf.Fig.4)	20	-0.230	-1.81×10^4	65.4	1.24	-1.4
	10	-0.505	-1.44×10^4	117.4	1.06	-2.9

$$f^{ent} = \frac{P}{1 - \frac{1}{M}}, \qquad \zeta = \frac{(N-1) f^{ent}}{4K}, \qquad S = \frac{-d_s}{f^{ent} M \left(1 - \frac{1}{M}\right)^4}$$

P = 350mm, K = 250mm

§ 6 Optical property of the einzel lens

To find the optimum shape of the einzel lens to be used in conjunction with the accel-lens, calculation has been made for various shapes of electrodes. A stringent requirement was placed for the high voltage holding capability since this einzel lens has to hold the same high voltage as that of the accel-lens. Fig.7 shows the arrangement with the best quality thus obtained. Values of δW_i for these lenses, computed with the same code as used for the accel-lens, are shown in Fig.8 as a function of M. Here different values of M are obtained by changing the object distance. Very steep increase of the value of δW_i occurs for the smaller values of M. This feature can be understood reasonably from the general nature of the spherical aberration which is expressed by a formula of the same form as formula (9) ($\delta W_i \propto M^{-3}$).

The values of S, determined by this formula are plotted in Fig.8. We see that Sis a considerably good constant which can be used as the measure of goodness ofthe lens property.

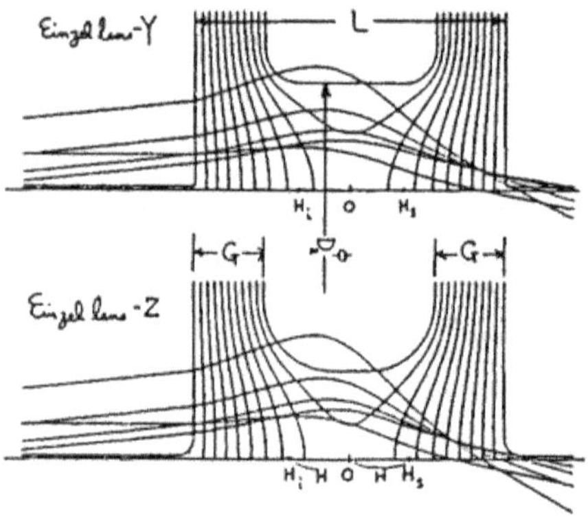

L = 450 mm G = 100 mm D_2 = 300 mm
f = 264 mm H = 74.5 mm $\alpha = 10^{-3}$

a	M	d_6	S	δW_i
1550 mm	-0.21	-3.24×10^5 mm	-5.3	-0.32 μm
1050	-0.34	-0.89×10^5	-4.1	-0.09
800	-0.49	-0.37×10^5	-3.4	-0.04
675	-0.64	-0.22×10^5	-3.0	-0.02

L = 450 mm G = 100 mm D_2 = 250 mm
f = 236 mm H = 86 mm

a	M	d_6	S	δW_i
1560 mm	-0.18	-3.15×10^5 mm	-4.0	-0.32
1060	-0.29	-0.95×10^5	-3.6	-0.10
810	-0.41	-0.39×10^5	-2.9	-0.04
685	-0.52	-0.23×10^5	-2.6	-0.02

a: distance from H_s to object point.
d_6 : spherical aberration coeff.

$$\delta W_i = d_6\,\alpha^3 = -S\alpha^3 f\,\frac{(M-1)^4}{M^3}$$

Fig.7 Property of the einzel lens

Steep rise of the spherical aberration, above mentioned, can be avoided of course, in the actual system by placing a defining slit in front of the einzel lens. But if we want to keep the acceptance angle not less than 10^{-3} radians, values of magnification can not be lowered much less than 0.5, as far as the lens dimension shown in Figs.3 and 4 is used.

If the dimension of the lens is reduced to the limit of high voltage holding capability, it may be possible to obtain smaller values of M. But in any case it does not seem easy to get the magnification less than 10^{-1} with keeping the spherical aberration less than 0.1μm.

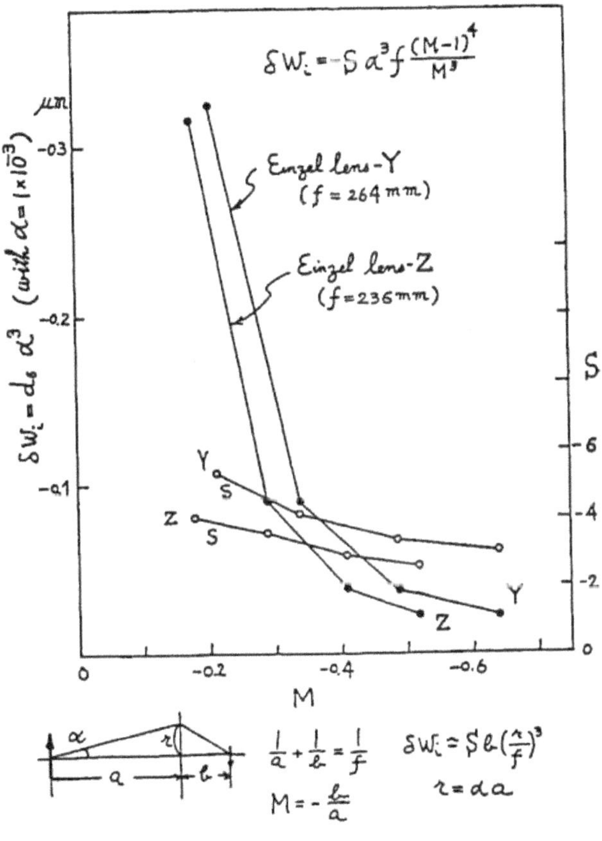

Fig.8 Spherical aberration of the einzel lens as a function of magnification M

§ 7 Proposed design of the optics of the microbeam apparatus capable of producing 0.1 μm beam

On the auguments in the last sections we conclude that ion lens systems, composed of accel and einzel lenses, can be used to obtain micro-beams less than 0.1 μm. Since the practical limit of contraction is about 0.02 for the accel-lens and about 0.3 for the einzel lens, assuming the acceptance angle of 10^{-3} radians for both lenses, overall magnification available becomes about 6×10^{-3}. This lens system should give a beam spot of 0.1 μm with an incident beam slit of 10^{-2} mm in diameter.

Large reduction of the magnification without deteriorating the spherical aberration can be achieved by using two accel-lenses in cascade. This possibility is shown in Fig.9. The overall spherical aberration of the combined lens system is mostly determined by the last lens, so the lens with the smaller N value has to be placed in the last. This figure shows that magnification of 1×10^{-3} can be achieved with the spherical aberration of 0.1 μm, assuming the incident acceptance angles of 10^{-3} radians for both lenses.

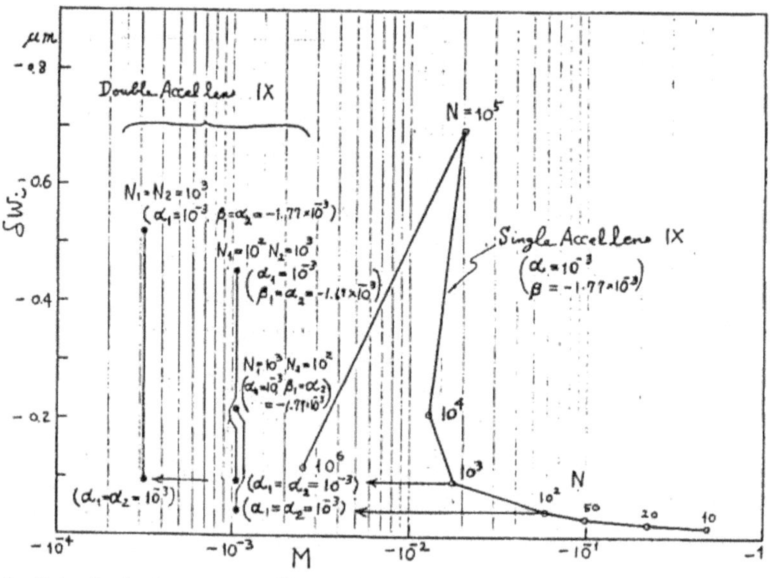

Fig.9 Spherical aberration δW_i of the two-stage accel-lens as a function of magnification M

One of crucial points to be considered carefully is that energy spread of incident low energy ion beams obtained from the ion source, might cause chromatic aberration. This effect occurs being associated with the longitudinal extension of the image. This magnitude can be evaluated by Elkind's formula. Results of calculation are shown in Table II. Here the longitudinal extension of the first image is 0.21mm, with a diverging angle of 1.77×10^{-3} radians. If the divergent angle is limited to 1×10^{-3} radians by placing a defining slit in front of the second lens, the effective lateral extension of the beam flux outgoing from this focus point is found to be 0.2μm. This is small enough compared to the expected image size at this point(~several μm)and hence has little effect. For the second accel-lens with the same N value as the first one, the longitudinal extension of the focused beam is less than that of the first lens, owing to the larger field strength E for the second lens. Thus we see that the chromatic aberration can be made much smaller than 0.1μm for the energy spread of the initial ion beam of 1eV. This value of energy spread is considered to be realizable with the duoplasmatron type ion source.

Table II Two-stage accel-lens system

	V_i	V_f	N	K	E	dL/dV_i	ΔV_i	ΔL
	V	V		mm	V/mm	mm/V	V	mm
1 st lens	10	10^3	100	10	99	0.21	1	0.21
2 nd lens	10^3	10^5	100	250	360	0.05	1	0.05

ΔL is the longitudinal extension of the image for each lens, caused by the energy spread ΔV_i of the incident beam, which is the same for both lenses. $dL/dV_i = (N^2/E) \times d(L/K)/dN$ can be evaluated by the Elkind's formula, giving $d(L/K)/dN = -0.0021$.

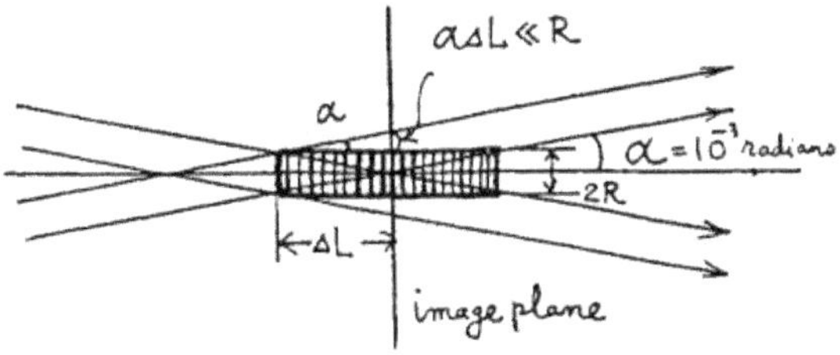

ON NEUTRON TRANSMUTATION DOPING OF SILICON

K. Dyrbye,
TOPSIL SEMICONDUCTOR MATERIALS,
DK-3600 Frederikssund, Denmark

ABSTRACT

For the application of silicon for high power semiconductor devices precision doping is required. Doping silicon crystals with phosphorus by neutron irradiation is the most common technique. Here, the neutron transmutation doping technique is briefly presented. The discussion focusses on the outstanding uniformity of the doping profile, essential for the performance of high power devices. Simple estimates of the significance of the factors limiting the uniformity are made.

INTRODUCTION

Since the early 1970's the application of large mono-crystalline silicon ingots has increased very fast and tremendous efforts have been invested in developing techniques for the growth of perfect Si-crystals with ever increasing dimensions. These crystals are the basic material for a wide range of electric and electronic devices. For most applications various elements have to be added to the pure silicon to taylor the electric characteristics to be appropriate for the devices in mind. Depending on the element added, and hence on the dominating charge carrier, silicon crystals are commonly divided into n-type and p-type. The most widely used dopants are phosphorus for n-type and boron for p-type silicon. When talking about neutron transmutation doping (NTD)[1,2] of silicon we are considering the production of n-type silicon doped with phosphorus by transmutation of

^{30}Si into ^{31}P by neutron irradiation. TOPSIL was pioneer in the production of NTD-silicon and has been involved in developing this technique since the very beginning in the early nineteen seventies[1,3,4] where production of sufficiently clean crystals to benefit from the outstanding doping characteristics obtained by NTD became possible. The NTD-technique is used when the application requires very low doping levels of high precision and uniformity. Typical dopant levels for NTD silicon and the corresponding resistivities are indicated in Table 1. The major market for silicon of this extremely clean and uniform quality is power semiconductor devices[5]. For the performance of thyristors and rectifiers uniformity is essential for obtaining high and welldefined breakdown voltages. Also, the market for the silicon based particle detectors is important[6].

Table 1: *In the typical resistivity range of NTD-silicon corresponding phosphorus doping concentrations and required neutron doses to introduce the phosphorus are listed.*

Res. (ohm cm)	Dopant conc. (10^{13} atom/cm^3)	ppba Phos.	Neutron dose (10^{16} /cm^2)
30	14.5	2.9	86
100	4.3	0.85	25
200	2.1	0.42	12
300	1.4	0.28	8

The NTD-technique exists along with a range of chemical doping methods, however, for high resistivity silicon doping, i.e. in the range of table 1, NTD is by far the most precise and uniform way to introduce phosphorus. In making comparisons the resistivity and the resistivity distribution, which reflect the carrier concentration, are often the key parameters. I will comment on the connection between the dopant concentration and the resistivity later, however,

roughly the resistivity is proportional to the reciprocal dopant concentration.
In the present paper the technique will be discussed, focussing on the factors limiting the homogeniety of silicon crystals doped by irradiation. In the range of lower resistivities the irradiation precision is the dominating uncertainty, at higher resistivities the requirements for the starting material become very restrictive, if we want to avoid the material properties to become the dominating uncertainty.

THE NEUTRON TRANSMUTATION TECHNIQUE.

For the conventional chemical doping methods, e.g. incorporation of impurities from the melt during crystal growth, the thermodynamics play an important role. Since nearly all impurities of interest have distribution coefficients much less than unity, they prefer to stay in the melt. This implies a non-uniform impurity distribution in the as grown crystals. In addition, lack of control of the melt and the feedrate of the doping impurities enhances the non-uniformity. This is in particular true for very low dopant concentrations, i.e. in the application range of NTD. For NTD the role of thermodynamics is negligible. The homogeneity is determined by the uniformity of the neutron flux over the volume of the Si-crystal and by the homogeneity (purity) of the as grown crystals.
The irradiation induced P-atoms are created via neutron capture by a ^{30}Si atom. Hereby the unstable ^{31}Si isotope is formed. ^{31}Si subsequently transmutesto ^{31}P by ß-emision, with a halflife of 2.6 hours (see Table 2). In addition to this reaction a number of other reactions take place, of which the most important are shown in Table 2.

Table 2: *Nuclear reactions*

$^{30}Si(n,\gamma)^{31}Si \xrightarrow{2.6h} {}^{31}P + \beta^-$
$^{28}Si(n,\gamma)^{29}Si$
$^{29}Si(n,\gamma)^{30}Si$
$^{31}P(n,\gamma)^{32}P \xrightarrow{14.3h} {}^{32}S + \beta^-$

Simple calculations, however, show that these reactions need not be considered, as their production of new isotopes is negligible for reasonable doping levels. The crystals are exposed to the desired dose of neutrons, placing the crystal close to the core of a nuclear reactor. In practice, of course, such dopings are not perfect. Neither is the neutron flux homogene, nor is the background impurity levels in the as grown crystals negligible. Both give rise to non-uniformities in the final doping profile. Also, the irradiation partly damages the crystal structure, mainly by the introduction of pointlike defects. Prior to application these must be annealed out. The resistivities of irradiated crystals referred to later in this paper are all measured after an annealing. The defect annealing aspect of the NTD will not be discussed here.

IRRADIATION INHOMOGENEITIES

In the present chapter we disregard effects caused by impurities in the grown crystal. This approximation is indeed good in the low resistivity crystals (the most heavily doped ones). Taking into account this assumption, the doping profile is determined by the irradiation alone. We often distinguish between two types of homogeneity, the doping uniformity along the crystal axis, which plays no role for the individual wafers cut from the crystal, and the radial variation, essential for each wafer.

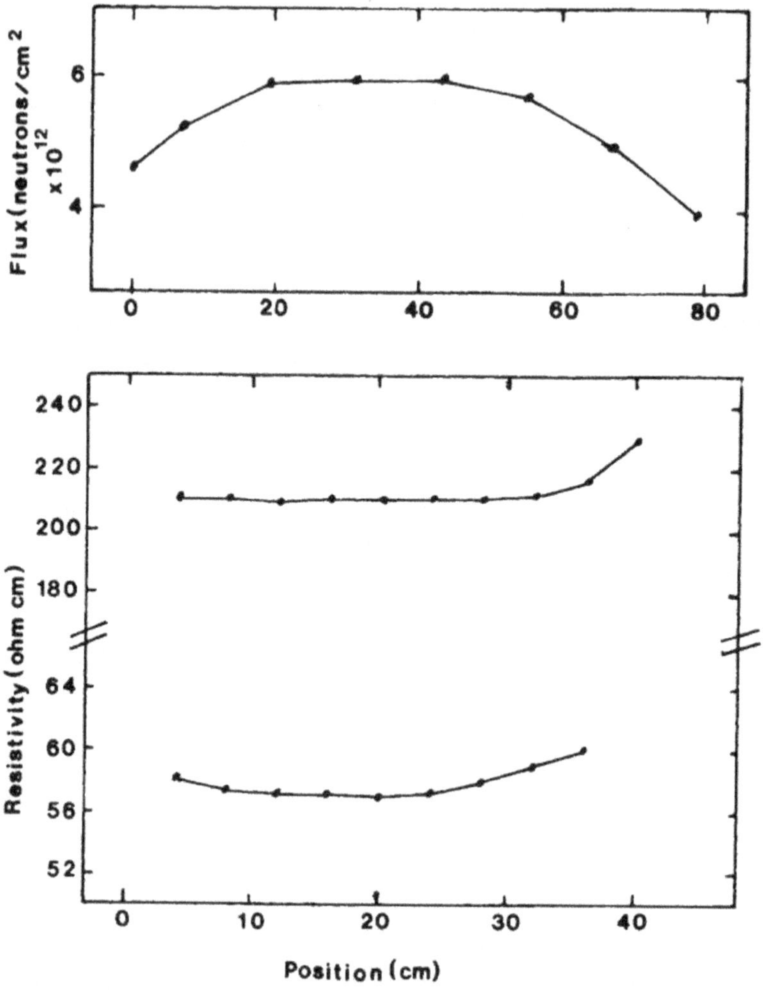

Fig 1: a) Neutron flux versus position in a vertical irradiation facility. b) Resistivities of two crystals irradiated in the channel of fig 1a versus the axial position on the crystals. The flux profile is reflected.

In general Si irradiation is carried out in vertical channels beside the reactor core. The crystals have diameters up to 5" and lengths typically up to 60 cm during irradiation. Since the neutron transmutation is carried out at small research reactors the flux is rather inhomogene over the crystal length if no precautions are taken. The effect is seen on Fig. 1, showing the neutron flux along the channel axis, (a), and resistivities of crystals irradiated in this facility versus the position on the crystal (b). Clearly the flux profile is reflected in the doping profile. The solutions to the problem are either flattening of the flux on the dispense of intensity or moving the crystal during the irradiation.

The radial resistivity profile is determined by the attenuation of the neutron flux when penetrating the silicon crystal. The decay length in silicon is about 19 cm[3], hence for stationary crystals the flux inhomogeneity is rather large, as seen from Table 3.

Table 3: *Radial flux variation (%)*

Crystal dia.	Stationary crystal	Rotating crystal
3"	33	1.0
4"	41	1.7
5"	48	2.5

To minimize this effect the ingots must rotate during irradiation. If we consider the cylindrical crystals surrounded by material with the same decay length as silicon itself we can calculate the non-uniformity of dose by integrating the time average of the flux at a distance from the crystal axis over a rotation period[3]. To first order the result is

$$\frac{\Phi(a)}{\Phi(0)} \approx 1 + \frac{a^2}{4\ b^2} \qquad (1)$$

where Φ is the average flux at a distance from the crystal axis, a is the radius of the crystal and b the decay length. Results are shown in table 3. The rotation is seen to be a considerable improvement compared to stationary irradiations. In practice the situation is sligthly different, since the material surrounding the ingots normally has decay lengths longer than 19cm. In most irradiation facilities the radial flux inhomogeneity is much less than the axial.

RESISTIVITY VARIATIONS

To convert the inhomogeneities in the flux to resistivity variations the relations between resistivity and neutron dose must be considered. First, however, we consider the relations between resistivity and doping levels taking into account that both acceptor and donor impurities are present. The resistivity is given by

$$\delta = \frac{1}{ne\mu_e + pe\mu_p} \qquad (2)$$

where e is the charge unit, n and p are the electrons and hole concentrations and μ_e and μ_p is the electron and hole mobilities, respectively. In the pure type of silicon crystals, we are discussing here, μ_e and μ_p can be treated as constants, hence we only need to estimate n and p. Assuming that the only electric active impurity is B as acceptor and P as donor, a very good estimate is given by[7]

$$n = \tfrac{1}{2}\sqrt{(N_P-N_B)^2+4n_i^2} + \tfrac{1}{2}(N_P-N_B) \qquad (3)$$

where N_P and N_B are the phosphorus and boron concentration, and n_i is the intrinsic carrier concentration commonly taken to be 1.45 $10^{10}cm^{-3}$ at 300K. p is known from the relation

$$pn = n_i^2 \tag{4}$$

We can now make assumptions about the B-impurity level, and calculate the silicon resistivity from equations (2), (3) and (4). Some results are shown on Fig. 2, where the resistivity is shown versus P-concentration for B-concentrations in the range 0.01-0.025 ppba.

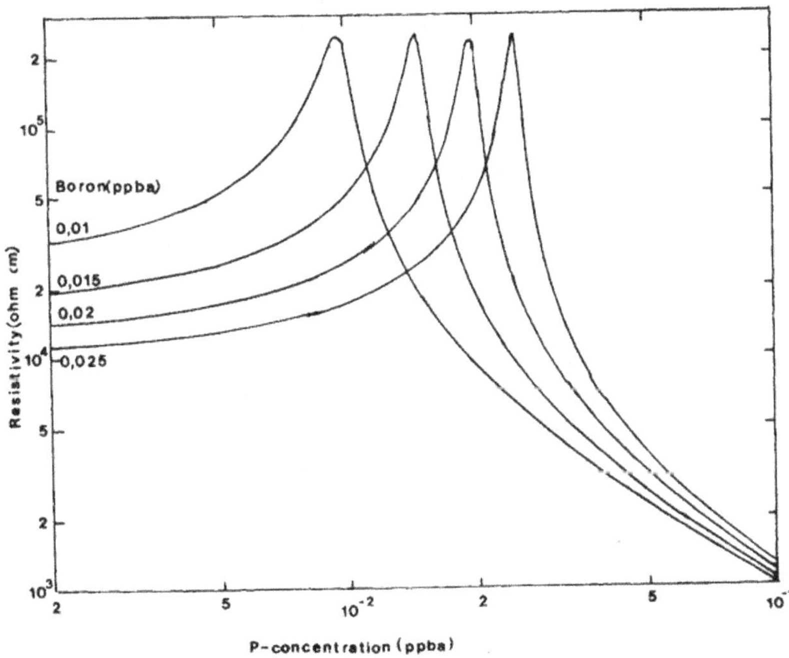

Fig. 2: *Estimated resistivities in near intrinsic silicon at 300K versus phosphorus concentration for various background levels of boron.* μ_n *and* μ_p *are chosen to* $1396 cm^2/Vs$ *and* $492 cm^2/Vs$.

For all B-concentrations, at lower P-concentrations the resistivity increases with P-concentration due to compensation of the holes. It peaks as the B and P concentration is about equal. The different mobilities of the holes and electrons move the peak position sligthly away from exactly equal concentration of P and B. At the peak the silicon type changes from p- to n-type. At higher P-concentrations the resistivity decreases, still following separate curves for each B-concentration, however, the curves approach each other at lower resistivities even on the logaritmic axis used on the figure. This feature is important since we want to dope to a certain resistivity adding phosphorus uniformly, and since precision control of these rest impurities is impossible and not better than conventional doped silicon. The nearly universal behaviour of the resistivity at the low resistivities inspires us to characterize the starting material for the NTD by its resistivity only. If we do so, and assume that the phosphorus donor levels introduced by the irradiation are fully ionized and uniformly distributed, we can get an easy overview of the homogeneity of the doped crystals. The amount of extra P introduced is given by

$$N_{P,NTD} = k\Phi, \qquad k = d\ c(^{30}Si)\sigma \tag{5}$$

where Φ is the neutron dose, d the atomic density of silicon, $c(^{30}Si)$ the relative presence of the ^{30}Si isotope and σ the cross section for the transmutation reaction. From (2) and (5) the following relation between the neutron dose and target resistivity is obtained

$$\frac{1}{\delta_{target}} = k\Phi + \frac{1}{\delta_{initial}} \tag{6}$$

Where δ_{target} and $\delta_{initial}$ are the resistivity of the doped and as grown crystal respectively. On Fig. 3 this approximate universal relation between the neutron dose and the silicon resistivity is displayed for various resistivities of the starting material.

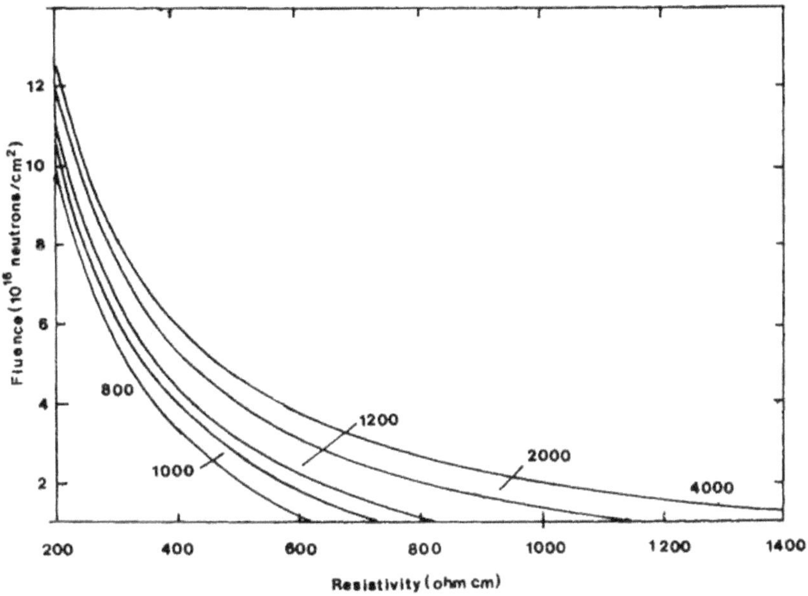

Fig. 3: *Neutron dose versus the resistivity obtained for crystal of various resistivities indicated on the figure in ohmcm. Only target resistivities higher than 200 ohmcm are displayed.*

Consider an ideal irradiation represented by a horizontal line on Fig. 3. Evidently the resistivity of the starting material, and therefore the homogeneity of the as grown crystals, is important for the final result. As in Fig. 2 the curves in this figure approach each other at higher doses, i.e. at lower resistivities, illustrating the increasing influence at higher resistivities from the as grown crystal on the final result. For example it is seen that with undoped crystal resistivities between 2000 and 4000 ohmcm and a

doping with 3 10^{16} neutrons/cm^2 we obtain silicon of about 650 ohmcm ±8%, whereas doses of 1.2 10^{17} neutrons/cm^2 result in 200 ohmcm with a variation of only ±4%.
We have illustated the influence of the non-uniformity of the irradiation and of the starting material on the NTD-crystal resistivity. Fig. 4 is included to evaluate their relative importance. The curves on Fig. 4 are based on equation (6) the resistivity variation over the whole crystal is shown versus the target resistivity. Making assumptions on the inhomogeneity of the irradiation and on the resistivity variation of the crystal the curves are obtained, estimating the maximun and minimun resistivities of the crystal. As in the previous chapters we disregard offset effects from problems in measuring the absolute neutron dose. Both in part a and b of the figure the upper curve represent the total variation, the dotted is the contribution from the irradiation (obtained assuming a homogene starting material), the dashed is due to the material properties.
In Fig. 4a the irradiation inhomogeniety is chosen to ±4% and the crystal in the resistivity range 2000-4000 ohmcm. The relative variation is seen to increase from 8% at low resistivities to about 40% at 1000ohmcm. In the low resistivity range the uncertainty is primarily due to the irradiation uncertainty, however, the irradiation contribution is nearly independent of the target resistivity, whereas the influence from the starting material increases with resistance and becomes the dominating part above 300 ohmcm. In Fig. 4b the variation is limited to 4% for lower resistivities by improving the quality of the irradiation to ±2% flux variation. At 1000 ohmcm the total variation is 16%, in this range the improvement with respect to part a) is primarily due to better purity of the material. Note that the relative homogeneity of the material in Fig. 4a and Fig. 4b are the same. A third route to more uniform NTD crystal is of course more uniform crystals.

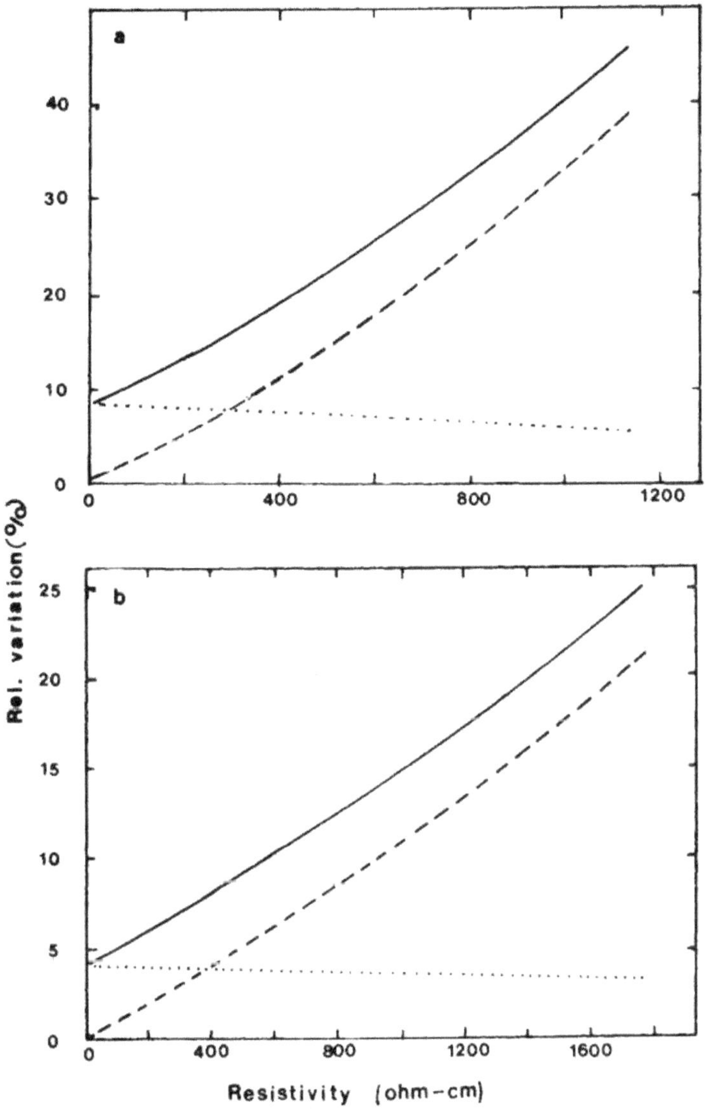

Fig. 4: The relative resistivity variations are shown versus the target resistivity. The full line is the total variation, dashed linie is contribution from the as grown crystal and the dotted is the irradiation contribution. In part a) crystal resistivity between 2000 and 4000 ohmcm and an irradiation variation of ±4% are arbitrary chosen. In part b) the parameters are 5000-10000 ohmcm and ±2% flux variation.

CONCLUSION

We have discussed the outstanding homogeneity of the doping profiles obtained by NTD, focussing on the factors limiting the uniformity. The importance of the as grown crystal and the irradiation have been compared. The increasing need for extremely pure crystals, when higher resistivities are desired, has been pointed out. For lower resistivities the major uncertainty factor is the irradiation.

ACKNOWLEDGEMENTS

Thanks are due to G.F. Eriksen, K.O. Nielsen and T. Visler, TOPSIL SEMICONDUCTOR MATERIALS, for helpfull discussions.

REFERENCES

1: Neutron Transmutation Doped Silicon, ed. J. Guldberg (Plemum Press, New York, 1981)

2: Silicon Transmutation Doping Techniques and Practices, IAEA-TECDOC-456, Vienna, 1988.

3: H.M. Janus and O. Malmros, IEEE Transac. on Elec. Dev., 8, 797 (1976).

4: N. Kaltenborn, O. Malmros in Neutron Transmutation Doping of Semiconductor Materials ed. R.D. Larrabee (Plenum Press, New York, 1984), 139.

5: H. Herzer in Semiconductor Processing ed. D.C. Gupta (ASTM Publication, Philadelphia, 1984) 533.

6: W.v. Ammon, J. Kemmer in Semiconductor Processing ed. D.C. Gupta (ASTM Publication, Philadelphia, 1984) 546.

7: J.M. Meese, P.J. Glairon in Neutron Transmutation Doping in Semiconductors ed. J.W. Meese (Plenum Press, New York, 1979) 109.

STRUCTURAL CHARACTERIZATION OF II-VI SEMI-CONDUCTORS BY ION CHANNELING

B. DAUDIN, E. LIGEON, J. FONTENILLE and A.C. CHAMI*
CENTRE D'ETUDES NUCLEAIRES DE GRENOBLE, DRF-G/SBT, BP 85 X, 38041 GRENOBLE CEDEX, F.
* INSTITUT DE PHYSIQUE, USTHB
BP 31, EL ALIA, ALGER, ALGERIA.

ABSTRACT :
We present a survey on the structural characterization of II-VI semiconductors using Rutherford backscattering combined to ion channeling. The possibility of determining the polarity and the bonding direction of atoms in the zinc-blende structure is discussed. The dechanneling by structural defects such as twins or misfit dislocations is described. An application to quantum well characterization is given.

INTRODUCTION :

A great interest in the II-VI material $Hg_{1-x}Cd_xTe$ has grown during recent years in relation to detector applications, due to the possibility of covering the entire spectral range from 0.8 μm (CdTe) to over 30 μm by adjusting the x value.

In addition, it has been found that Molecular Beam Epitaxy (MBE) could be successfully used to produce heterostructures such as quantum wells (QWs) or superlattices (SLs) composed of very thin (CdTe/ZnTe, ...) layers with abrupt interfaces. The quality of the resulting devices depends on the electronic properties of the semiconductor materials as well as on the crystalline quality : accurate monitoring of the heterostructures period and quality of the interfaces. The first point implies that the growth rate is rigorously controlled. To achieve the second goal, it is necessary that no interdiffusion takes place at the interface and that no structural defects are present. Measurement of chemical

composition, thickness, lattice strain and structural defects density in these layers are therefore of primary importance for the understanding of their physical properties. High resolution electron microscopy, X-rays diffraction, and Rutherford backscattering (RBS) with ion channeling are commonly used for this purpose. These techniques provide complementary informations.
Nevertheless, it is worth pointing out that ion beam techniques, in contrast to electron microscopy, require no sample preparation. On the other hand, in contrast to X-rays diffraction techniques, they can be successfully applied to the analysis of very thin layers. For these reasons, RBS and ion channeling are a very powerful tool for structural characterization of thin films.

The aim of this paper is to illustrate the above statement through significant examples. Results of crystal polarity determination by ion channeling as well as interface and growth structural defects characterization will be presented. An application to quantum wells will also be discussed.

POLARITY DETERMINATION :

The face polarity in zinc blende compound semiconductors is of great importance for their chemical and physical properties. In particular, it has been shown that epitaxial growth exhibited significant differences when initiated on A or B {111} faces [1].

In the case of CdTe, the (111) B face (the Te face) corresponds to the so-called "best face for epitaxy" and several methods have been developed in the past in order to determine it, ranging from chemical etching [2] and X-ray diffraction [3, 4] to, more recently, Auger electron spectroscopy [5] and electron diffraction [6]. Most of these methods require sample preparation. On the other hand, no particular surface processing is necessary for ion channeling. This method, which is related to the asymetry of the channeling dips in zinc-bende structure, provides direct polarity determination of the {111} face.

From figure 1, it is clear that (111) surface polarity is related to the CdTe (or TeCd) stacking sequence of inclined $(11\bar{1})$ planes. Such a sequence can be deduced from the backscattering yield at critical angle ψ_c in $(11\bar{1})$ planar channeling [7]. As a matter of fact, when the incident ion beam impinges on $(11\bar{1})$ planes at critical angle ψ_c, the channeled particles trajectory is tangent to the Cd or Te planes at a depth $\lambda/4$, where λ is the oscillation length [8]. As a result, a maximum in the backscattering yield (see figure 2) corresponding to a preferential diffusion on Cd planes is expected at a depth of $\lambda/4$ for an incident angle of $-\psi_c$ with respect to the $(11\bar{1})$ plane direction, whereas the symmetrical situation at $+\psi_c$ gives rise to a preferential diffusion on Te planes at the same depth. Such a situation, corresponds to a B face polarity.

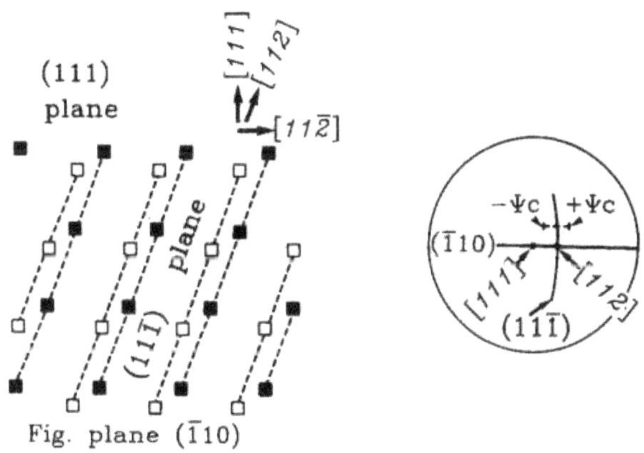

Figure 1 : $(\bar{1}10)$ plane view of (111) B face CdTe crystal. Te atoms are represented by ■ and Cd atoms by □. The circle shows the stereographic [111] projection, indicating $\pm\ \psi_c$ angle used in channeling experiments.

The above method has also been applied to epitaxially grown (111) CdTe on (001) GaAs [7]. The experimental procedure was to grow a 4 μm thick (111) CdTe layer and then prepare a step by chemical etching in order to perform channeling analysis on both CdTe and GaAs substrate (see figure 3). Angular scans through (11$\bar{1}$) CdTe plane revealed that the deposited CdTe was (111) B type. On the other hand, scanning through (111) GaAs planes allowed to determine the stacking sequence of the (111) planes and the direction of the Ga and As dangling bond.

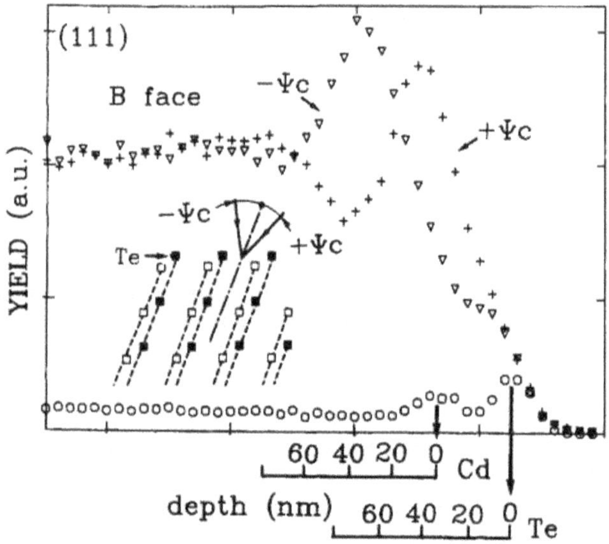

Figure 2 : *Channeling experiments on (111) $Cd_{0.96}Zn_{0.04}Te$ crystal. $E_{^4He}$ = 2.4 MeV. The depth scale for each element is given. The spectra were recorded at critical angle $-\psi_c$ and $+\psi_c$ (ψ_c=0.3°)*

Channeling also allowed to check the following epitaxial relationships :

a) [111] CdTe parallel to [001] Ga As.
b) One {110} plane ($\bar{1}$10) is common to CdTe and GaAs.
c) In direction perpendicular to the ($\bar{1}$10) plane, the (11$\bar{2}$) CdTe plane is parallel to the (110) GaAs plane.

From the above considerations, it was therefore possible to suggest a model of the (111) CdTe/(001)GaAs epitaxial growth basically stating that As sites can be filled with Te atoms, as shown in figure 3, [7].

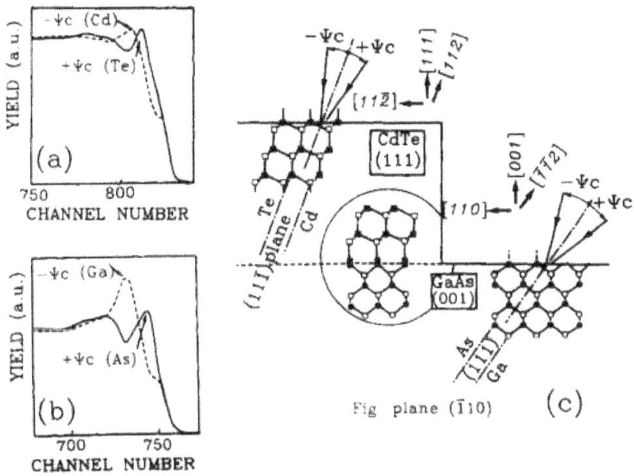

Figure 3 : (a) Spectra recorded for the (111) epitaxial grown CdTe on (001) GaAs. Stacking sequence of $(11\bar{1})$ plane indicates a B face (Te face). (b) Spectra recorded at critical angles $\pm\,\psi_c$ with respect to the (111) GaAs substrate after CdTe layer removal, giving the sequence of the (111) GaAs planes. (c) Schematic representation of channeling experimental results. We have represented the (111) CdTe polarity face, stacking sequence in (001) GaAs, and a possible epitaxial relation between GaAs and CdTe at the interface (encircled). Representation : (O) Ga, (●) As, (□) Cd, and (■) Te.

In conclusion of this section, it is worth stressing that the stacking sequence determination by ion channeling is only possible for polar and asymetrically spaced planes or axes. If twins are present, it is expected that the maximum in the backscattering yield shown in figure 2 will be drastically smoothed. This was actually observed for (111) CdTe grown on (001) GaAs (see figure 3a).

STRUCTURE OF (111) CdTe EPILAYERS ON MISORIENTED (001) GaAs :

GaAs is a good quality as well as cheap material, making it an attractive substrate for epitaxy of II-VI compounds. Nevertheless, characteristic of the (111) CdTe growth on (001) GaAs substrate is the appearance of a high twin and stacking fault density near the interface [9, 10]. On the other hand, it has recently been shown that growth on misorientated GaAs can drastically decrease microtwin formation [11 - 13]. Consequently, the structural and optical qualities are greatly improved, which is of primary importance for practical applications.

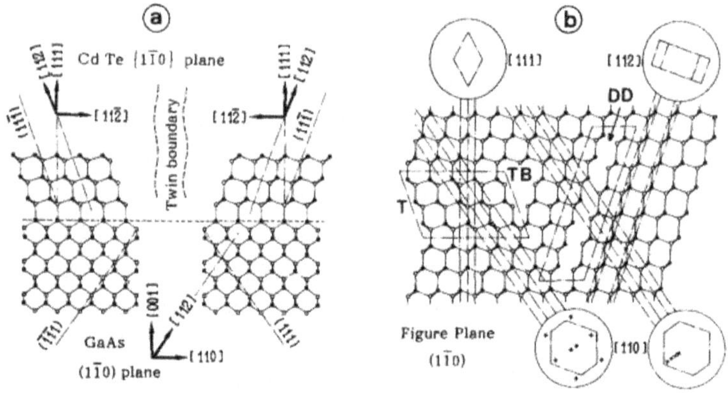

Figure 4 : *(a) Representation of the (111) CdTe epitaxy on nominal (001) GaAs. The {1̄10} plane is common to (001) GaAs and (111) CdTe. The two growth directions of the CdTe are equivalent and a twin boundary takes place at the junction of these two regions. (b) Extended defects in a (111) CdTe layer - T = twinned region ; TB = twin boundary ; DD = dissociated dislocation [for simplification the stacking fault lies only on three (111) planes]. Introduction of a twinned region does not change the [111] axis but displaces lattice atoms when observed in [110] channels. For dissociated dislocations the lattice atoms appear displaced along the [110] axis (little [112] changes are observed).*

Figure 4-a shows the two equivalent growth directions of (111) CdTe deposited on nominal (001) GaAs, responsible for the formation of twins. The resulting extended defects are shown in figure 4-b, from which it is clear that the dechanneling yield will depend on the observation direction.

An illustration of twin characterization by channeling is shown in figure 5. The sample was (111) CdTe grown on nominal (001) GaAs. Near the surface, channeling is as good as on bulk material. The twins density decreases from the interface to the surface. An important consequence is that one particular orientation dominates far from the interface. On the other hand, close to the interface, the large peak in the [110] aligned spectrum is a direct evidence of twins, since only a weak peak appears on the [111] spectrum, likely due to the presence of twin boundaries.

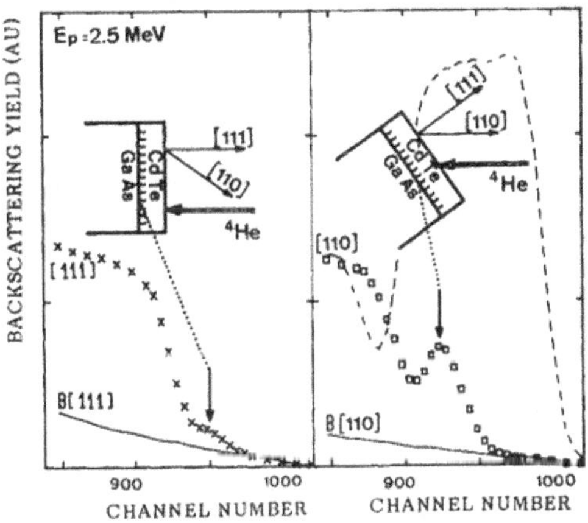

Figure 5 : *Evidence of twins in (111) CdTe/(001) GaAs by protons channeling (2.5 MeV, T = 20 K, layer thickness about 2 μm). (X) Aligned spectrum along the [111] axis of the CdTe layer. ( ) Aligned spectrum along the [110] axis of the CdTe layer. (-) Aligned spectra along the [111] and [110] axes for a CdTe bulk sample (B[111] and B[110]) and random spectrum (---) recorded near the [110] axis.*

As mentioned above, the growth of CdTe on a slightly misoriented GaAs substrate allows to reduce the twin density. Starting from (001) GaAs tilted around $[1\bar{1}0]$ exposes Ga dangling bonds at the step edge, while tilting around [110] exposes As bonds. The nature of steps (Ga or As) is commonly controlled by RHEED but can also directly be deduced from polarity determination, as described in the preceeding section. Figure 6 shows channeling spectra corresponding to (111) CdTe grown on the same run on two GaAs substrates which were 4° misoriented around $[1\bar{1}0]$ and [110] in order to display, respectively, Ga or As steps. Both were mounted side by side on the same sample holder. It is clear from figure 6 that a consequence of growing on As steps is a bad structural quality correlated to a large direct scattering peak in the channeling spectrum. On the other hand, the low backscattering yield observed for the sample tilted around $[1\bar{1}0]$ (Ga steps) is comparable to the value observed for bulk CdTe, which means that the cristallinity is good.

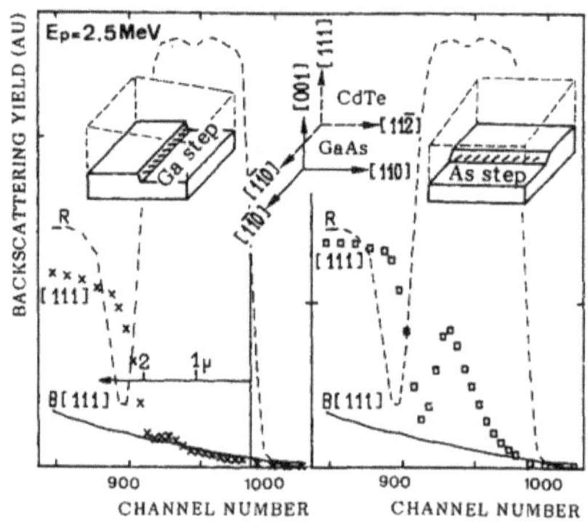

Figure 6 : *Comparison of aligned [111] CdTe layer on 4° misoriented [001] GaAs towards [110], [Ga steps (X)] and towards $[1\bar{1}0]$, [As steps (□)]. Random (---) and [111] aligned (-) spectra (B [111]) corresponding to bulk CdTe.*

Additional experiments not discussed here have shown

that the microscopic growth process is closely related to the value of the misorientation angle [14].

CHARACTERIZATION OF MISFIT DISLOCATIONS BY ION CHANNELING :

Another kind of structural defects present in epitaxial layers is related to the elastic strain relaxation. Extensive theoretical work has been devoted to this problem, leading to the determination of a critical thickness at which misfit dislocations appear at the interface in order to relax strain energy accumulation in the layer [15, 17].

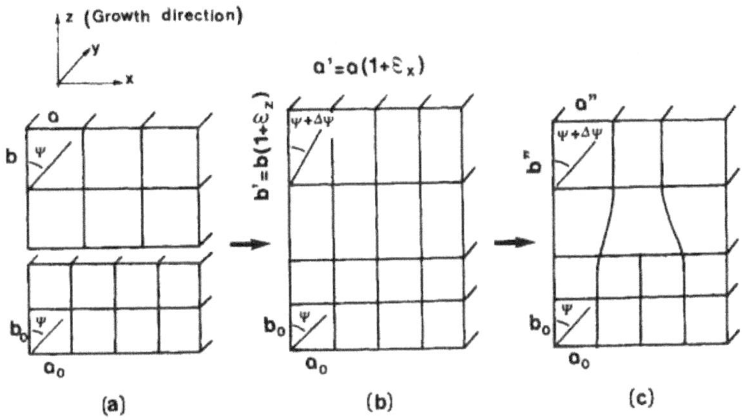

Figure 7 : *(a) Representation of independent layer and substrate lattices. (b) After coherent epitaxy : fully strain accommodated mismatch. (c) Incoherent epitaxy : mismatch accommodated by misfit dislocations and residual strain.*

If the layer is sufficiently thin, the difference in lattice constants is entirely accomodated by elastic deformation of the crystal structure of the material in each layer so as to obtain an uniform lattice spacing in the plane of the interface (figure 7-b). As a consequence of the Poisson effect, this leads to a tetragonal deformation (ϵ_T) of the epitaxial material. Such a deformation can be directly detected by ion channeling, by measuring the angular scan shift between layer and substrate (see figure 7-b). One can therefore derive the relations [18].

$$\mathrm{tg}\ \psi = \frac{a}{b} \text{ and } \frac{2\Delta\psi}{\sin 2\psi} = \frac{\Delta a}{a} - \frac{\Delta b}{b} = \epsilon_T, \qquad (1)$$

where a and b are the sizes of the cell submitted to the strain in the plane and normal to the plane, respectively, $\epsilon_x = \epsilon_y$ is the in-plane strain, ϵ_z the normal strain and $\epsilon_T = \epsilon_x - \epsilon_z$ the tetragonal distortion. Using the elastic coefficients c_{11} and c_{12} (defined for the <001> orientation) and assuming no stress in the normal direction, one finds

$$\epsilon_T = \frac{2\Delta\psi}{\sin 2\psi} = (1 + 2\ c_{12} / c_{11})\ \epsilon_x \qquad (2)$$

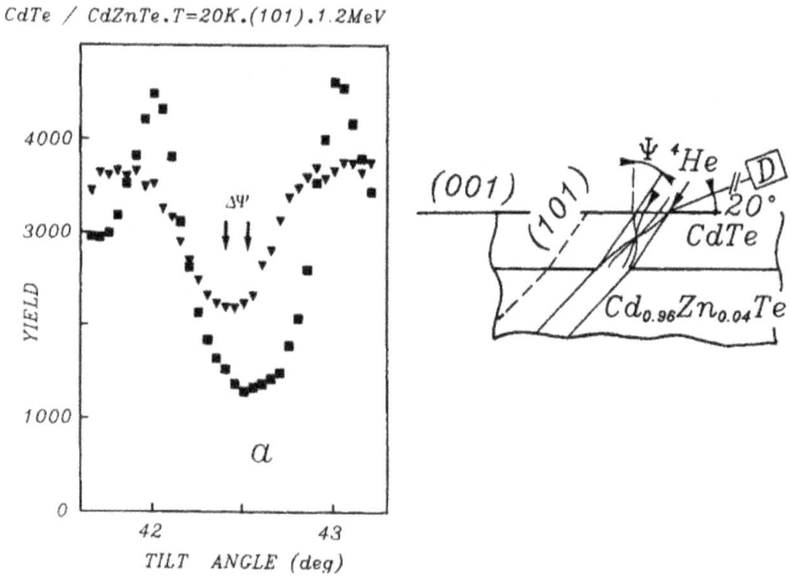

Figure 8 : *Angular scan yield recorded at T = 20 K through (101) plane for CdTe(001)/$Cd_{0.96}Zn_{0.04}Te$ (001). Layer thickness = 50 nm. $E_{^4He}$ =1.2 MeV. The detector grazing exit angle is kept constant (~20°) during the angular scan related to layer (■) and substrate (▼). Angular scan shift between layer and substrate corresponds to about 0.20°, as indicated by arrows.*

For a fully strain-accommodated mismatch, $\epsilon_x = 2.3\times10^{-3}$ corresponds to the relative lattice difference between CdTe and $Cd_{0.96}Zn_{0.04}Te$. Since for [001] CdTe, $C_{11}=5.60\times10^{10}$ N/m^2, and $C_{12} = 3.84\times10^{10}$ N/m^2 [19], we deduce $2\Delta\psi/\sin2\psi = 2.4\epsilon_x$. For the measurement along the (101) plane ($\psi = 45°$), the calculated $\Delta\psi$ is 0.16°. Channeling measurement of $\Delta\psi$ along the (101) plane is presented in fig. 8. Angular scan yield indicates a misorientation of about 0.20° ± 0.02° between layer and substrate. According to relation (2), we deduce that $\epsilon_x \sim 2.9\ 10^{-3}$ which is close to the lattice mismatch between layer and substrate. This result indicates that for a 50-nm-thick CdTe layer on $Cd_{0.96}Zn_{0.04}Te$, the mismatch is fully strain accomodated [18]. It is worth pointing out that absolute $\Delta\psi$ values are obtainable only if all channeled particles reach the interface with parallel velocity vectors [18, 20]. This condition is achieved when $\lambda/2$ (or $n\lambda/2$) is matched to the interface depth. In other cases, it appears that the $\Delta\psi$ value is energy dependent and that the value of the tetragonal distortion cannot be properly determined. This lattice distortion increases the strain energy of the layer. When the layer thickness increases up to a critical value (t_c), misfit dislocations appear at the interface (see fig. 7-c) in order to relax the strain energy accumulation in the layer [15-17]. The lattice misfit between substrate and layer is then accomodated by a dislocation array. For the system (001) CdTe/(001) $Cd_{0.97}Zn_{0.03}Te$, which presents a low misfit $\left(\frac{\Delta a}{a} = 1,8\ 10^{-3}\right)$, full strain relief corresponds to an average distance between dislocations of 127 nm and the elastic interaction between dislocations can be neglected. Channeled particles, crossing the dislocation strain field, are dechanneled with a probability $P_D = N_D\sigma_D$ where N_D is the dislocation length per unit surface at the interface area and σ_D the dechanneling cross section per unit length of dislocation. The σ_D value related to ^{4}He particles channeling in the (110) CdTe plane is about 20 nm, so that the dechanneling radius surrounding the dislocation line is about 10 nm. Aligned spectra in Fig. 9 reveal a step due to an abrupt dechanneling increase when channeled particles cross the

$CdTe/Cd_{0.97}Zn_{0.03}Te$ interface. Identical interface dechanneling is found for (110) and $(1\bar{1}0)$ planes, indicating symmetrical distribution of misfit dislocations in the two <110> perpendicular directions in the interface plane. Depth scale determination indicates that the position of the dechanneling step corresponds fairly well to the interface location (750 nm in this case). The dechanneling probability P_D at the interface, as deduced from fig. 9, is obtained through the relationship :

$$P_D = (\chi_D - \chi_\nu)/(1 - \chi_\nu),$$

where $(1 - \chi_\nu)$ and $(1 - \chi_D)$ represent the amount of channeled particles before and after the interface, respectively. The contribution of the interface dislocations to the dechanneling yield can therefore be deduced, independently of the growth defects concentration within the layer. Nevertheless, it is clear from the spectra shown in figure 9 that the growth defects concentration is higher in the layer than in the substrate as the dechanneling slope related to the layer is higher than the corresponding one to the substrate.

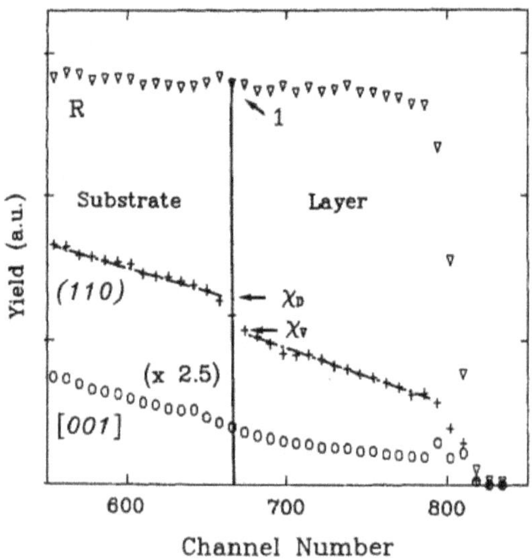

Figure 9 : Rutherford backscattering spectra ($E_{^4He}$ =2.4 MeV) for $CdTe(001)/Cd_{0.97}Zn_{0.03}Te$. (O) Aligned spectrum through [001] axis ; (+) Aligned spectrum through (110) plane ; (▽) random spectrum.

Figure 10 represents the dechanneling probability P_D at the interface versus the thickness of the deposited layers. For a given channeled particle energy ($E_{^4He}$ = 2.4 MeV), σ_D is constant and ~ 20 nm. So, for any layer thickness, P_D is proportional to N_D (dislocation length/unit surface at the interface). For thicknesses lower than 390 nm, no interface dechanneling is observed (N_D = 0) ; the CdTe layer is fully strain accomodated (see fig. 7-b). For higher thicknesses, interface dechanneling appears (N_D = 0). This boundary value corresponds to the critical thickness t_c (390 nm) of this epitaxial system. For layer thicknesses $t > t_c$, the dislocation length N_D at the interface first increases and then tends towards a limit. Calculations of the critical thickness are reported in [21].

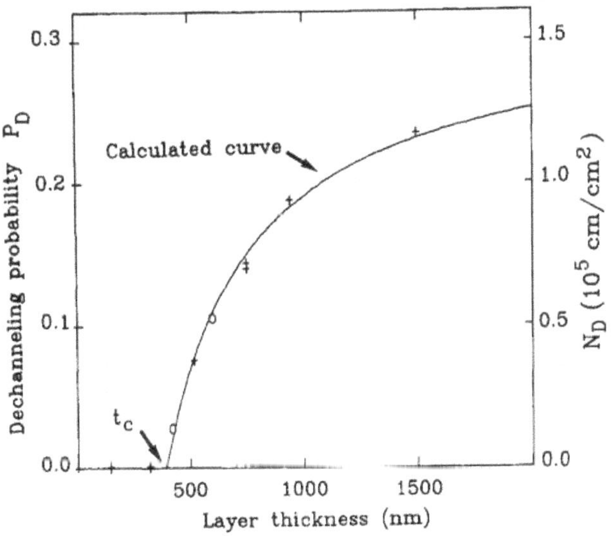

Figure 10 : *Dechanneling probability of 2.4 MeV ^{4}He through (110) plane of CdTe(001)/$Cd_{0.97}Zn_{0.03}Te$ (001) for various layer thicknesses. (+) P_D experimental values obtained in the case of $Cd_{0.97}Zn_{0.03}Te$ substrate ; (O) P_D normalized experimental values obtained in the case of $Cd_{0.96}Zn_{0.04}Te$ substrate. The full line is a calculated curve (after[21])*

The sensitivity of the misfit dislocation density measurements determined from planar channeling then can be estimated. For a mean distance d between two successive misfit dislocations in a square array, the dechanneling probability is $P_D \sim 2\sigma_D/d$. Assuming that a step corresponding to a minimum dechanneling ratio, P_d, of 5%, can be detected, it appears that the minimum distance between two dislocations must be d_{min} = 800 nm. This corresponds to the misfit dislocation density of a fully relaxed layer having a lattice misfit of about 3×10^{-4} with respect to its substrate.

STRUCTURAL CHARACTERIZATION OF QUANTUM WELLS (QW) :

Quantum wells are a current application of semiconductor epitaxial layers. They basically consist of a semiconductor layer confined between two barriers, for instance a 2.5 nm thick CdTe layer confined between two $Cd_{0.1}Zn_{0.9}$ Te barriers. The experimental and calculated RBS spectra corresponding to such a system are shown in figure 11. In order to increase the analysed thickness of the QW, the sample is titled (by 56°) from the beam direction and a grazing exit angle detection (18°) is used. The small peak appearing between the Te and Zn fronts corresponds to the

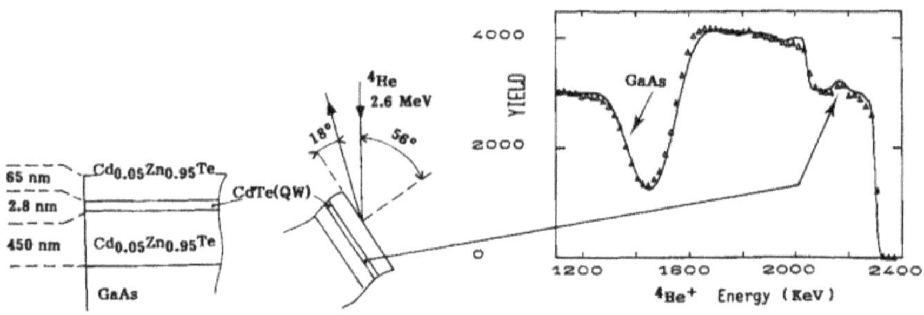

Figure 11 : *Backscattering spectrum for 2.6 MeV ^{4}He ions incident on the CdZnTe/CdTe/CdZnTe/GaAs quantum well system (Δ). Experimental elements concentrations and layer thicknesses are indicated. The full line is the RBS calculated spectrum.*

presence of the CdTe QW. (The GaAs is observed at lower energy). The best fit of the experimental data corresponds to a thickness of 2.8 nm for the QW and of 65 nm and 450 nm for the two barriers (Expected values were 2.5 nm, 50 nm and 400 nm). The chemical composition of the barrier is $Cd_{0.05}Zn_{0.95}Te$, which is slightly different from the expected composition $Cd_{0.1}Zn_{0.90}Te$.

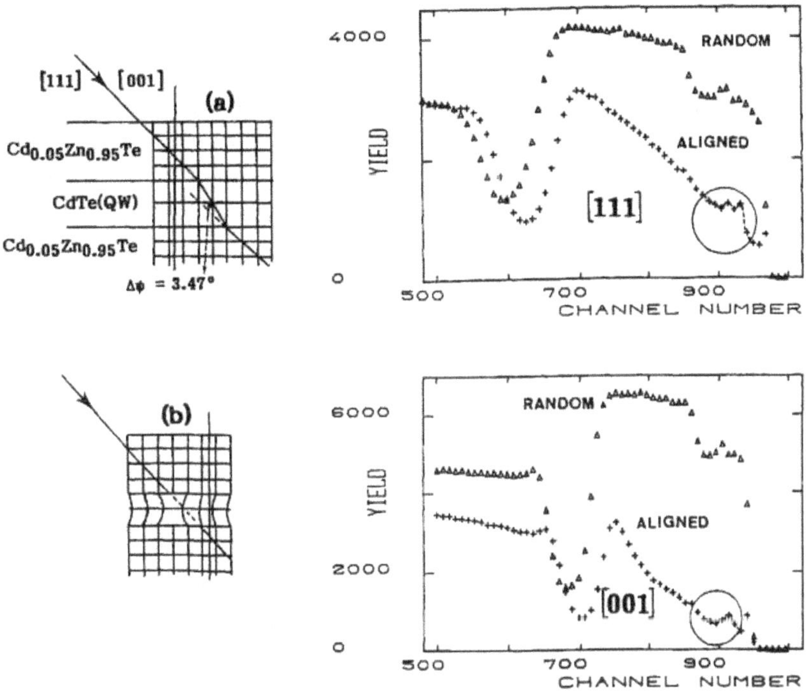

Figure 12 : *Channeling spectra, $E_{^4He}$ =2.4 MeV, along [001] and [111] directions of a thin CdTe layer confined within two CdZnTe barriers. The encircled peaks show the direct scattering on CdTe due to the channel distortion in the thin CdTe layer as schematized in part (b), opposite to (a).*

Channeling spectra along the [001] and [111] directions are shown in fig. 12. The large dechanneling yield, observed on both [001] and [111] directions, varies with the incident beam energy as E_0. Two types of defects can induce such an energy dependence of the dechanneling yield :

(i) stacking faults of the dissociated dislocations created during the growth of the layers
(ii) distortion of the [001] and [111] channels induced by the presence of a dense misfit dislocation array at the GaAs - CdZnTe interface [22] (due to the large lattice mismatch (7,5%) between the materials). The defect density is maximum at the interface and decreases towards the surface. These assumptions are confirmed by HREM results. In addition channeling spectra show that the Te and Cd atoms of the QW give a noticeable rise of direct scattering in the [001] and [111] direction (Encircled peak in fig. 12). Direct scattering in the [001] direction suggests that the QW is not fully strain accomodated (part (a) of fig. 12) as initially expected. The large mismatch between the lattices of the QW and that of the two barriers (~ 7%) is probably accomodated by a misfit dislocation array taking place on each side of the QW. Thus this channeling experiment reveals that a CdTe layer confined between two $Cd_{0.05}Zn_{0.95}Te$ barriers has a critical thickness less than 2.8 nm.

In a different field, ion implantation is a potential method to obtain the two or three dimensional quantum well structure. Such a method, previously described [23, 24], is used to mix the quantum well barrier interfaces and induce an interdiffusion of the element in order to adjust the width of the quantum well. However the vacancies generated by the implanted ions trigger the misfit dislocations formation in the strained quantum well. On fig. 13, aligned spectrum through [001] axis of implanted system (0.8 nm of CdTe confined between two ZnTe barrier) are presented. Before ion implantation, the lattice mismatch (~ 7% between CdTe and ZnTe) is accomodated by strain and no dechanneling in [001] direction is reported. After ion implantation (10^{14} Cd^{+} cm^{-2}, 200 keV) the presence of misfit dislocations is put in evidence by a direct scattering peak on aligned spectra. This example show that layer thickness as small as 0.8 nm can be studied by channeling method.

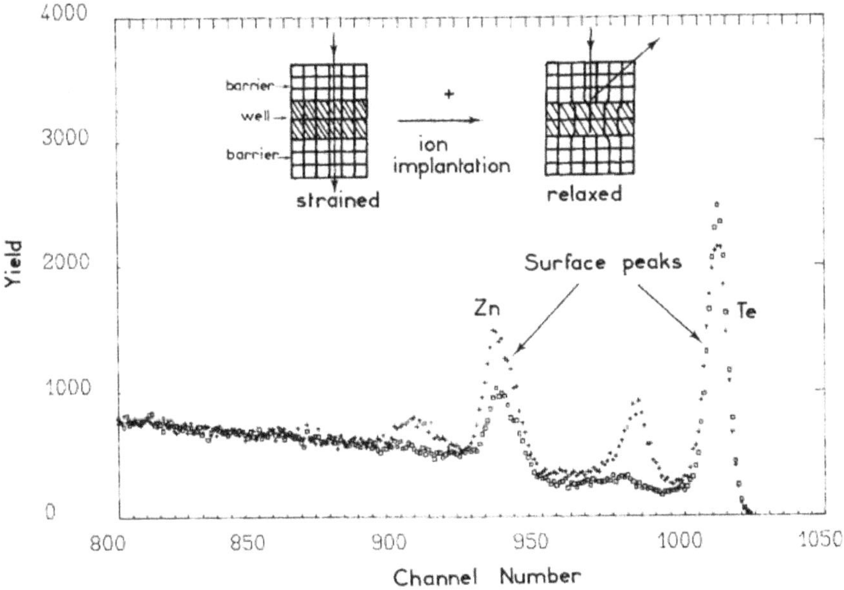

Figure 13 : *Rutherford backscattering spectra ($E\ ^4He$=2.4 MeV) for ZnTe/CdTe/ZnTe system, through [001] axis. (□) : before ion implantation. (+) : after implantation of 10^{14} $Cd^{+}.cm^{-2}$, at 200 keV. The extra peaks (Te and Zn) are due to the presence of misfit dislocations.*

CONCLUSION :

We have shown that ion beam analysis, namely RBS and channeling, provides valuable information on the structural properties of epitaxial layers. Compared to electron microscopy or X-rays diffraction, the ion beam techniques allow in-depth analysis to be performed. Depth resolution can reach about 3 nm near the surface (a few 10 nm at about 1 μm). Comparison of experimental and calculated spectra allow to determine the chemical composition of a layer. This was applied to the analysis of a 3 nm thick CdTe quantum well covered by a 65 nm thick $Cd_{0.05}Zn_{0.95}Te$ barrier. Provided that the surface element has a mass higher than the substrate one, a sensivity better than one monolayer can be reached.

The structural informations obtainable using the ion beam techniques range from surface polarity to the lattice strain, the presence of twins or the density of misfit dislocations. Lattice deformation in a quantum well having a large mismatch with respect to the barrier can be put in evidence by channeling. In addition, it is possible to determine whether such a well is relaxed or not.

REFERENCES :

[1] J.P. Faurie, J. Cryst. Growth, 81, 483, (1987).

[2] T.H. Myers, J.P. Schetzina, T.J. Magee, and R.D. Ormond, J. Vac. Sci. Technol. A 1, 1598, (1983).

[3] E.P. Warckois, M.C. Lavine, A.N. Mariano, and H.C. Gatos, J. Appl. Phys. 33, 690, (1962).

[4] P.F. Fewster and P.A.C. Whiffin, J. Appl. Phys. 54, 4668, (1983).

[5] T.C. Lu, C.M. Stahle, J. Morimoto, R.H. Bube, and R.S. Fiegelson, J. Appl. Phys. 61, 924, (1987).

[6] E.A. Hewat, L. Di Cioccio, A. Million, M. Dupuy, and J.P. Gailliard, J. Appl. Phys. 63, 1 May (1988).

[7] A.C. Chami, E. Ligeon, R. Danielou and J. Fontenille Appl. Phys. Lett. 52, 1502, (1988).

[8] J.H. Barrett, Phys. Rev. B 20, 3535, (1979).

[9] F.A. Ponce, G.B. Anderson, and J.M. Ballingall, Surf. Sci. 168, 564, (1986).

[10] N. Otsuka, L.A. Kolodziejsi, R.L. Gunshor, S. Datta, R.N Bichnell, and J.F. Schetzina, Appl. Phys. Lett. 46, 860, (1985).

[11] S.A. Dvoretsky, A.K. Gutakovsky, Y. Yu Karasev, N.A. Kiselev, I.V. Sabinina, Y. Sidorov, and S.I. Stenin, in *Eurem 88*, edited by P.J. Goodhew and H.G. Dickinson, Inst. Phys. Conf. Ser. No. 93 (Institute of Physics, Bristol, 1988), Vol. 2, p. 407.

[12] J. Cibert, Y. Gobil, K. Saminadayar, S. Tatarenko, A. Chami, G. Feuillet, Le Si Dang, and E. Ligeon, Appl. Phys. Lett. 54, 828, (1988).

[13] J.L. Reno, P.L. Gourley, F. Monfroy, and J.P. Faurie Appl. Phys. Lett. 53, 1747, (1988).

[14] E. Ligeon, C. Chami, R. Danielou, G. Feuillet, J. Fontenille, K. Saminadayar, A. Ponchet, J. Cibert, Y. Gobil and S. Tatarenko, J. Appl. Phys. 67, 2428, (1990).

[15] C.A.B. Ball and J.H. Van der Merwe, in *Dislocations in Solids*, edited by F.R.N. Nabarro (North-Holland, Amsterdam, 1983), p. 121.

[16] J.W. Matthews, in *Dislocations in Solids*, edited by F.R.N. Nabarro (North-Holland, Amsterdam, 1979), p. 462.

[17] R. People and J.C. Bean, Appl. Phys. Lett. 47, 322, (1985).

[18] A.C. Chami, E. Ligeon, J. Fontenille and R. Danielou J. Appl. Phys. 62, 3718, (1987).

[19] O. Madelung Ed., Landolt-Börnstein numerical data and functional relationships in Science and Technology, Springer, Berlin, 1982, Vol. 17.

[20] W.K. Chu, J.A. Ellison, S.T. Picraux, R.M. Biefeld and G.C. Osbourn, Phys. Rev. Lett. 52, 125, (1984).

[21] A.C. Chami, E. Ligeon, R. Danielou, J. Fontenille, G. Lentz, N. Magnea and H. Mariette, Appl. Phys. Lett. 52, 1874, (1988).

[22] A.C. Chami, E. Ligeon, J. Fontenille, G. Feuillet and R. Danielou, J. Appl. Phys. 64, 637, (1988).

[23] H. Okamoto, Jpn. J. Appl. Phys. 26, 315, (1987).

[24] J. Cibert, P.M. Petroff, D.J. Werder, S.J. Pearton, A.C. Gossard and J.H. English, Appl. Phys. Lett. 49, 223, (1986).

MUONIC X-RAYS AS A TOOL FOR NONDESTRUCTIVE MATERIALS RESEARCH

H. Daniel
Physik-Department
Technische Universität München
D-8046 Garching, Fed. Rep. of Germany

ABSTRACT

Negative muons in matter are slowed-down and Coulomb-captured. They then are part of a muonic atom, with one electron replaced by the muon. They cascade down under emission of Auger electrons and, preferably from low-lying states, muonic x-rays. These x rays are characteristic for the element in which they occur. Their energies are, roughly, higher than the energies of the electronic x-rays by a factor muon mass over electron mass. So the suitable instrument is the Ge detector. Every element, in practice with the exception of H, can easily be recognized. The sum over all Lyman transition intensities, almost equal to the sum over all Lyman quantum intensities, yields the number of all muons Coulomb-captured, except for H where transfer of the muon to heavier atoms is possible. With a suitable calibration a quantitative chemical elemental analysis is obtained. Problems connected with this calibration will be treated in detail. Scanning analysis is possible. Future applications will certainly be in the field of surface analysis, performed with very slow muons or (neutral) pμ atoms which are stopped in the first few layers; monolayer analysis will be feasible. Beam cooling, in one way or the other, will enlarge the range of applica tions. A fairly new way of cooling will be sketched, namely cooling with the help of a dc electric field in a gas. Not much hope can be set on the investigation of chemical bonding by muonic x-rays, as the chemical effects are not very pronounced. Although not yet done, analysis with x rays may be combined with imaging where the incoming muon track and the track of the outgoing decay electron are used to pin down the place of the muonic atom. The analysis with muonic x-rays is almost nondestructive; a very weak radioactivity is induced. In practice, however, it will be very hard to detect this activity after the irradiation. Many examples of analyses will be presented.

1. INTRODUCTION

A negative muon (μ^-) is a heavy electron, so it has no strong interaction. Its mass is $m_\mu = 207\ m_e$ where m_e is the electron mass. When stopped in matter the μ^- immediately forms a muonic atom. Due to the large μ^- mass the atomic orbit radii are small and the muonic x-ray energies high. Every negative muon stopped in matter will on the average produce several x-ray quanta.

The muon decays via weak interaction. The free particle mean life is 2.2 μs. Although muons are present in cosmic radiation artificial production is necessary for efficient use. The usual method of production is to bombard nuclei by protons at laboratory energies around or slightly above 600 MeV. The proton interacts mainly with a single nucleon:

$$p + n \rightarrow p + p + \pi^- \quad (1)$$

$$\pi^- \rightarrow \mu^- + \bar{\nu}_\mu . \quad (2)$$

The preference of the above given proton energy comes from the fact that the system p + π has a resonance at 160 MeV in the center of mass reference frame (m_π = 140 MeV).

Another way, again via π^-, is to shoot (negative) electrons (e^-) on nuclei:

$$e^- + n \rightarrow e^- + p + \pi^- . \quad (3)$$

Here the preferred e^- energy is around 300 MeV; the π^- are converted into μ^- as above [eq. (2)]. More efficient than the methods eqs. (1) and (3) would be to employ deuterons (d) or even tritons (t) as projectiles, as neutrons (part of d and t) can interact both with neutrons and protons of the target nuclei to produce π^-; however, there are disadvantages connected with this method, e.g. high n-induced contamination. The use of storage rings or similar devices has been discussed, but, probably due to technical difficulties, not yet tried.

The π^- are collected in a magnetic channel, nowadays mostly with a superconducting solenoid as the main component, where they decay to muons. The muons travel within the same channel until they reach their final destination, e.g. a target whose muonic x-rays shall be measured.

Analogous to the production of μ^- one can also produce μ^+ and use them for magnetic spin rotation experiments (μSR). However positive muons do not produce muonic x-rays and will not be treated here.

Medium energy accelerators mostly built for producing mesons in large quantities are called 'meson factories' . The three best known meson factories in the world are LAMPF at the Los Alamos National Laboratory, Los Alamos, NM, U.S.A.; PSI, Villigen, Canton of Aargau, Switzerland and TRIUMF, Vancouver, BC, Canada. However, there are many more places in the world with useful muon beams.

2. MUONS IN MATTER

2.1 Overview

In order to understand the analysis with muonic x-rays it is important to understand the slowing down and the Coulomb capture.

The curriculum of a negative muon in matter while still on its way is short. Produced at medium energies it is slowed down and captured by Coulomb interaction in a time of the order of 10^{-9} s. It then cascades down the atomic ladder in about 10^{-13} s until it reaches the 1s level. While sitting in this level it may either decay almost the same way as a free particle (predominant in low Z atoms) or may be captured via weak interaction into the nucleus [predominant for medium Z and high Z atoms, the divide being at Z = 12 (Mg)].

The principal muonic Lyman energies range from 2 keV in the case of H to 6.4 MeV in the case of U. Higher series such as the Balmer and Paschen series are also observable in the case of atoms with atomic numbers which are not too low. All these lines are characteristic of the element which emits them. Hence every element can be recognized rather easily by its muonic x-rays, as first proposed already 1969[1)].

2.2 Slowing Down

In order to characterize the mechanisms of energy loss, the quantity

$$W_B = \frac{M}{2}(\alpha c)^2 = \frac{M}{m}\,13.6 \text{ eV}, \qquad (4)$$

where M and m are the muon and electron rest masses, respectively, α the fine structure constant and c the velocity of light, is a useful number. For muon energies $W \gg W_B$ the Bohr mechanism of energy loss is dominant, characterized by the validity of the sudden approximation ("frozen" atomic electrons).

At very low energies, $W \lesssim W_B$, the energy loss is mostly treated being due to collisions with the electrons forming a Fermi gas. The treatment is similar to that of electron conductivity in a metal. The main feature at kinetic energies $W_{kin} \approx 0$ is an energy loss proportional to the particle velocity v,

$$- \frac{dW}{ds} \propto v. \tag{5}$$

At very small energies details of the electronic structure are important for the slowing down process. In condensed matter serious difficulties arise as to how to take the higher electron density and smaller atomic dimensions, both compared to the gaseous state, into account.

The energy loss at very low energies was calculated in closed form by the present author[2)]. Fig. 1 shows the (calculated) energy loss for gaseous Ar and a hypothetical Ar with the density of condensed matter[2)].

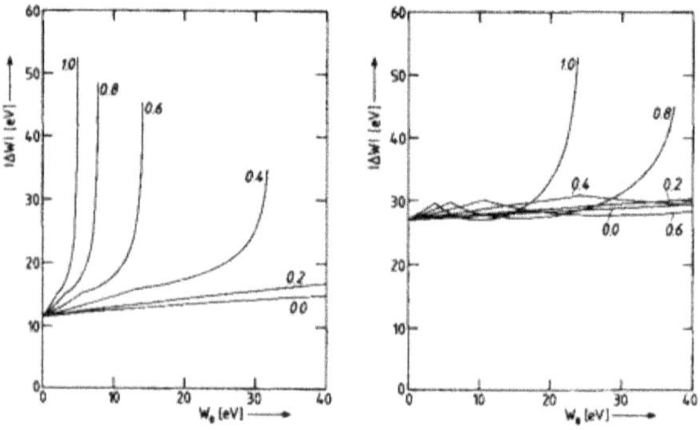

Fig. 1 Energy loss $|\Delta W|$ as function of incoming particle energy W_o. The parameter on the individual curves is the ratio impact parameter over electronic atomic radius r. Left: $r = 1.91 \times 10^{-10}$ m (gaseous); right: $r = 1.40 \times 10^{-10}$ m ("condensed").

In any theory of slowing down and Coulomb capture, it is essential to use correct transport theory:

a. particles captured already at higher energies are no longer available to be slowed down to lower energies;

b. even without this trivial effect there is an important relation between slowing down and spectral flux density of the muons.

Spectral flux density n(W) here means, as usual, the number of particles with energies between W and W + dW which enter a sphere of (small) radius r per unit time, divided by dW and πr^2. As can easily be shown[3], the following equation holds:

$$n(W)\ S(W) = n(W_o)\ S(W_o) = \text{const} \tag{6}$$

where W_o is the energy of the incoming beam. Fig. 2 shows an experimental n(W) distribution measured with a refined time-of-flight method[4]; the starting time is determined with a scintillation detector telescope, the stopping time with a Ge diode detecting muonic x-rays from a AuPd or Si target. Note that there are many muons even at very low energy (down to about one electron volt!).

2.3 Coulomb Capture

2.3.1 Coulomb capture treated quantum mechanically

About half of all papers on Coulomb capture which have appeared at any time treat the problem quantum mechanically, and about half treat it semiclassically. In the quantum mechanical treatment the muon "suddenly" jumps from a continuum state (traveling wave) into a bound state (Fig. 3). This jump is accompanied by the emission of either an Auger electron or an x ray. The treatment is difficult because there are large Coulomb effects on both incoming muon and outgoing electron. Hence almost all computations are for free atoms where one knows at least the initial electron states fairly well.

Older quantum mechanical treatments do not agree with experiment. In particular the Coulomb capture is calculated to take place already at high energies. Also, transport theory is often not properly taken

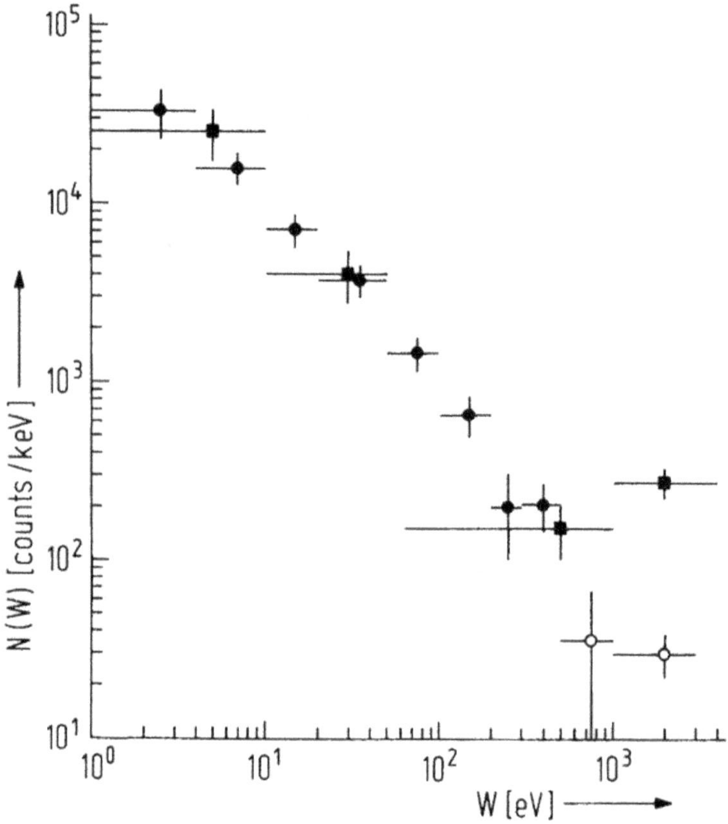

Fig. 2 N(W), over beam cross section and time integrated spectral flux density n(W), vs. energy W for muons emerging from CH_2. Circles: $Au_{80}Pd_{20}$ target; squares: Si target; open circles: energy too high for complete stopping in target.

into account.

Newer results are mostly for free hydrogen atoms; so they are not of real help in our task here.

2.3.2 Coulomb capture treated semiclassically

Semiclassical treatment here means that the muon follows a classical trajectory while the electrons are treated quantum mechanically. The muon is mostly assumed to orbit in the potential of a Thomas-Fermi

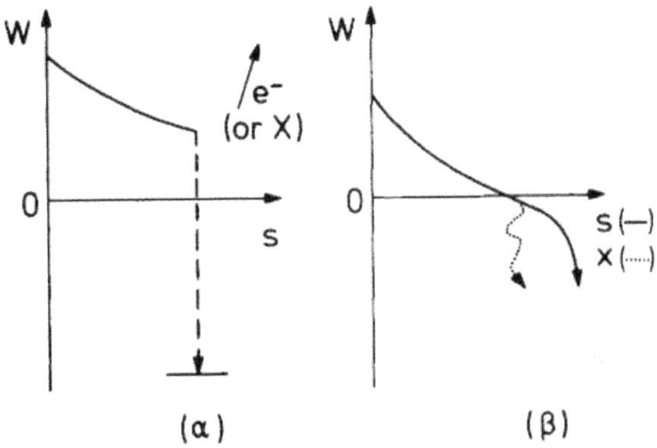

Fig. 3 Symbolized Coulomb capture, quantum mechanically treated (α), left, and semiclassically treated (β), right. s denotes the trajectory length, x in (β) a space coordinate, and W the muonic energy (kinetic plus potential).

atom. As in section 2.2, serious difficulties arise if one wants to treat condensed matter.

An important simplification results from treating the energy loss, which actually is due to individual collisions with a limited number of electrons, as arising from a continuously effective frictional force. The gross features of the per-atom Coulomb capture ratio between elements Z_1 and Z_2 not taking individual atomic data into account, are fairly well represented by[5]

$$A(Z_1,Z_2) = \frac{Z_1^{1/3}}{Z_2^{1/3}} \frac{\ln 0.57\ Z_1}{\ln 0.57\ Z_2}. \tag{7}$$

The formula is refined by taking the electron density change due to the condensed state into account[2]. This yields

$$A(Z_1,Z_2) = \frac{Z_1^{1/3}\ln(0.57Z_1)r_2}{Z_2^{1/3}\ln(0.57Z_2)r_1}, \tag{8}$$

where r_1 and r_2 are the atomic radii of atoms Z_1 and Z_2, respectively, for the respective valence states. Fig. 4 shows a comparison[6] of this formula with experiment. Note that eq. (8) does not contain any adjustable parameter; the discrepancy in the rare earth region is basically understood[7].

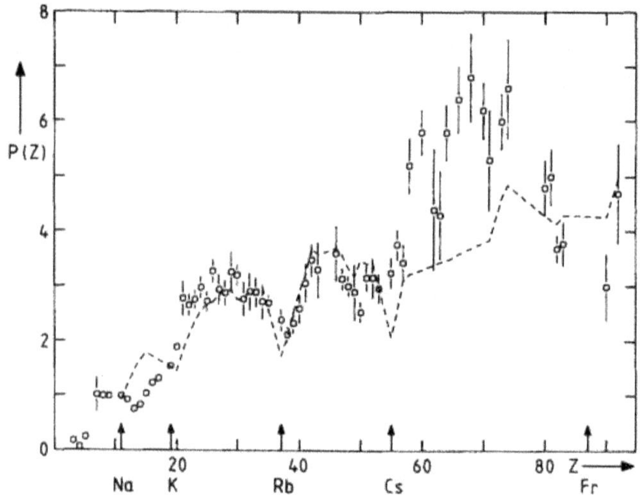

Fig. 4 Experimental values P(Z) with $P(Z_1)/P(Z_2) = A(Z_1,Z_2)$ and theoretical curve from eq. (8) both obtained with setting $Z_2 = 8$ and P(8) = 1. The P(Z) values are averages over all available data[6].

In the (still) simplified model leading to eq. (8) the spectral flux density n(W) at low energies turns out to be a horizontal straight line ("white spectrum"), mainly due to a steep potential at the atomic radius which describes the condensed state fairly well (but not the gaseous). The effect of variations in n(W) was calculated and, in the case of a strong deviation from a horizontal straight line, found to be strong[8].

In all the above treatment the per-atom capture ratio has been taken to be independent of the individual concentrations. This is true for solid matter[9] but not for gases[10].

2.4 Muonic Atoms And Muonic X-Ray Spectra

In matter not containing hydrogen a muon captured into an individual atom will stay in this atom until it decays or reacts with the nucleus. First, it usully cascades down to lower levels by Auger transitions and later by radiative transitions, emitting mesonic x-rays. The investigation of this cascade is usually done with Ge detectors. The experiment is so much standard that it seems not necessary to describe it here in detail. Fig. 5 shows a set-up[11] and Fig. 6 the Lyman series[12] of Mg; the Lyman series of heavier elements can be even better resolved in many more lines.

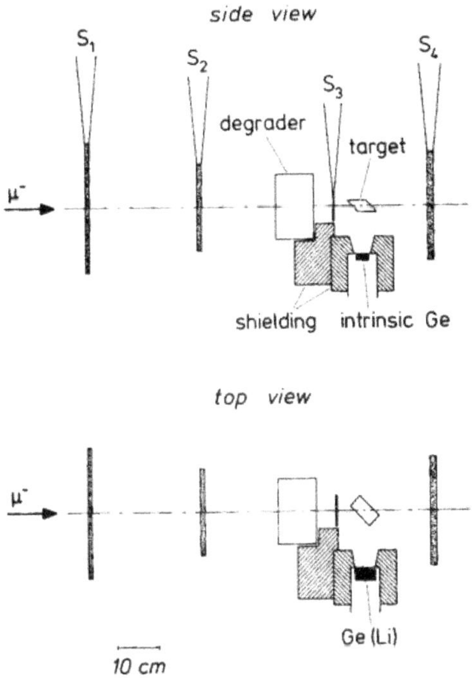

Fig. 5 Set-up for muonic x-ray measurements. (S_1, S_2, S_3, $\bar{S}_4$) scintillation counter telescope. Simultaneous use of two Ge detectors [Ge(Li) withdrawn in side view, Ge withdrawn in top view]. Target tilted.

In hydrogen-containing matter an effect not discussed before in this paper may occur. Muonic hydrogen is electrically neutral. Hence, it can penetrate into other atoms. It then experiences an attractive polarization potential yielding a closer approach to the nucleus of the penetrated atom which is assumed to be heavier. The atomic levels of the two systems cross (actually it is a pseudocrossing), and the muon

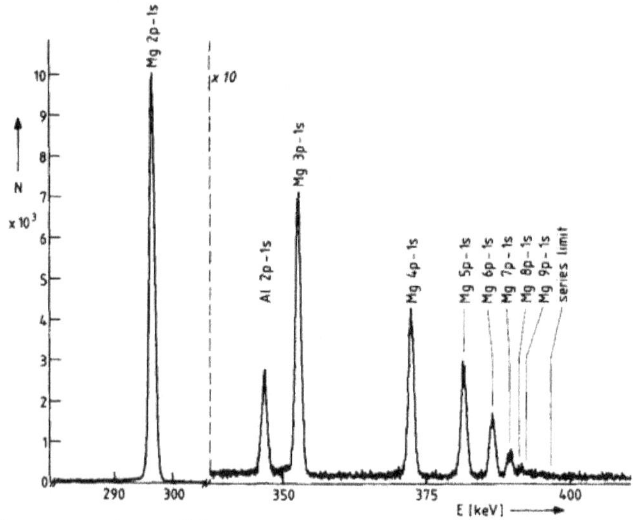

Fig. 6 Lyman series of Mg.

may be transferred to the heavier atom. The transfer cross section σ is large, of the order of the geometric cross section of the electronical atom. σ has been calculated[13] to be proportional to Z.

3. ANALYSIS OF SAMPLES

3.1 X-Ray Intensities And Elemental Composition

Since muonic x-rays are characteristic of every element, the natural application for the analysis of samples is to determine the elemental composition. Although there is some influence of the chemical form, the chemical effects are not very pronounced.

The greatest advantage of the analysis with muonic x-rays is: the analysis is (almost) noninvasive and nondestructive. The radiation risk is very low[1]. Another advantage is that the method easily yields a bulk analysis. If desired, however, we can select special parts of the specimen with scanning analysis or even combined with thoroughly three-dimensional imaging, with lateral resolutions of the order of 1 mm close to the surface and 5 mm well within the specimen and depth resolution of about 0.1 mm. With suitable set-ups we can perform surface analyses (down to below monolayers, where one sees the first atom and almost only the first one).

The quantities directly observed are intensity ratios of x-rays. Usually, one sums over the total Lyman series intensities and takes this number as the number of muons captured in the element. What one wants to know, however, are not the x-ray intensities but the abundances. Hence, one needs a calibration. A formula for the capture ratio into elements with atomic numbers Z_1 and Z_2 is easily obtained from the per-atom capture ratio $A(Z_1,Z_2)$:

$$R(Z_1,Z_2) = \frac{k_1}{k_2} A(Z_1,Z_2); \tag{9}$$

here k_1 and k_2 are the respective atomic abundances. For $A(Z_1,Z_2)$ we may take the value from eq. (8). Higher accuracy, however, is achieved by an experimental calibration. Table 1 lists the composition of various glass samples and Table 2 summarizes calibration data ob-

Table 1 Composition of the measured glass samples (atom%)

Element	E 15[a]	E 25[a]	E 39[a]	WH[a]	EOMAG[a]	Duran[b]	Kalk-natron[b]
O	60.93	60.98	61.89	60.33	59.86	64.96	60.14
Na	8.24	7.12	3.84	9.08	9.54	2.70	9.46
Mg	2.04	2.07	2.13	1.85	2.60	0.01	2.05
Al	0.48	0.49	0.50	0.26	0.69	1.07	0.51
Si	25.70	25.23	26.05	25.00	24.12	30.90	24.70
S	0.06	0.07	0.06	0.05	0.06	[c]	0.07
K	0.22	0.22	2.28	0.13	0.22	0.32	0.19
Ca	2.20	3.71	2.29	3.23	2.86	0.03	2.85
Ti	0.01	0.01	0.27	0.03	0.01	[c]	0.01
Fe	0.03	0.03	0.27	0.03	0.03	0.01	0.02
Ba	0.07	0.07	0.42	0.003	0.003	[c]	[c]

[a] Flachglas AG

[b] Schott & Gen

[c] No data given by the manufacturer

Table 2 Per-atom capture ratios of muons in modern glasses (Si=1.00)

Element	E 15[a]	E 25[a]	E 39[a]	WH[a]
O	1.23 (2)	1.29 (1)	1.14 (2)	1.15 (1)
Na	1.26 (2)	1.30 (2)	1.30 (3)	1.27 (2)
Mg	1.02 (5)	1.05 (5)	1.05 (6)	1.01 (4)
Al	1.12 (14)	1.17 (14)	0.83 (10)	f
Si	1.00	1.00	1.00	1.00
K	1.73 (38)	1.28 (21)	1.85 (11)	1.64 (39)
Ca	2.36 (8)	2.23 (5)	2.16 (11)	2.11 (7)
Ti	7 (7)	5 (5)	3.01 (20)	6 (3)
Fe	7 (4)	7 (3)	3.69 (29)	6 (3)
Ba	5 (2)	3.0 (13)	3.91 (39)	40 (40)

Element	EOMAG[a]	Duran[b]	Kalknatron[b]	Mean
O	1.27 (1)	1.28 (3)	1.19 (1)	1.22 (2)[c]
Na	1.32 (2)	1.34 (6)	1.27 (2)	1.29 (1)[c]
Mg	1.03 (4)	e	1.07 (17)	1.03 (1)[c]
Al	0.93 (7)	1.21 (7)	0.98 (8)	1.04 (6)[c]
Si	1.00	1.00	1.00	1.00
K	1.91 (24)	1.68 (18)	1.82 (19)	1.75 (4)[c]
Ca	2.15 (7)	1.5 (6)	2.12 (14)	2.20 (2)[c]
Ti	8 (5)	g	1.2 (12)	3.01 (20)[d]
Fe	7 (3)	5.5 (21)	3.4 (12)	3.69 (29)[d]
Ba	20 (20)	g	g	3.91 (39)[d]

[a] Flachglas AG
[b] Schott & Gen.
[c] External error
[d] Value taken from E 39 only
[e] Superimposed by the muonic Na(3-1) line
[f] Correction for background aluminum lines not possible
[g] No concentration data given by the manufacturer

tained from these glasses[14]. As can be seen, the accuracy amounts to 1 % of the atomic abundance of the element in question in favorite cases. Other calibration data, for elements occuring in tissue, was obtained from "tissue equivalent plastic" (Shonka plastic) and "tissue equivalent liquid" (TE liquid)[15]. Table 3 summarizes the results. "Z law" means that the capture is proportional to Z, an expression which yields about the same capture ratio as eq. (8) in the case of light elements, and "modified Z law" means that the muons captured in H are transferred to the geometrically nearest atom with Z > 1 and hence show up as if captured by that nearest atom.

Table 3 Comparison of calculated and measured muonic x-ray yields

Sample element (atomic %) with H	(atomic %) without H	Yield of Lyman x-rays (%) modified Z-law	measured
Shonka plastic			
H (58.28 ± 0.58)	...	...	
C (37.38 ± 0.13)	89.61 ± 0.31	87.99 ± 0.30	87.78 ± 0.29
N (1.46 ± 0.02)	3.50 ± 0.06	3.76 ± 0.06	3.67 ± 0.13
O (2.43 ± 0.16)	5.81 ± 0.37	6.65 ± 0.42	6.48 ± 0.25
F (0.30 ± 0.08)	0.72 ± 0.18	0.85 ± 0.21	0.93 ± 0.20
Ca(0.15 ± 0.04)	0.36 ± 0.09	0.94 ± 0.24	0.46 ± 0.15
TE liquid			
H (63.47 ± 0.63)	...	...	
C (6.35 ± 0.06)	17.56 ± 0.18	14.70 ± 0.15	14.12 ± 0.55
N (1.56 ± 0.02)	4.29 ± 0.04	5.00 ± 0.05	3.94 ± 0.22
O (28.38 ± 0.31)	78.15 ± 0.80	80.30 ± 0.80	81.94 ± 0.59
S (0.06 ± 0.01)	0.16 ± 0.03	...	unobserved
K (0.06 ± 0.01)	0.16 ± 0.03	...	unobserved

The question of a concentration dependence of $A(Z_1,Z_2)$ was explicitly checked for single-phase metallic alloys[9]. No dependence was found (Fig. 7). A more indirect but nevertheless significant check is provided by the above glass calibration, as the abundances of the several elements vary strongly between the samples.

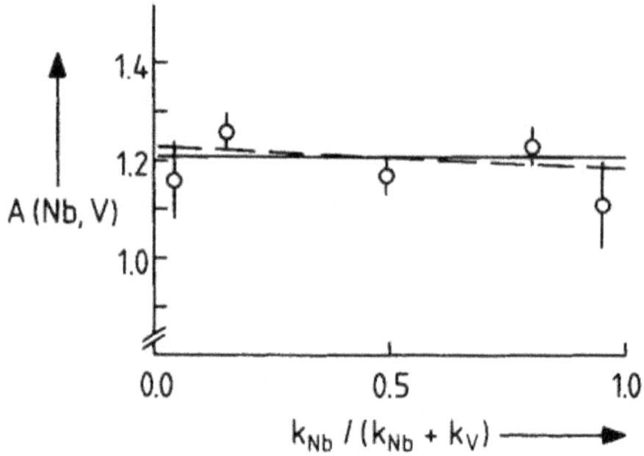

Fig. 7 Per-atom capture ratio in single phase alloys of Nb and V vs. Nb concentration. Solid line: one-parameter fit (horizontal straight line); dashed line: two-parameter fit (tilted straight line). The one-parameter fit yields a lower normalized χ^2!

3.2 Results On Bulk Analysis

The first experiments on analysis done at all with muonic x-rays were performed on organic materials from animals. The Heidelberg group, performing their experiments at CERN, investigated samples of fat and muscle tissue as well as bones from healthy and diseased animals[16]. They found differences between pairs of samples that were selected as identical as possible, except one kind was from a healthy animal and the other from a diseased (mostly cancer). Table 4 gives an example.

Table 4 Experimental ratios of summed up Lyman intensitiy ratios (double ratios)

Double ratio[a]	calf liver	horse muscle
$(\Sigma C/\Sigma O)_p/(\Sigma C/\Sigma O)_n$	0.64(10)	0.77(3)

[a] the subscripts p and n denote pathological and normal samples, respectively

Although the differences are beyond the errors, the results have to be interpreted with care, because the number of samples was too small to draw definite conclusions, as stated by the authors.

Later experiments, with the principal goal to explore the quantitative elemental bulk analysis of parts of the human body, were performed at Los Alamos[17]. The samples were put into liquid nitrogen immediately after separation and kept frozen until the end of the x-ray runs. Although the runs were a few hours each, the data-taking time that would be needed nowadays with better beams and detectors, would be of the order of half an hour for each sample, including mounting, and so on. Part of the difficulties in comparing the numbers obtained by conventional analysis (microanalysis with gas chromatography for C and N, absorption analysis of O) and muonic x-ray analysis consist in representative sampling for the conventional analysis. Nevertheless, the agreement between the two methods is satisfactory. Figs. 8 and 9 show examples. One easily recognizes the difference between muscle (small C content) and fat (large C content).

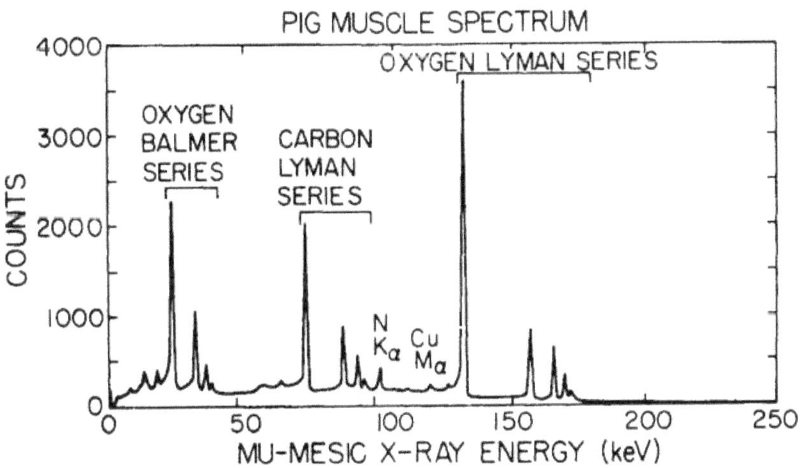

Fig. 8 Muonic x-ray spectrum from pig muscle.

The sensitivity of the method was obtained to about 10^{-4} atomic abundance.

Fig. 10 shows a typical object analyzed in a study of ancient glasses (Celtic and Roman, all before J.C.). It is a bracelet, wonder-

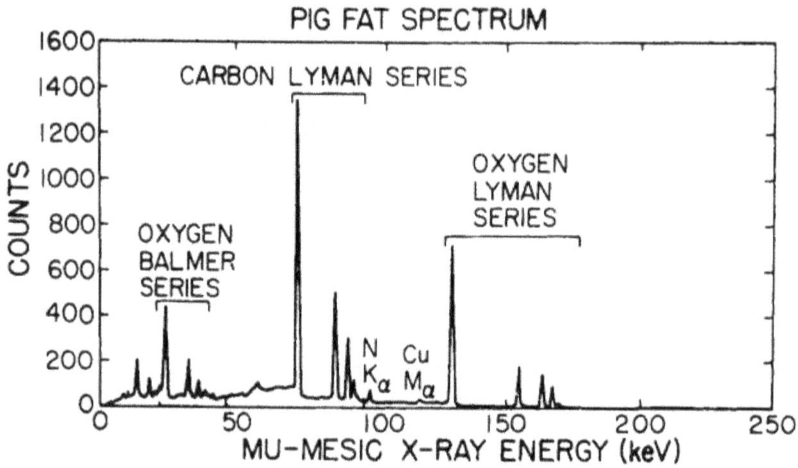

Fig. 9 Muonic x-ray spectrum from pig fat.

Fig. 10 Celtic bracelet from Bavaria (Manching).

fully colored. Fig. 11 depicts the muonic x-ray spectrum. All pieces investigated[18] were provided by Prähistorische Staatssammlung, Munich. Table 5 summarizes the results. An immediate consequence was to recognize that sample inventory no. 1974/2451 was not genuine but only about 200 y old! As seen in Table 5 apart from some exceptions, the glasses from Manching are of an astonishing homogeneity. Soda was used exclusively as a flux since sodium fully dominates, with concentrations

Table 5 Results on Celtic and Roman glasses in at.% (error in parentheses). Data normalized to 100%. (#1-19 Celtic,#20-30 Roman)

#	Inventory No.	origin	O	Na	Al
1	1974/2026+2034	Manching	60.3(3)	9.9(3)	0.7(2)
2	1974/419	Manching	60.4(5)	11.4(5)	0.0(3)
3	1974/1860	Manching	59.9(4)	10.9(5)	0.0(3)
4	1974/1607	Manching	59.9(4)	10.9(5)	0.92(15)
5	1974/1202	Manching	60.0(4)	10.6(5)	0.0(3)
6	1974/1517	Manching	60.1(4)	11.0(5)	0.0(3)
7	1974/1838	Manching	61.7(9)	11.2(5)	0.0(3)
8	1974/2110	Manching	58.8(4)	12.9(5)	0.39(12)
9	1974/2197	Manching	59.6(4)	11.0(5)	0.0(3)
10	1974/2451	Manching	59.0(6)	11.2(5)	4.7(5)
11	1903/14,1	Manching	62.4(13)	10.2(5)	0.8(2)
12	1893/82,6	Manching	60.9(10)	11.1(4)	0.8(3)
13	1903/19,2	Manching	61.9(8)	8.6(3)	0.3(3)
14	1893/79,3	Manching	62.5(9)	11.3(4)	0.3(2)
15	IV 389	Mamming	61.1(7)	11.3(3)	0.0(3)
16	EM 995	Roseninsel	57.9(12)	9.5(4)	0.0(3)
17	1951, 489	Neufahrn	60.6(10)	10.9(4)	0.3(8)
18	1978, 765	Sachsendorf	60.4(9)	10.4(3)	0.0(3)
19	1963, 458	Nassenfels	61.1(8)	9.7(3)	0.6(8)
20	1958, 1920	Lorenzberg	57.1(6)	13.8(3)	1.0(2)
21	1971, 630 g	Auerberg	58.6(6)	12.8(3)	1.06(15)
22	1971, 665b(1)	Auerberg	59.0(7)	6.1(2)	5.2(3)
23	1972, 629	Auerberg	58.9(7)	12.7(3)	0.93(9)
24	1973, 592	Auerberg	58.9(6)	11.7(3)	1.02(8)
25	1973, 736aa	Auerberg	59.1(8)	11.7(4)	1.1(3)
26	1975, 440 a	Auerberg	58.3(7)	12.6(3)	1.9(11)
27	1975, 472 a	Auerberg	58.3(5)	12.8(3)	0.86(10)
28	1975, 479	Auerberg	58.3(6)	13.3(3)	0.98(13)
29	1975, 1175b	Auerberg	58.2(7)	12.0(3)	0.92(7)
30	1971, 655 g	Auerberg	59.7(8)	10.3(3)	1.09(12)

Table 5 continued

#	Si	K	Ca	Mn	Fe	Cu
1	25.0(2)	0.27(5)	3.1(3)	0.59(4)	0.10(2)	
2	26.7(2)	0.33(8)	1.1(3)	0.00(5)	0.00(3)	
3	25.7(2)	0.71(12)	2.8(3)	0.00(5)	0.00(3)	
4	24.8(2)	0.33(5)	2.6(2)	0.34(3)	0.21(2)	
5	25.1(2)	0.43(7)	3.0(3)	0.54(5)	0.24(3)	
6	25.7(2)	0.13(3)	2.7(2)	0.20(4)	0.20(3)	
7	24.0(5)	0.48(7)	2.5(2)	0.00(5)	0.14(3)	
8	23.3(2)	0.31(4)	3.1(3)	1.10(7)	0.79(12)	
9	24.3(2)	0.37(7)	3.6(3)	0.72(5)	0.33(3)	
10	21.0(2)	0.28(7)	2.9(2)	1.01(8)	0.00(3)	
11	23.9(10)	0.43(3)	2.12(11)	0.10(4)	0.063(13)	
12	23.6(7)	0.27(4)	2.54(10)	0.52(7)	0.19(2)	0.09(8)
13	25.3(6)	0.180(12)	3.41(11)	0.00(5)	0.23(3)	0.12(11)
14	23.8(7)	0.12(2)	1.83(7)	0.00(5)	0.19(7)	0.09(7)
15	24.5(5)	0.12(2)	2.68(8)	0.00(5)	0.21(2)	
16	28.4(10)	0.36(4)	3.46(15)	0.10(5)	0.23(3)	
17	25.1(7)	0.19(4)	2.54(9)	0.00(5)	0.32(3)	0.1(3)
18	26.4(6)	0.22(4)	2.12(8)	0.00(5)	0.14(4)	0.3(5)
19	25.2(5)	0.23(3)	2.58(8)	0.12(5)	0.26(3)	0.3(3)
20	23.9(4)	0.62(9)	2.55(13)	0.70(9)	0.19(8)	0.1(2)
21	24.3(5)	0.30(6)	2.68(13)	0.00(3)	0.06(4)	0.26(15)
22	22.1(4)	1.9(1)	5.2(2)	0.00(3)	0.43(7)	
23	24.4(5)	0.13(8)	2.78(11)	0.086(15)	0.07(3)	
24	25.2(4)	0.26(4)	2.88(9)	0.00(3)	0.10(2)	
25	24.9(6)	0.26(11)	2.8(2)	0.00(3)	0.22(11)	
26	24.6(5)	0.18(5)	2.79(13)	0.10(2)	0.25(4)	0.11(9)
27	24.4(4)	0.24(4)	3.13(12)	0.20(3)	0.13(3)	
28	23.8(4)	0.35(6)	2.47(10)	0.66(8)	0.13(5)	0.02(12)
29	25.5(5)	0.23(3)	2.95(9)	0.11(2)	0.07(2)	0.06(3)
30	24.6(6)	0.27(5)	3.22(13)	0.45(5)	0.30(4)	0.11(8)

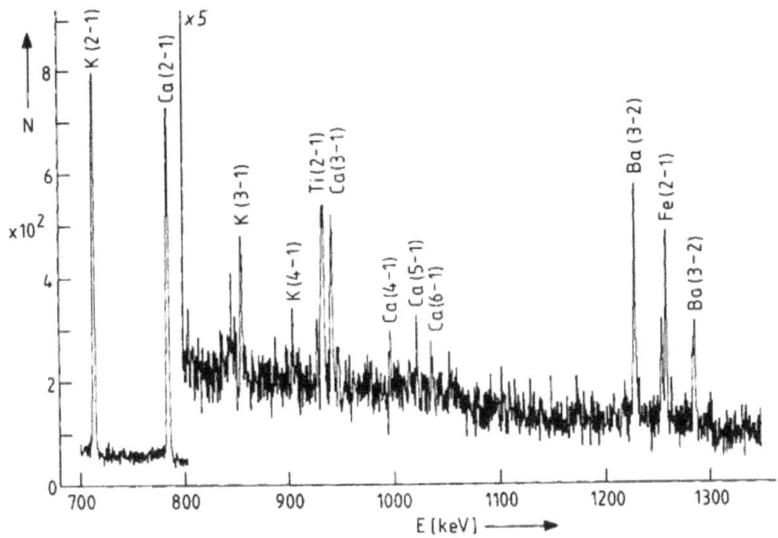

Fig. 11 Spectrum from the bracelet shown in Fig. 10.

of 8-11 %, over potassium, which is below 0.5 % in almost all samples. Calcium with a normal concentration between 1 % and 3 % was used as a stabilizer, while aluminum is below 1 % or below the detection limit of 0.3 %. The concentration of iron, manganese and copper vary considerably, in line with the different colours of the glasses. Iron is responsible for yellowish, greenish and light bluish hues, according to the concentrations of other elements and the kiln atmosphere. Manganese causes a deep purple colour in higher concentrations, while in medium concentrations it makes the greenish colour of glass stained by iron disappear and a completely colourless glass results. Copper in smaller concentrations is responsible for the more or less intense bluish colour of the glass.

3.3 Results On Scanning And Surface Analysis

Scanning analyses have up to now been performed with scanning via the muon energy only (no spatial selection of the incoming muons).

An investigation of several Roman coins (sesterces) of different age made a devaluation of currency showing up, an effect not unknown in present times[18].

A suitable application of the scanning method is also the analysis of glazes on ceramic objects. Fig. 12 shows an example. Since in history there are considerable differences in the type of flux, such as sodium carbonate, potassium carbonate and lead oxide, and in the kind of stabilizer, such as natural calcium or aluminum compounds, the analysis of the glaze promises information about differences in the technique of glazing.

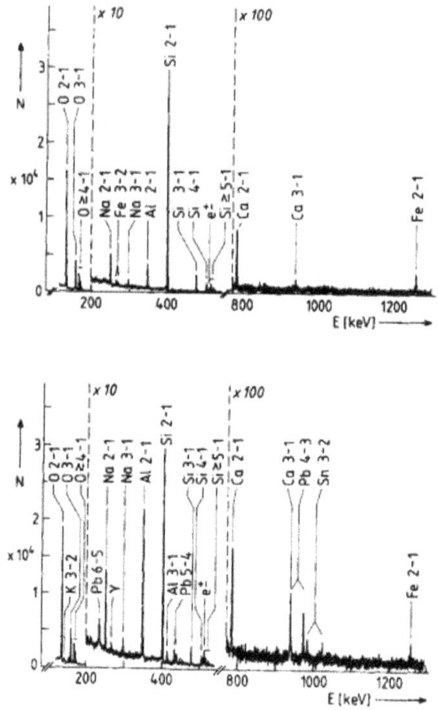

Fig. 12 Spectra from an Islamic tile, Cairo, fourteenth century A.D. Top: base material (fired clay); bottom: glaze

Two Islamic bowls and one Islamic tile were analyzed with muonic x-rays[18]. As shown in Table 6 an Egyptian bowl from the fourteenth century is characterized by a lead glaze with an aluminum stabilizer, while for the tile of the same period a soda glaze with a very low amount of stabilizers was used. The results are in agreement with the general experience that glazes containing higher amounts of potassium and tin, which are usual in Europe, were not used in Islamic countries.

Table 6 Scanning analysis results on Islamic bowls and tiles. Composition in at.% (error in parentheses). Data normalized to 100 %.

Element	Bowl, Cairo, 14th century		Tile, Cairo, 14th century	
	Glaze	Body	Glaze	Body
O	59.3(11	57.2(7)	55.0(7)	59.2(6)
Na	0.08(8)	0.21(4)	6.1(2)	2.44(9)
Al	8.7(11)	11.7(7)	1.9(6)	1.4(2)
Si	22.8(8)	24.8(5)	32.0(6)	34.6(6)
K	1.18(16)	1.61(9)	0.84(6)	0.67(6)
Ca	2.5(2)	2.25(11)	2.31(10)	1.26(6)
Fe	0.91(15)	0.67(8)	0.30(0)	0.25(5)
Sn			0.55(7)	0.06(3)
Pb	4.6(4)	1.6(2)	1.06(14)	0.16(3)

Surface analysis in solid state physics in the strict sense of the word, namely analysis of the very first layer, has been tried at SIN: The Munich group, applying their slow-muon technique[19], measured the muonic x-ray spectra from a Si target in high vacuum after capture of very slow muons and from the same surface after a short "shower" with O_2 gas. They did observe oxygene in the latter case but not in the first. The results, however, were not very promising[20].

Another way of surface analysis appears much more advantageous: detection of muonic x-rays after muon transfer from hydrogen to a heavier element in the gaseous phase. Due to the large transfer cross section the muons are preferably transferred to the outermost atoms of a chemical compound. The technique was a long time ago applied at CERN in a large 1000 atm hydrogen target[21]. Measurements were done with SF_6, with the six F atoms symmetrical around the S center, at 19 atm and with a mixture of protium and SF_6, SF_6 concentration 0.02 at%, at 1000 atm, both at room temperature[22]. Table 7 summarizes the experimental results, labeled "exp", and the theoretical results, labeled "the". We see a pronounced effect when comparing experimental and theoretical values for the mixture, i.e. for transfer.

Table 7 Per-atom capture and transfer ratios, respectively, F : S

pure SF_6,exp	pure SF_6,the*	H_2+SF_6,exp	H_2+SF_6,the
1.23±0.17	0.95	1.80±0.22	0.56

*based on eq. (8)

A visualization experiment with two-dimensional imaging making use of the decay-electron track only[1)] was performed at Los Alamos [23)]. Fig. 13 shows object and "muonograph" (image obtained with muons).

4. DISCUSSION AND OUTLOOK

4.1 Discussion

The elemental analysis with muonic x-rays has been shown to be widely applicable. If we do not take a spurious induced radioactivity, in practice hardly to detect at all, into account, the method is completely nondestructive. It is, however, very expensive and, moreover, restricted to set-ups at a medium-size or big accelerator. So it is certainly not the method for the small company that wants to routinely analyze its samples. The sensitivity is limited and cannot compete with, for example, accelerator mass spectrometry. On the other hand it offers a whole bunch of peculiar advantages such as, perhaps with the exception of hydrogen, sensitivity for every element, the possibility of scanning and even looking into the interior well through layers above. There are several opportunities even not looked at seriously. In the first place we think of "feeling" the shape of a molecule with the transfer technique; it is understood that one would nowadays not need a (somewhat inconvenient!) pressure of 1000 atm within a volume of well above a liter! In the second place we think of three-dimensional imaging with simultaneous x-ray analysis.

4.2 Outlook

If one tends to take advantage of the non-standard applications mentioned in this paper such as detailed scanning analysis, maybe combined with truly three-dimensional imaging[1)], and shape "feeling" in

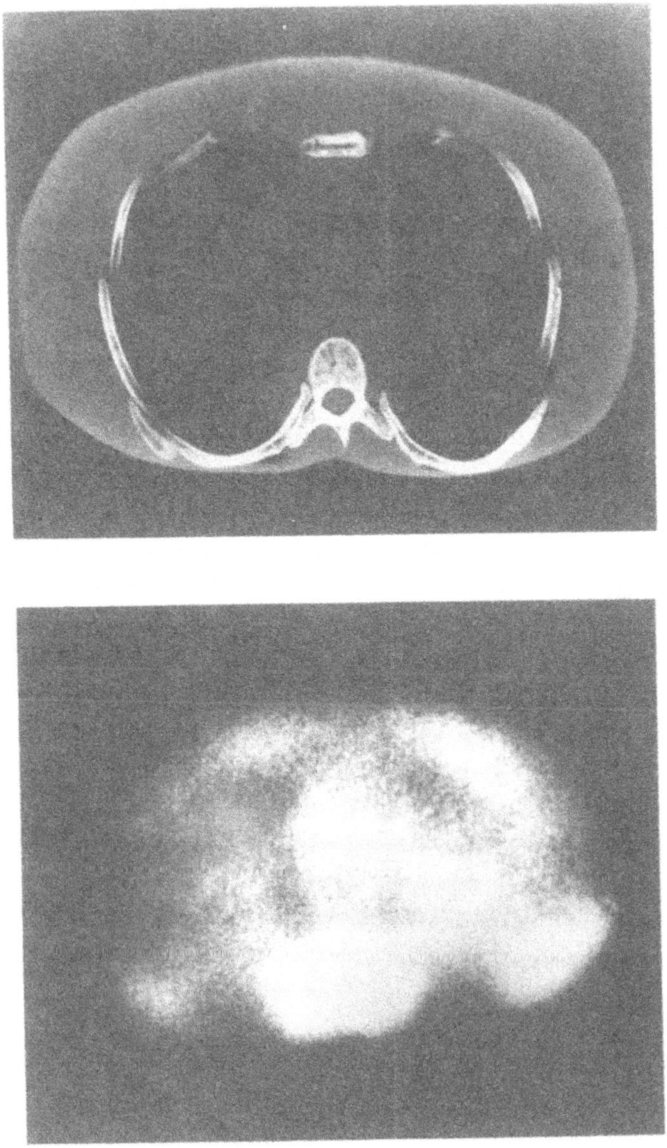

Fig. 13 Top: x-ray photograph of Rando phantom slice. Bottom: Pion stopping density distribution in the slice, not corrected for inhomogeneous illumination by the pion beam.

gases one would like to have higher intensities of the muon beam, and in particular a higher phase space density at low muon energies. Unfortunately standard beam cooling methods, namely stochastic cooling and electron cooling, do not seem to work in the case of the primarily low phase-space density muon beams. A cybernetic beam has been proposed, and progress was reached[24]; the final fate of the project is, however, uncertain. A new method, namely frictional cooling with simultaneous acceleration either by an rf field[25] or by a dc field[26], has very recently been proposed. It is based on energy loss by ionization in the energy range where the stopping power S decreases with decreasing energy, in practice between 0 and about 6 keV for muons in hydrogen. In this range S effects cooling[3] while multiple scattering yields heating. So there will be an equilibrium. If one uses gas one may avoid discharges by applying a magnetic field $\underline{B}$ perpendicular to the accelerating electric field $\underline{E}$. However, this method has to demonstrate its feasibility in the first place!

We add in parentheses: nuclear γ-ray spectra from the μ capture by the nucleus are very different for differing isotopes of the same element and hence allow a nondestructive isotopic analysis of the sample. No application to analysis, however, has been reported.

REFERENCES

1) Daniel, H., Nucl. Med. 8, 311 (1969).
2) Daniel, H., Ann. Phys. (N.Y.) 129, 303 (1980).
3) Daniel, H., Nucl. Inst. Meth. 147, 297 (1977).
4) Fottner, G., Daniel, H., Ehrhart, P., Hagn, H., Hartmann, F.J., Köhler, E. and Neumann, W., Z. Phys. A304, 333 (1982) and unpublished results.
5) Daniel, H., Phys. Rev. Lett. 35, 1649 (1975).
6) von Egidy, T. and Hartmann, F.J., Phys. Rev. A26, 2355 (1982).
7) Daniel, H. and Naumann, R.A., Muon Cat. Fus. 4, 121 (1989).
8) Daniel, H., Z. Phys. A302, 195 (1981).
9) Bergmann, R., Daniel H., von Egidy, T., Hartmann, F.J., Reidy, J.J. and Wilhelm, W., Phys. Rev. A20 , 633 (1979).

10) Ehrhart, P., Hartmann, F.J., Köhler, E. and Daniel, H., Z. Phys. A311, 259 (1983).

11) Daniel, H., Biol. Trace Elem. Research 13, 301 (1987).

12) Hartmann, F.J., Bergmann, R., Daniel, H., Pfeiffer, H.-J., von Egidy, T. and Wilhelm, W., Z. Phys. A305, 189 (1982).

13) Gershtein, S.S., Soviet Physics JETP 16, 501 (1963) [J. Exptl. Theoret. Phys. (U.S.S.R.) 43, 706 (1962)].

14) Daniel, H., Hartmann, F.J. and Köhler, E., Fresenius' Z. Anal. Chemie 321, 65 (1985).

15) Reidy, J.J., Hutson, R.L., Daniel, H. and Springer, K., Anal. Chem. 50, 40 (1978).

16) Daniel, H., Pfeiffer, H.-J. and Springer, K., Biomed. Tech. 18, 222 (1973) and Phys. Med. Biol. 20, 1035 (1975).

17) Hutson, R.L., Reidy, J.J., Springer, K., Daniel, H. and Knowles, H.B., Radiology 120, 193 (1976).

18) Daniel, H., Hartmann, F.J., Köhler, E. and Beitat, U., Archaeometry 29 (1), 110 (1987); Daniel, H., Hartmann, F.J. and Riederer, J. in Prokovskii, V.H. et al. (eds.) Proc. Int. Symp. on Muon and Pion Interactions in Matter, Dubna 1987, p. 451.

19) Daniel, H., Fottner, G., Hagn, H., Hartmann, F.J., Stoeckel, P. and Wilhelm, W., Phys. Rev. Lett. 46, 720 (1981).

20) Schott, W., Daniel, H., Hartmann, F.J. and Neumann, W., Proc. Int. School of Phys. of Exotic Atoms "Ettore Majorana", Erice 1989, ed. by L. Simons, to be published.

21) Pfeiffer, H.-J., Springer, K. and Daniel, H., Nucl. Phys. A254, 433 (1975).

22) Pfeiffer, H.-J. and Daniel, H., Z. Phys. A275, 313 (1975).

23) Daniel, H., Reidy, J., Hutson, R., Bradbury, J. and Helland, J., Rad. Res. 68, 171 (1976).

24) PSI Jahresbericht 1988, Villigen 1989, p. 14.

25) Daniel, H., Muon Cat. Fus. 4, 109 (1989).

26) Daniel, H., invited paper given at the 18th INS Int. Symp. on Physics with High-Intensity Hadron Accelerators, Tokyo, Mar 14 - 16, 1990; to be published in the Conf. Proc.

The Design of the Israeli X-ray Telescope.

B. Rosner, Department of Physics, Technion, Haifa.

Abstract

The properties of an x-ray telescope, based on a matrix of 421 solid state detectors with a unique collimation system, for the investigation of celectial objects emitting x-rays in the 3-30 KeV energy range are described. It has a 1.7 KeV energy resolution, 10 µsec temporal resolution and extremely low background. The essensial features of the an x-ray telescope were designed by a multidepartmental group. They meet the constraints dictated by the Israeli Space Agency namely, less than 20 kg mass, less than 20 watt power and a modest cost.

Introduction

Many astrophysical objects emit x radiation. The x rays are generally emitted by charged particles accelerated to high energies. Most frequently the accelerations take place in very strong gravitational fields of compact objects, for example, accretion processes onto a neutron star are strong sources of hard radiation. Furthermore the intensity of many of those objects is variable in time which gives a clue on their sizes and interaction mechanisms. Observations of celestial x rays is done mainly by instruments flown on satellites or balloons because of the strong absorption of this radiation by the atmosphere.

Details of the telescope

The technical details of the telescope which is shown in Fig. 1 are divided into 4 parts:

1. The collimator.
2. The structure or the telscope.
3. The matrix of x ray detectors.
4. The electronics for data aquisition and preprocessing.

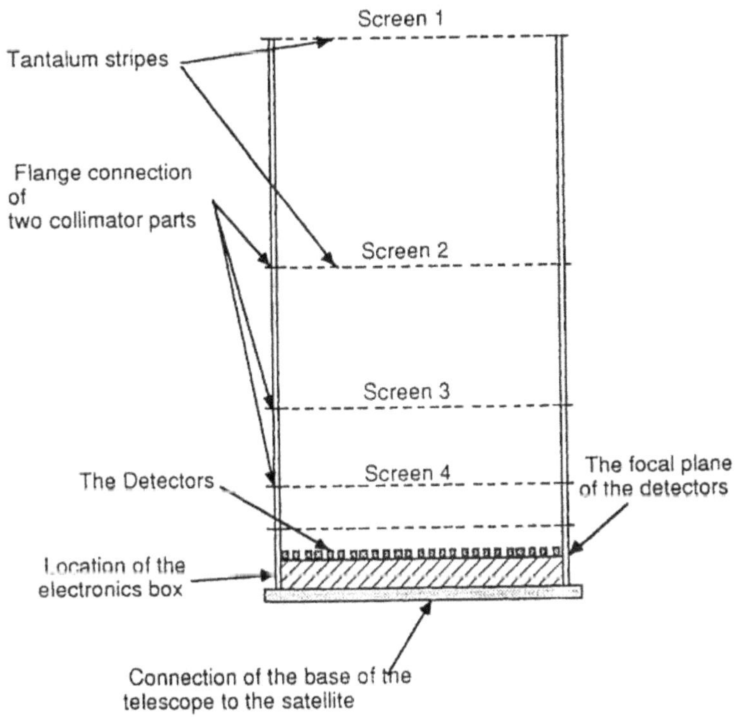

A Schematic View of the Telescope

Figure 1 Schematic view of the telescope and the main parts. The total length of the telescope is 100 cm and the diameter is 50 cm. The screens area made of Tantalum.

The collimator

The most massive part of any x ray telescope is the collimator. It serves to define the solid angle viewed by each of the 421 detectors and to attenuate the radiation arriving from other directions. The constraints on the length of the payload is one meter, the size of a each detector is 1 cm^2, this defines a spatial resolution of 0.57°. The collimator will contain several screens made of tantalum and a honeycomb structure at the far end. The total weight of the collimator is about 9 kg. It will attenuate radiation in the 3-30 KeV range at least by a factor of 10^{-4}.

The basic idea of the operation of the screens is shown in Fig. 2. The apertures in the first screen define the solid angle viewed by each detector. The second and successive screens are positioned so that radiation entering through the apertures not in front of a detector, will not reach it. As the distance between screens decreases and many are needed, it is better to replace them by an "egg-crate" structure preferably at the far end, as remote as possible fron the detectors.

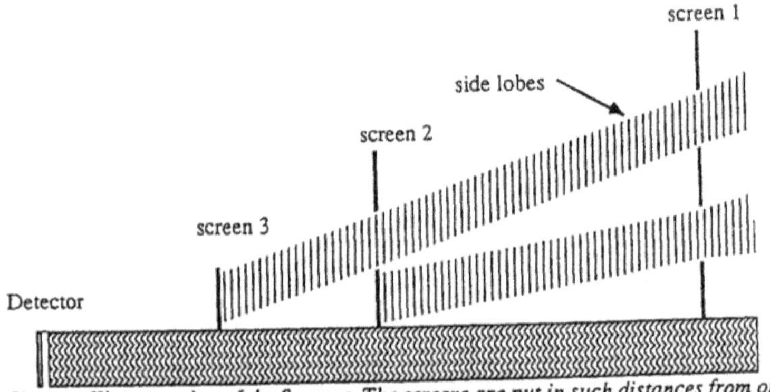

Figure 2. The operation of the Screens. The screens are put in such distances from one another as to shade the detector from photons that may enter through the 'wrong' opening. The distances between the screens are determined from the theory. However, as can easily be verified from the figure, there is quite a large 'tolerance' in the relative location of the screens. The wavy area indicates the spatial resolution of the telescope.

A computer program to assist in the design of the collimating system and to test its performence, using sophisticated computer graphic hardware was written. It calculates the spatial response of the telescope and provides also the possibility to design a telescope with a "Lobster Eye" type of imaging. Such a design will reduce source confusion and help in the estimation of the backgroud intensity.

The supporting structure

The screens are held by a supporting cylindrical structure. The basic units of the telescope are the rectangular detectors, but demands for stiffness of the entire structure favors a cylindrical shape. Beside supporting the screens, the cylinder must shield the detectors from the side. The material used for the main structure is aluminum. The full length of the telescope is divided into several small cylindrical sections which are connected with flanges at the location of the screens. The thin wall cylindrical parts obtain their stiffness from "integral stringers" prepared by a process developed in our aerospace engineering department. Laboratory tests have verified the structural properties of this shape and assure that it will successfully stand the launch load. The weight of the cylindrical frame without the thermal blanket, passive coolers, nuts and bolts is estimated to be about 2 kg.

The detectors unit

The detectors unit will be constructed from a matrix of 421 x ray detectors operating at -20C°. Each element of the matrix is a PIN silicon photodiode and an AMPTEK 250 preamplifier built in a modular way. Most of the detectors are of the HAMAMATSU 1723-09 type with an active area of 1 cm^2 and 0.5 mm thick. The energy resolutin is a function of the temperature. At -20°C which is the operating temperature, we have 1.7 KeV FWHM. This is due mainly to the electronic noise of the preamplifier. In view of the measured efficiency, the total effective areas of the detectors are 408 cm² at 10 KeV, 147 cm² at 20 KeV and 54 cm² at 30 KeV. A thin aluminazed mylar will be placed in front of the detectors to protect their active area of the from light.

Energy calibration of the detectors will be done by two methods. A continuous calibration mode, using a very weak Am-241 source with an alpha-gamma coincidence system. In the second mode, a Cd-109 electron capture source emitting Ag characteristic x rays will be used. It will be placed in small container to be opened only for fast calibration needs.

Electronic circuit architecture

Two architectures were experimented with analog circuitry following the charge preamplifier stage and the synchronous node was preferred. As a first implememtation, a working circuit with 8 channels was built. Based on CMOS amplifier LMC660, the power consumption and amplification bandwidth requirements are satisfied. On the basis of these measurments the parameters that were chosen as satisfactory are given:
Minimal signal output of the preamplifier - 300 microvolts.
Signal amplification before analog memory, A=10.
Discriminator time delay - 0.2 microseconds.
Signal amplification before ADC, A=30.
ADC with full scale of 2.5 volts, AD 7576.

Estimate of background radiation

A Monte Carlo computer program was used to estimate the intensity of background radiation reaching the detectors. The satellite and telescope are represented by a cylinder which consists of three regions made of different materials as can be seen in Fig. 3. The outer part is made of 200 kg aluminum representing the satellite. The silicon detectors with their natural shape and density are positioned as a planar disk. The region above and below the silicon disk is filled with 15 kg tantalum, representing the x ray shielding material. Photons that reach the detectors in the 3-30 KeV energy range after passing through the structure and shielding material are considered as background counts.

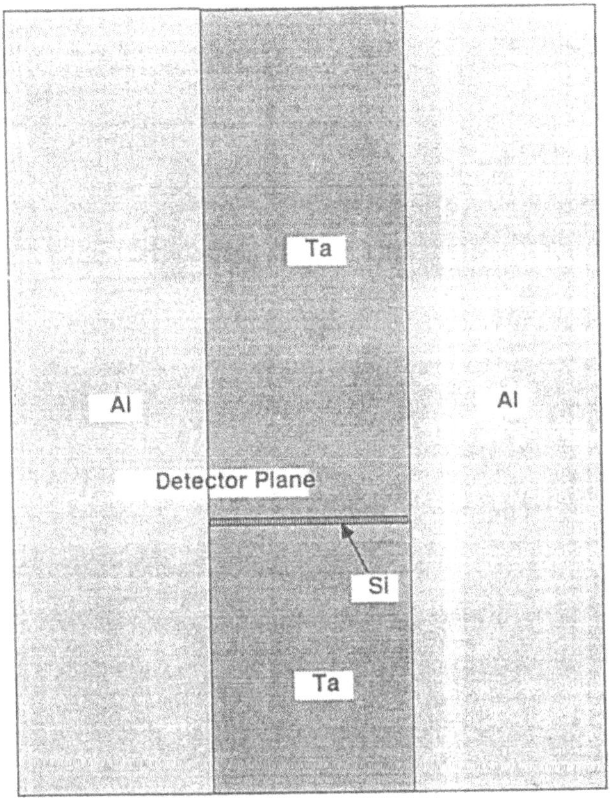

Figure 3 A schematic view of the satellite and the telescope used in the Monte Carlo simulations.

The background has the following sources: x and gamma rays, and high energy particles predominantly protons and electrons. The background in the detectors arises from radiation hitting the detector directly through the apertures in the screens and indirectly due to Compton scattering, pair production and bremsstrahlung in the structure and shielding material. The most interesting result of the calculations is the very low background count of $3.5*10^{-5}$ counts/cm^{2}sec. This low value is due mainly to the small mass of the telescope.